New Applications and Developments in Synthetic Peptide Chemistry

New Applications and Developments in Synthetic Peptide Chemistry

Editors

Giovanni N. Roviello
Rosanna Palumbo

Basel • Beijing • Wuhan • Barcelona • Belgrade • Novi Sad • Cluj • Manchester

Editors
Giovanni N. Roviello
Institute of Biostructure and
Bioimaging (IBB)
Italian National Council
for Research (CNR)
Napoli
Italy

Rosanna Palumbo
Institute of Biostructure and
Bioimaging (IBB)
Italian National Council
for Research (CNR)
Napoli
Italy

Editorial Office
MDPI
St. Alban-Anlage 66
4052 Basel, Switzerland

This is a reprint of articles from the Special Issue published online in the open access journal *Pharmaceuticals* (ISSN 1424-8247) (available at: www.mdpi.com/journal/pharmaceuticals/special_issues/peptide_chem).

For citation purposes, cite each article independently as indicated on the article page online and as indicated below:

Lastname, A.A.; Lastname, B.B. Article Title. *Journal Name* **Year**, *Volume Number*, Page Range.

ISBN 978-3-0365-9601-3 (Hbk)
ISBN 978-3-0365-9600-6 (PDF)
doi.org/10.3390/books978-3-0365-9600-6

© 2023 by the authors. Articles in this book are Open Access and distributed under the Creative Commons Attribution (CC BY) license. The book as a whole is distributed by MDPI under the terms and conditions of the Creative Commons Attribution-NonCommercial-NoDerivs (CC BY-NC-ND) license.

Contents

About the Editors vii

Preface ix

Rosanna Palumbo, Hayarpi Simonyan and Giovanni N. Roviello
Advances in Amino Acid-Based Chemistry
Reprinted from: *Pharmaceuticals* **2023**, *16*, 1490, doi:10.3390/ph16101490 1

Rosanna Palumbo, Daniela Omodei, Caterina Vicidomini and Giovanni N. Roviello
Willardiine and Its Synthetic Analogues: Biological Aspects and Implications in Peptide Chemistry of This Nucleobase Amino Acid [†]
Reprinted from: *Pharmaceuticals* **2022**, *15*, 1243, doi:10.3390/ph15101243 4

Muhammad Luqman Nordin, Ahmad Khusairi Azemi, Abu Hassan Nordin, Walid Nabgan, Pei Yuen Ng and Khatijah Yusoff et al.
Peptide-Based Vaccine against Breast Cancer: Recent Advances and Prospects
Reprinted from: *Pharmaceuticals* **2023**, *16*, 923, doi:10.3390/ph16070923 20

Minji Kim, Kush Savsani and Sivanesan Dakshanamurthy
A Peptide Vaccine Design Targeting KIT Mutations in Acute Myeloid Leukemia
Reprinted from: *Pharmaceuticals* **2023**, *16*, 932, doi:10.3390/ph16070932 46

Sana Khajeh pour, Arina Ranjit, Emma L. Summerill and Ali Aghazadeh-Habashi
Anti-Inflammatory Effects of Ang-(1-7) Bone-Targeting Conjugate in an Adjuvant-Induced Arthritis Rat Model
Reprinted from: *Pharmaceuticals* **2022**, *15*, 1157, doi:10.3390/ph15091157 69

Radoslav Chayrov, Tatyana Volkova, German Perlovich, Li Zeng, Zhuorong Li and Martin Štícha et al.
Synthesis, Neuroprotective Effect and Physicochemical Studies of Novel Peptide and Nootropic Analogues of Alzheimer Disease Drug
Reprinted from: *Pharmaceuticals* **2022**, *15*, 1108, doi:10.3390/ph15091108 85

Xiaoxiao Fu, Jing Wang, Huaying Cai, Hong Jiang and Shu Han
C16 Peptide and Ang-1 Improve Functional Disability and Pathological Changes in an Alzheimer's Disease Model Associated with Vascular Dysfunction
Reprinted from: *Pharmaceuticals* **2022**, *15*, 471, doi:10.3390/ph15040471 104

Sara La Manna, Marilisa Leone, Flavia Anna Mercurio, Daniele Florio and Daniela Marasco
Structure-Activity Relationship Investigations of Novel Constrained Chimeric Peptidomimetics of SOCS3 Protein Targeting JAK2
Reprinted from: *Pharmaceuticals* **2022**, *15*, 458, doi:10.3390/ph15040458 125

Monica Iavorschi, Ancuța-Veronica Lupăescu, Laura Darie-Ion, Maria Indeykina, Gabriela Elena Hitruc and Brîndușa Alina Petre
Cu and Zn Interactions with Peptides Revealed by High-Resolution Mass Spectrometry
Reprinted from: *Pharmaceuticals* **2022**, *15*, 1096, doi:10.3390/ph15091096 140

Carlo Diaferia, Elisabetta Rosa, Giancarlo Morelli and Antonella Accardo
Fmoc-Diphenylalanine Hydrogels: Optimization of Preparation Methods and Structural Insights
Reprinted from: *Pharmaceuticals* **2022**, *15*, 1048, doi:10.3390/ph15091048 160

Diana I. S. P. Resende, Marta Salvador Ferreira, José Manuel Sousa-Lobo, Emília Sousa and Isabel Filipa Almeida
Usage of Synthetic Peptides in Cosmetics for Sensitive Skin
Reprinted from: *Pharmaceuticals* **2021**, *14*, 702, doi:10.3390/ph14080702 . . . . **174**

About the Editors

Giovanni N. Roviello

Dr. Giovanni Roviello graduated with honors in chemistry from Federico II University (2002, Naples, Italy) and received his PhD in biotechnology from the same university in 2006. He works as a senior researcher at the Institute of Biostructure and Bioimaging (IBB) of the Italian National Research Council (CNR) in Naples, Italy. He was a visiting researcher in Germany (Institute of Molecular Biotechnology (IMB), Jena 2003, Georg-August University of Goettingen 2005, FAU University, Erlangen 2018), the UK (University of Greenwich, Medway Campus in Chatham Maritime 2022), and Ireland (UCD, Dublin 2017) working in the field of biorganic chemistry, and a visiting professor in Poland at Adam Mickiewicz University (AMU), Poznan (September 2023) in the frame of the "Excellence Initiative—Research University" program. He has strong scientific and teaching connections with Yerevan State University (Armenia), with which he works on two funded research projects; AMU University (Poland), for which he served as a lecturer for the nucleic acids course for Ph.D. students in 2021; and Geomedi University (Georgia), having been appointed the honorary professorship in medicinal chemistry for this university in 2012. He is the principal investigator for CNR for different international research activities, including the EU-funded project NobiasFluors (ID 872331; 2020–2024), the Royal Society (UK)-funded project 'Design, synthesis, and in vitro evaluation of TMPRSS2 inhibitor analogues as potential novel anti-COVID-19 drugs' (2022–2024), the project code 21T-1D057 funded by the Republic of Armenia (2021–2024), and the MESRA (Armenia)-CNR-funded bilateral project 'Synthetic alpha-amino acids and peptide-based systems for anticancer strategies' (2023–2025). He is a full member of Sigma Xi, the Scientific Research Society (United States), serves as an academic editor for numerous international scientific journals, and is the author of more than 110 scientific articles.

Rosanna Palumbo

Rosanna Palumbo received her degree cum laude in biological sciences in 1985 from the Federico II University of Naples, Italy. In 1989, she was a visiting scientist at the Cold Spring Harbor Laboratory (NY, USA). In 1991, she obtained a specialization degree in microbiology and virology from the Medical School of the II University of Naples, Italy. From 1995 to 2003, she worked as a research scientist at the Institute of Food Science of the Italian National Research Council (CNR), and currently she is a researcher at the Institute of Biostructure and Bioimaging (IBB) at CNR, Italy. From 1999 to 2000, she was scientifically responsible for the operative unit seen in the concerted action "Dietary Exposure to Vegetal Estrogen and Related Compounds and Effect on Skeletal Tissue", funded by the European Union. From 2000 to 2001, she served as a scientific advisor for the research operative unit of the scientific project funded by the Italian Ministry of Agricultural and Forestry Policies. She was also appointed as a component of the scientific teams for several Italian research projects funded by the Italian Association for Cancer Research (AIRC), Regione Campania, etc. Her research activities are mainly focused on the development and preclinical validation in vitro of innovative approaches for tumor diagnosis and therapy using molecular targets and labeled probes that are conducted thanks to her outstanding ability to coordinate the activities of different research teams aimed at developing biologically active compounds, together with her capability of working in highly multidisciplinary environments involving organic chemists and physicists, collaborating with scientists having the most diverse backgrounds in order to develop strategies for the implementation of multidisciplinary projects. Finally, she is the author of about 60 scientific articles in international journals and more than 60 scientific contributions presented at international and national conferences.

Preface

'*New Applications and Developments in Synthetic Peptide Chemistry*' is a reprint for anyone who is interested in peptide science, an attractive and successful interdisciplinary field whose study can promote enrichment in different areas of chemistry and biomedicine. Peptides are of great interest in biomedicine for the treatment of diseases of high economic and social impact, including cancer and neuropathies, and as cosmetics, as we show in this reprint. We would like to highlight that peptides have a pivotal role in the fight against viral diseases, such as the pathologic condition caused by the coronavirus SARS-CoV-2, which was at the origin of the recent COVID-19 pandemic. In this context, future research could investigate synthetic peptide inhibitors of coronavirus enzymes that are of utmost relevance in the fight against possible future coronavirus pandemics, considering that these pathogens might reemerge in the near future. Peptides are able to interact with biological macromolecules involved in key biological pathways, such as nucleic acids and proteins. Moreover, after functionalization by nucleobases, they become powerful tools in oligonucleotide targeting, as testified by the scientific reports on peptide nucleic acids (PNAs) and nucleobase-containing peptides (nucleopeptides), with attractive therapeutic and diagnostic potentials. Not less importantly, peptide-based derivatives are endowed with self-assembly properties at the origin of nanometric-scale materials, which are attracting particular attention, especially in nanomedicine and technology. Novel experimental techniques (i.e., synthetic, spectroscopy-based, and biological) and theoretical efforts based on molecular docking, molecular dynamics, and machine learning for peptide design and development are expected in the near future, with particular attention to be paid to the study of natural amino acid derivatives and peptides. In fact, some structures, almost neglected in our opinion, such as plant nucleoamino acids, polydiamino acids, and nucleopeptides, deserve new investigative work as the comprehension of the role played by these molecules in their natural hosts could disclose new scenarios for biomedical applications.

Giovanni N. Roviello and Rosanna Palumbo
Editors

Editorial
Advances in Amino Acid-Based Chemistry

Rosanna Palumbo [1], Hayarpi Simonyan [2] and Giovanni N. Roviello [1,*]

1 Institute of Biostructures and Bioimaging, Italian National Research Council (IBB-CNR), Via P. Castellino 111, 80131 Naples, Italy; rosanna.palumbo@cnr.it
2 Institute of Pharmacy, Yerevan State University, 1 Alex Manoogian Str., Yerevan 0025, Armenia; hayarpi.simonyan@ysu.am
* Correspondence: giovanni.roviello@cnr.it; Tel.: +39-081-220-3415

Numerous applications of amino acid-based compounds and peptide derivatives in different biomedicine- and nanotechnology-related fields were described in the recent scientific literature [1]. For example, glycine derivatives including glycyl-glycyl-glycine, glycyl-glycine, sarcosine, dimethylglycine, all of which were functionalized with memantine (Figure 1) moieties, were found to exert a neuroprotective effect, improving cell viability against copper- and glutamate-induced neurotoxicity [2].

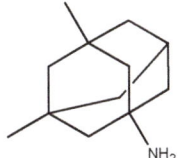
3,5-dimethyladamantan-1-amine

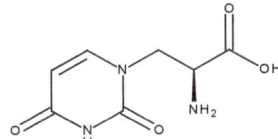
(2*S*)-2-amino-3-(2,4-dioxopyrimidin-1-yl)propanoic acid

Figure 1. Structural representation of memantine (**left**) and L-Willardiine (**right**) with respective International Union of Pure and Applied Chemistry (IUPAC) names.

A synthetic octapeptide, derived from activity-dependent neuroprotective protein (ADNP), that is able to bind to Cu^{2+} and Zn^{2+} showed peculiar crystallization properties that were influenced by the metal ions. It also exerts a neuroprotective effect due to both its metal chelating properties and its ability to interact with amyloid beta (Aβ) peptide, whose abundant deposition in the brain is famously linked to the Alzheimer's disease [3,4].

Synthetic peptides in conjunction with growth factors can show neuroprotective properties, making them potential candidates as innovative neurodrugs. The synthetic dodecapeptide C16 administered together with the growth factor angiopoietin-1 improved functional disability and reduced neuronal cell death in animal models by protecting vascular endothelial cells, thereby inhibiting inflammatory cell infiltration and maintaining blood–brain barrier (BBB) permeability [5].

Novel cyclic peptidomimetics of the protein suppressor of cytokine signaling 3 (SOCS3) [6] were designed and synthesized for the development of novel therapeutic strategies involving the ternary protein complex formed by SOCS3 with Janus Kinase 2 and glycoprotein 130 [7].

Signs of the potential anti-metastatic activity of sugar–amino acid derivatives were discovered by the collaborative efforts of Armenian and Italian chemists who used the Amadori reaction to obtain novel synthetic conjugates and also discovered novel molecules with therapeutic potential [1].

Peptides are also useful in the field of prophylactics for the realization of vaccines. Among the others, peptide-based vaccines have recently been attracting a growing attention in the prevention and recurrence of breast cancer [8,9].

Citation: Palumbo, R.; Simonyan, H.; Roviello, G.N. Advances in Amino Acid-Based Chemistry. *Pharmaceuticals* **2023**, *16*, 1490. https://doi.org/10.3390/ph16101490

Received: 5 October 2023
Accepted: 18 October 2023
Published: 19 October 2023

Copyright: © 2023 by the authors. Licensee MDPI, Basel, Switzerland. This article is an open access article distributed under the terms and conditions of the Creative Commons Attribution (CC BY) license (https://creativecommons.org/licenses/by/4.0/).

Chimeric compounds whose structures include both nucleobases and amino acid residues are known in the scientific literature as nucleoamino acids, which in turn form larger structures that are often labeled as nucleopeptides [10,11]. L-Willardiine (Figure 1) is one of several examples of a nucleoamino acid that occurs in nature, and in particular, it functions as a neurotransmitter in the human organism. Synthetic nucleoamino acids and the corresponding nucleopeptides can also be used in several biomedical and nanotechnological applications [12].

Amino acid-based materials have also been found to be capable of forming biocompatible hydrogels as new nanomaterials that can be employed in biomedical strategies. For example, synthetic derivatives of diphenylalanine were shown to form hydrogels whose structural arrangement and behavior in terms of matrix porosity, stiffness, and stability is influenced by the different formulation strategy [13].

Interestingly, peptides are also useful in cosmetics and can be used as active ingredients on sensitive skin due to their ability to interact with skin cells with high potency at low dosage and to penetrate the stratum corneum [14].

In conclusion, amino acid-derivatives and peptides are molecular tools with a vast number of applications in the field of human health, ranging from therapy to disease prophylaxis, but also find use in cosmetics and nanotechnology, as we mentioned in this work.

List of Contributions

1. Chayrov, R.; Volkova, T.; Perlovich, G.; Zeng, L.; Li, Z.; Štícha, M.; Liu, R.; Stankova, I. Synthesis, Neuroprotective Effect and Physicochemical Studies of Novel Peptide and Nootropic Analogues of Alzheimer Disease Drug. Pharmaceuticals 2022, 15(9), 1108; https://doi.org/10.3390/ph15091108.
2. Iavorschi, M.; Lupăescu, A.; Darie-Ion, L.; Indeykina, M.; Hitruc, G.; Petre, B. Cu and Zn Interactions with Peptides Revealed by High-Resolution Mass Spectrometry. Pharmaceuticals 2022, 15(9), 1096; https://doi.org/10.3390/ph15091096.
3. Fu, X.; Wang, J.; Cai, H.; Jiang, H.; Han, S. C16 Peptide and Ang-1 Improve Functional Disability and Pathological Changes in an Alzheimer’s Disease Model Associated with Vascular Dysfunction. Pharmaceuticals 2022, 15(4), 471; https://doi.org/10.3390/ph15040471.
4. La Manna, S.; Leone, M.; Mercurio, F.; Florio, D.; Marasco, D. Structure-Activity Relationship Investigations of Novel Constrained Chimeric Peptidomimetics of SOCS3 Protein Targeting JAK2. Pharmaceuticals 2022, 15(4), 458; https://doi.org/10.3390/ph15040458.
5. Nordin, M.; Azemi, A.; Nordin, A.; Nabgan, W.; Ng, P.; Yusoff, K.; Abu, N.; Lim, K.; Zakaria, Z.; Ismail, N.; Azmi, F. Peptide-Based Vaccine against Breast Cancer: Recent Advances and Prospects. Pharmaceuticals 2023, 16(7), 923; https://doi.org/10.3390/ph16070923.
6. Palumbo, R.; Omodei, D.; Vicidomini, C.; Roviello, G. Willardiine and Its Synthetic Analogues: Biological Aspects and Implications in Peptide Chemistry of This Nucleobase Amino Acid. Pharmaceuticals 2022, 15(10), 1243; https://doi.org/10.3390/ph15101243.
7. Diaferia, C.; Rosa, E.; Morelli, G.; Accardo, A. Fmoc-Diphenylalanine Hydrogels: Optimization of Preparation Methods and Structural Insights. Pharmaceuticals 2022, 15(9), 1048; https://doi.org/10.3390/ph15091048
8. Resende, D.; Ferreira, M.; Sousa-Lobo, J.; Sousa, E.; Almeida, I. Usage of Synthetic Peptides in Cosmetics for Sensitive Skin. Pharmaceuticals 2021, 14(8), 702; https://doi.org/10.3390/ph14080702.

Author Contributions: All authors have contributed equally to this work. All authors have read and agreed to the published version of the manuscript.

Acknowledgments: The editorial board would like to acknowledge and thank the contributions of the authors as well as all the reviewers whose efforts, expertise and constructive comments have contributed significantly to the quality of this Special Issue. G.N. Roviello and H. Simonyan would also like to thank Italian National Research Council (CNR) and Armenian Science Committee of the Ministry of Education and Science (MESRA) for their support to their collaboration [Armenian-Italian bilateral research project 23SC-CNR-1D002; CNR/MESRA (Armenia) scientific cooperation (CNR Prot. N. 19140 del 20230125 (2023-CNR0A00-0019140)].

Conflicts of Interest: The authors declare no conflict of interest.

References

1. Fik-Jaskółka, M.A.; Mkrtchyan, A.F.; Saghyan, A.S.; Palumbo, R.; Belter, A.; Hayriyan, L.A.; Simonyan, H.; Roviello, V.; Roviello, G.N. Spectroscopic and SEM evidences for G4-DNA binding by a synthetic alkyne-containing amino acid with anticancer activity. *Spectrochim. Acta Part A Mol. Biomol. Spectrosc.* **2020**, *229*, 117884. [CrossRef] [PubMed]
2. Vieira, A.D.C.; Medeiros, E.B.; Zabot, G.C.; de Souza Pereira, N.; do Nascimento, N.B.; Lidio, A.V.; Scheffer, Â.K.; Rempel, L.C.T.; Macarini, B.M.N.; de Aguiar Costa, M. Neuroprotective effects of combined therapy with memantine, donepezil, and vitamin D in ovariectomized female mice subjected to dementia model. *Prog. Neuro-Psychopharmacol. Biol. Psychiatry* **2023**, *122*, 110653. [CrossRef] [PubMed]
3. Choi, S.H.; Tanzi, R.E. Adult neurogenesis in Alzheimer's disease. *Hippocampus* **2023**, *33*, 307–321. [CrossRef] [PubMed]
4. Wang, C.; Shao, S.; Li, N.; Zhang, Z.; Zhang, H.; Liu, B. Advances in Alzheimer's Disease-Associated Aβ Therapy Based on Peptide. *Int. J. Mol. Sci.* **2023**, *24*, 13110. [CrossRef] [PubMed]
5. Cornelissen, F.M.; Markert, G.; Deutsch, G.; Antonara, M.; Faaij, N.; Bartelink, I.; Noske, D.; Vandertop, W.P.; Bender, A.; Westerman, B.A. Explaining Blood–Brain Barrier Permeability of Small Molecules by Integrated Analysis of Different Transport Mechanisms. *J. Med. Chem.* **2023**, *66*, 7253–7267. [CrossRef] [PubMed]
6. Yin, Y.; Liu, W.; Dai, Y. SOCS3 and its role in associated diseases. *Hum. Immunol.* **2015**, *76*, 775–780. [CrossRef] [PubMed]
7. Berishaj, M.; Gao, S.P.; Ahmed, S.; Leslie, K.; Al-Ahmadie, H.; Gerald, W.L.; Bornmann, W.; Bromberg, J.F. Stat3 is tyrosine-phosphorylated through the interleukin-6/glycoprotein 130/Janus kinase pathway in breast cancer. *Breast Cancer Res.* **2007**, *9*, 1–8. [CrossRef] [PubMed]
8. Fatima, G.N.; Fatma, H.; Saraf, S.K. Vaccines in Breast Cancer: Challenges and Breakthroughs. *Diagnostics* **2023**, *13*, 2175. [CrossRef] [PubMed]
9. Zhu, S.-Y.; Yu, K.-D. Breast cancer vaccines: Disappointing or promising? *Front. Immunol.* **2022**, *13*, 828386. [CrossRef] [PubMed]
10. Ghosh, S.; Ghosh, T.; Bhowmik, S.; Patidar, M.K.; Das, A.K. Nucleopeptide-coupled injectable bioconjugated guanosine-quadruplex hydrogel with inherent antibacterial activity. *ACS Appl. Bio Mater.* **2023**, *6*, 640–651. [CrossRef] [PubMed]
11. Hoschtettler, P.; Pickaert, G.; Carvalho, A.; Averlant-Petit, M.-C.; Stefan, L. Highly Synergistic Properties of Multicomponent Hydrogels Thanks to Cooperative Nucleopeptide Assemblies. *Chem. Mater.* **2023**, *35*, 308. [CrossRef]
12. Palumbo, R.; Omodei, D.; Vicidomini, C.; Roviello, G.N. Willardiine and its synthetic analogues: Biological aspects and implications in peptide chemistry of this nucleobase amino acid. *Pharmaceuticals* **2022**, *15*, 1243. [CrossRef]
13. Li, X.; Bian, S.; Zhao, M.; Han, X.; Liang, J.; Wang, K.; Jiang, Q.; Sun, Y.; Fan, Y.; Zhang, X. Stimuli-responsive biphenyl-tripeptide supramolecular hydrogels as biomimetic extracellular matrix scaffolds for cartilage tissue engineering. *Acta Biomater.* **2021**, *131*, 128–137. [CrossRef] [PubMed]
14. Mazurowska, L.; Mojski, M. Biological activities of selected peptides: Skin penetration ability of copper complexes with peptides. *J. Cosmet. Sci.* **2008**, *59*, 59–69. [PubMed]

Disclaimer/Publisher's Note: The statements, opinions and data contained in all publications are solely those of the individual author(s) and contributor(s) and not of MDPI and/or the editor(s). MDPI and/or the editor(s) disclaim responsibility for any injury to people or property resulting from any ideas, methods, instructions or products referred to in the content.

Review

Willardiine and Its Synthetic Analogues: Biological Aspects and Implications in Peptide Chemistry of This Nucleobase Amino Acid [†]

Rosanna Palumbo, Daniela Omodei, Caterina Vicidomini and Giovanni N. Roviello *

Institute of Biostructures and Bioimaging-Italian National Council for Research (IBB-CNR), Via De Amicis 95, 80145 Naples, Italy
* Correspondence: giroviel@unina.it; Tel.: +39-081-2203415
† Dedicated to the Memory of Professor Ulf Diederichsen.

Abstract: Willardiine is a nonprotein amino acid containing uracil, and thus classified as nucleobase amino acid or nucleoamino acid, that together with isowillardiine forms the family of uracilylalanines isolated more than six decades ago in higher plants. Willardiine acts as a partial agonist of ionotropic glutamate receptors and more in particular it agonizes the non-N-methyl-D-aspartate (non-NMDA) receptors of L-glutamate: ie. the α-amino-3-hydroxy-5-methyl-4-isoxazole-propionic acid (AMPA) and kainate receptors. Several analogues and derivatives of willardiine have been synthesised in the laboratory in the last decades and these compounds show different binding affinities for the non-NMDA receptors. More in detail, the willardiine analogues have been employed not only in the investigation of the structure of AMPA and kainate receptors, but also to evaluate the effects of receptor activation in the various brain regions. Remarkably, there are a number of neurological diseases determined by alterations in glutamate signaling, and thus, ligands for AMPA and kainate receptors deserve attention as potential neurodrugs. In fact, similar to willardiine its analogues often act as agonists of AMPA and kainate receptors. A particular importance should be recognized to willardiine and its thymine-based analogue AlaT also in the peptide chemistry field. In fact, besides the naturally-occurring short nucleopeptides isolated from plant sources, there are different examples in which this class of nucleoamino acids was investigated for nucleopeptide development. The applications are various ranging from the realization of nucleopeptide/DNA chimeras for diagnostic applications, and nucleoamino acid derivatization of proteins for facilitating protein-nucleic acid interaction, to nucleopeptide-nucleopeptide molecular recognition for nanotechnological applications. All the above aspects on both chemistry and biotechnological applications of willardine/willardine-analogues and nucleopeptide will be reviewed in this work.

Keywords: willardiine; glutamate receptor; kainate receptor; nucleoamino acid; nucleopeptide; peptide; nucleobase; synthetic compounds; amino acids

Citation: Palumbo, R.; Omodei, D.; Vicidomini, C.; Roviello, G.N. Willardiine and Its Synthetic Analogues: Biological Aspects and Implications in Peptide Chemistry of This Nucleobase Amino Acid. *Pharmaceuticals* 2022, 15, 1243. https://doi.org/10.3390/ph15101243

Academic Editor: Rakesh Tiwari

Received: 9 September 2022
Accepted: 28 September 2022
Published: 10 October 2022

Publisher's Note: MDPI stays neutral with regard to jurisdictional claims in published maps and institutional affiliations.

Copyright: © 2022 by the authors. Licensee MDPI, Basel, Switzerland. This article is an open access article distributed under the terms and conditions of the Creative Commons Attribution (CC BY) license (https://creativecommons.org/licenses/by/4.0/).

1. Introduction

Willardiine was first identified by Rolf Gimelin in 1959 from the extracts of seeds of Acacia willardiana [1]. Structurally it corresponds to (2S)-2-amino-3-(2,4-dioxopyrimidin-1-yl)propanoic acid (**1**, Figure 1) and carrying an uracil moiety it can be ascribed to the category of nucleoamino acids [2–4]. Willardiine is synthesized by the single specific enzyme uracilylalanine synthase, and the N–heterocyclic moiety uracil obtained by the orotate pathway [5] proved to be an effective bioisostere for the distal carboxyl group of L-glutamate [6].

This is the main excitatory neurotransmitter found in the central nervous system (CNS) and mediates its actions through a variety of metabotropic (G-protein coupled) and ionotropic (ligand-gated cation channels) receptors [7]. Among the others, ionotropic glutamate receptors are proteins present in almost all mammalian brain structures at the excitatory synapses that facilitate signal transmission in the central nervous system mediating the fast excitatory neurotransmission, and are involved in human nervous system development and function, with their dysfunction being correlated with several CNS disorders [8–13]. Mechanistically, antagonists of ionotropic glutamate receptors prevent the closure of the bilobed ligand-binding domain while full agonists close it. In fact, extracellular cleft closure of ionotropic glutamate receptors is associated with receptor activation; however, the mechanism underlying partial agonism is not necessarily linked to the blockade of the full receptor cleft closure; instead, partial agonists more likely destabilize the cleft closure as suggested by in silico studies [14]. There are three types of ionotropic glutamate receptors: the N-methyl-D-aspartate (NMDA) receptor, a glutamate receptor and ion channel present in neurons, and the α-amino-3-hydroxy-5-methyl-4-isoxazole-propionic acid (AMPA) [15] and kainate receptors [16–18]. It was demonstrated that in mammalians **1** agonizes non-NMDA glutamate receptors acting as a partial agonist of ionotropic glutamate receptors [19] and in particular activates the AMPA receptors [20–22].

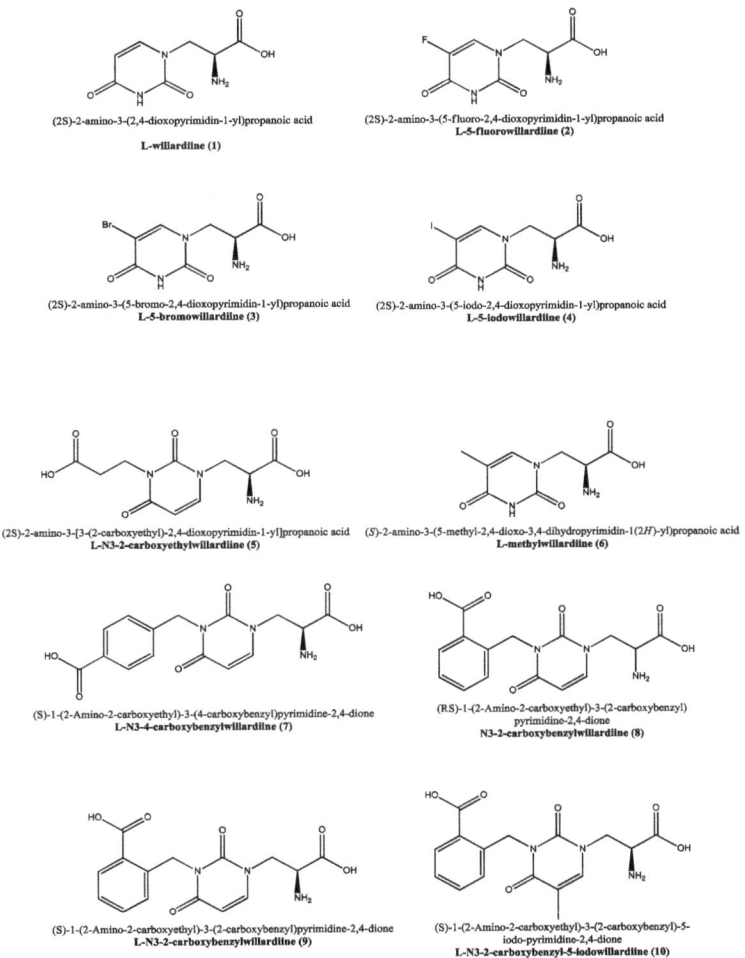

Figure 1. Molecular representation of the structures of willardiine and some of its analogues. IUPAC and use names are reported alongside with the numbering used in this manuscript.

Similar to willardiine, its analogues (Figure 1) often act as agonists of AMPA and kainate receptors, but due to their different binding affinities for each receptor, they have been employed not only to investigate the structural and functional consequences of activation of AMPA/kainate receptors, but also to determine the structural modifications required in order to convert **1** into an antagonist at AMPA and kainate receptors [23].

Thus, in this review we describe some of the main ligands, including competitive agonists and antagonists of glutamate receptor family members developed within the willardiine family alongside with the reported structure–activity hypotheses that can serve as starting point for the discovery of new analogues of neurotherapeutic importance with increased selectivity and/or potency properties as binders of kainate or AMPA receptors. Moreover, the role of willardiine and its analogues like the L-methylwillardiine (AlaT) in the peptide chemistry will be discussed in the light of the potential applications of the derived nucleopeptides [24] in diagnostics and biotechnology [25].

2. Willardiine and Its Analogues

2.1. Disease Relevance and Potential Pharmaceutical Role

Willardiine analogues may bind specifically to kainate and AMPA receptors, which are implicated in a variety of neurological disorders characterized by alterations in glutamate signaling. For example, the selective antagonists for glutamate receptor subtypes disclose therapeutic importance for a variety of neurological disorders characterized by aberrant activation of kainate or AMPA receptors. Though far from being exhaustive, in the following section we report on some neurological disorders characterized by dysregulation in either AMPA or kainate receptor activation.

2.1.1. AMPA Receptors in Neurological Disorders

AMPA receptors are particularly implied in neurodegenerative disorders through their connection with synaptic plasticity, which plays a fundamental role in many physiological aspects of cognitive abilities but also in neural development [26]. Many neuropathies are associated with the cognitive decline, which in turn is linked to changes in AMPA-mediated plasticity [27,28]. In animal models of Parkinson's disease, enhanced levels of AMPA receptors have been revealed in affected regions, while AMPA antagonists have been proposed as potential therapeutics. However, the benefits of inhibiting AMPA receptors with respect to curing the symptoms of Parkinson's disease do not always justify the use of this strategy that can cause severe off-target effects [29,30]. On the other hand, AMPA receptor deficits have been correlated with Huntington's disease, that causes the progressive degeneration of neural cells in the brain [31], as demonstrated in studies conducted in human postmortem tissues and animal models. In particular, modulating AMPA receptors in animal models led to a decrease of degeneration in both the striatum and memory deficits [29]. Moreover, genetic alterations leading to AMPA receptor GluA2 subunit defects were correlated with autism spectrum disorders [32]. The role of AMPA receptors in autism is also supported by the effectiveness of the pharmacological modulation of this type of receptors which rescued social impairments in animal models [33]. Among the many hypotheses about the biological basis of the major depressive disorder, the glutamate hypothesis was corroborated by the evidence that NMDA receptor antagonists blocking the activation of NMDA receptors, and thus causing the AMPA receptor activation to compensate for the decrease in glutamate signaling, have antidepressant effects in animal models [34]. Interestingly, AMPA receptors mediate also the dysregulated sleep, that is a major symptom of major depressive disorder [35].

2.1.2. Kainate Receptors in Neurological Disorders

Recent biochemical and behavioral studies suggest a key role for the kainate receptor GluK2 in controlling abnormalities related to the behavioral symptoms of mania, such as psychomotor agitation, aggressiveness and hyperactivity [36], while the kainate receptor GluK3 was associated with recurrent major depressive disorder [37]. Kainate receptors are expressed in regions of spinal cord implied in the transmission of sensory stimulation and pain, and their functional dysregulation was linked to pain [38,39]. Several studies led to a variety of linkages between epilepsy and kainate receptors including different and sometimes contradictory mechanisms that clearly demonstrate the difficulty in studying kainate receptors in this respect [39–41].

2.2. Molecular Insights on Bioactivity of Willardiine analogues

As anticipated, willardiine is a partial agonist of ionotropic glutamate receptors [14] and specifically, agonizes non-NMDA glutamate receptors in particular activating the AMPA receptors [15–17]. In analogy to willardiine, its most common analogues are also agonists of kainate/AMPA receptors. The addition of a halogen to the uracil moiety affects the binding affinities and stability of the analogues that are generally more stable than 1 and show higher binding affinity for AMPA receptors. Extensive research has been conducted on the 5-fluorowillardiine (2, Figure 1) that showed limited effects at the kainate receptor but acted as a selective agonist of the AMPA receptor and is widely employed in vitro to selectively stimulate AMPA receptors [42–48]. The activity of the halogen-based analogues 2–4 (Figure 1) at AMPA/kainate receptors was also assayed using mouse embryonic hippocampal neurons [42]. In particular, 5-bromowillardiine (3) acted as a potent agonist, and led to rapid but incomplete desensitizing responses. As for the 5-iodowillardiine (4) it resulted a selective kainate receptor agonist [49]. On the other hand, 2 with an 1.5 µM EC_{50} was seven times more potent than AMPA (EC_{50} = 11 µM) and 30 times more potent than 1 (EC_{50} = 45 µM); as for the potency sequence within the explored halogen derivatives the observed trend was F(2) > Br(3) > I(4) > willardiine (1, Table 1) [42].

Table 1. Steady-state EC_{50} [1] for AMPA/kainate receptor activation by willardiines.

Compound	EC_{50} (µM)	±SD
1	44.8	15.0
2	1.47	0.39
3	8.82	1.29
4	19.2	1.92

[1] adapted from data reported in [42].

Moreover, cross-desensitization experiments revealed that halogen-bearing willardiines bind with different affinity to desensitized receptors that are the same receptors activated by kainate and AMPA, and the rapidly desensitizing and equilibrium responses to the analogues under investigation were mediated by the same receptors. However, they produced different degrees of desensitization with the desensitization sequence being F(2) > willardiine (1) > Br(3) > I(4). More in detail, the Iodine-containing derivative (4) blocked the activation of the desensitizing response elicited by 1 and 2, while 1 and 2 blocked the equilibrium response to 4. It was possible to conclude that simple changes in the molecular structure of willardiine can lead to marked differences in the ability of agonists to produce desensitization of AMPA/kainate receptors [42].

The thermodynamics connected with the interaction of willardiine and its analogues with the ionotropic receptors was studied in order to obtain useful insights exploitable for new drug design [38]. For example, analogues of willardiine modified at position 5 with F, Cl, I, H and NO_2 led to deeper insights on the thermodynamics of the interaction of willardiine with AMPA receptors [50]. The binding of willardiine analogues to subtypes of the AMPA receptor was, in some cases, driven by increases in entropy. The major part of the studied analogues were partial agonists whose charged state was found in direct connection with the enthalpic contribution to the interaction [38]. In particular, the binding of the charged forms to AMPA receptor is driven essentially by enthalpy, while the interaction of the uncharged form is largely dominated by the entropic contribution due to changes in the hydrogen-bonding network within the binding site, with these findings providing clues for further neurodrug development [38].

Interestingly, the carboxyethyl derivative of willardiine (**5**) was found to act as antagonist of the AMPA receptor and more in detail, the crystal structure of the GluR2 binding domain of this receptor, which is crucial for mediating its calcium permeability, in complex with **5** was described in the literature (Figure 2a,b) [51].

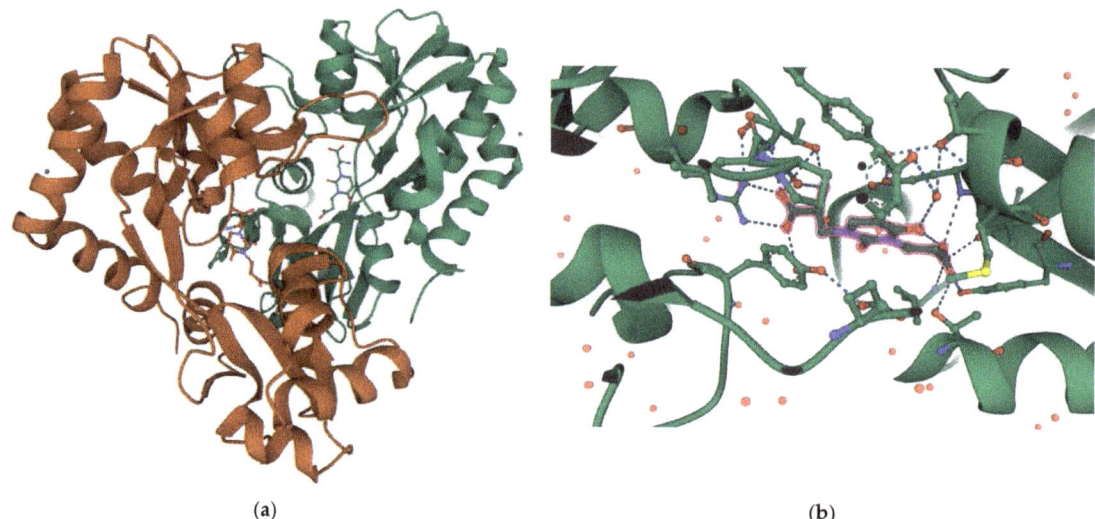

Figure 2. (**a**) 3D view of the crystal structure of the complex between the binding domain of the AMPA subunit GluR2 and **5** (PDB DOI: 10.2210/pdb3H03/pdb PDB ID: 3H03 https://www.rcsb.org/3d-sequence/3H03?assemblyId=1, accessed on 29 August 2022); (**b**) Detailed view of the ligand interaction in the same complex (for more details on protein residues involved in the binding see the link https://www.rcsb.org/3d-view/3H03?preset=ligandInteraction&label_asym_id=E, accessed on 29 August 2022). Note how the amino acid COOH forms a double H-bond with Arg93, while the ethyl COOH forms H-bonds with Tyr187 and Thr171.

This compound binds to one lobe of the protein with interactions similar to agonists, while the binding with the second lobe produces a stable lobe orientation that is similar to the apo state. The carboxyethyl substituent in the N(3) position of **5** keeps the protein lobes separated and the internal dynamics are minimal compared to the protein bound to the reference antagonist 6-cyano-7-nitroquinoxaline-2,3-dione, which minimizes the contacts with one of the two lobes. This latter complex is less stable than that observed with **5** despite a 100-fold higher affinity [45]. Overall, the study with **5** suggests that the antagonism of willardiine analogues is associated with the overall orientation of the lobes of the AMPA receptor rather than with specific interactions [45]. Another study [23] conducted on a range of novel willardiine analogues as antagonists acting at AMPA and kainate receptors suggested that for a derivative to act as AMPA/kainate receptor antagonist, an N3-substituent bearing a carboxylic acid (like in **5**) side-chain is needed, the S-stereochemistry should be present in the derivative and a iodine added to the 5-position of the uracil moiety enhances antagonism at kainate receptors. Moreover, in the same study the 3-methyl analogue **6** (Figure 1), was found to be a weak agonist, indicating that merely blocking ionisation of the uracil ring of willardiine even though decreases the agonist activity is not sufficient to convert the analogue into an antagonist [23]. Aiming at developing new antagonists of kainate/AMPA receptors also N3-carboxybenzyl substituted willardiine analogues were synthesized [49]. The N3-4-carboxybenzyl derivative (**7**) proved to be equipotent at AMPA and kainate receptors in the rat spinal cord. The racemic N3-2-carboxybenzyl analogue (**8**) was found to be a potent and selective kainate receptor antagonist in experiments conducted on native rat and human recombinant kainate and AMPA receptors. More in detail, the kainate receptor antagonist activity was demonstrated to reside essentially in the S enantiomer (9). 5-Iodo substitution of the uracil ring of **9** gave **10**, which proved to have enhanced selectivity and potency for the kainate receptor [49].

3. Willardiine in Peptide Structures

In general, synthetic nucleobase-containing molecules [52–59] as well as modified nucleosides [60–66] are interesting compounds with several biomedical properties [67–69]. Introducing single nucleobase-bearing amino acids into polyamino acid chains led to nucleopeptides able to stabilize certain protein and peptide structures, enhancing their function [70–72]. These DNA analogues are known not only for their ability to bind complementary nucleic acids [55,57,73] and permeate through cell membranes [55,56,74–77] but also to self-assemble [78,79], and are endowed with other biomedically-relevant properties [72,80,81]. In this regard, nucleoamino acids [56], made up of amino acids carrying DNA or RNA bases, are useful building blocks of modified peptides with improved biomolecular binding properties. Examples of natural nucleoamino acids are given not only by **1**, but also lathyrine, a non-proteinogenic amino acid from *Lathyrus* species, and discadenine, a plant cytokinin isolated from *Dictyostelium discoideum*. Refs. [82–85] Nonetheless, nucleoamino acid moieties are found in the natural peptidyl nucleosides [86], molecules that play a key role in biology and therapy, while a plethora of synthetic nucleoamino acids were developed as building blocks of nucleopeptides [87–92].

Examples of natural nucleopeptides containing **1** in their structures are given by γ-glutamylwillardiine **11** and γ-glutamylphenylalanylwillardiine **12** (Figure 3) that were isolated from vegetal sources and whose structures were established by hydrolytic procedures to give the constituent amino acids, and for the tripeptide, the constituent dipeptides, and spectroscopy [93]. The compound **1** in its Nα-acetyl-modified form was obtained synthetically as a racemic mixture and subsequently deacylated enzymatically to the free L-willardiine by treatment with an acylase enzyme [94]. The molar ellipticity at 260 nm (-2.0 $M^{-1}cm^{-1}$) was in agreement with the expected value for the pure enantiomer. After these steps, several peptides containing **1** were synthesized through solid phase peptide strategy using an oxim resin. In particular, they were obtained alternating **1** and glycine and placing a dansyl fluorophore at the N-terminus as a fluorescence probe. Remarkably, the willardine-based peptides showed β-sheet structures in aqueous solution as revealed by CD analysis [94].

The utilization of analogues of **1** in building up enzyme-resistant peptides in which specific amino acid residues of the native sequence were replaced by β-nucleo-α-alanines was reported in the literature [46]. Research efforts focused on this type of studies were devoted especially by the group led by prof. Ulf Diederichsen of University of Goettingen in Germany. They introduced the analogues of **1** into modified αvβ3-integrin-inhibiting RGD cyclopeptides, combining, thus, the properties of the peptide backbone with the properties inherent the heterocyclic nucleobases [2]. They also designed alanyl-nucleopeptide/DNA chimeras for DNA diagnostics whose solid phase synthesis protocol required the introduction of nucleotides as phosphoramidites, while the nucleoamino acids as monomers protected by the acid labile monomethoxytrityl (MMT) group on the α-amino groups, and by acyl protecting groups on the exocyclic amino nucleobase groups. In this context, they synthesized the four MMT/acyl-protected nucleo alanines, including the MMT-protected willardiine analogue AlaT (**13**, Figure 3), that are useful monomers for the obtainment of nucleopeptide/DNA chimeras under conditions that are compatible with the standard phosphoramidite DNA synthesis strategy [95]. Interestingly, nucleopeptides containing analogues of **1** such as AlaT, are endowed with cell-membrane permeability and were also able to reach the cell nucleus without exerting any cytotoxic effects [75].

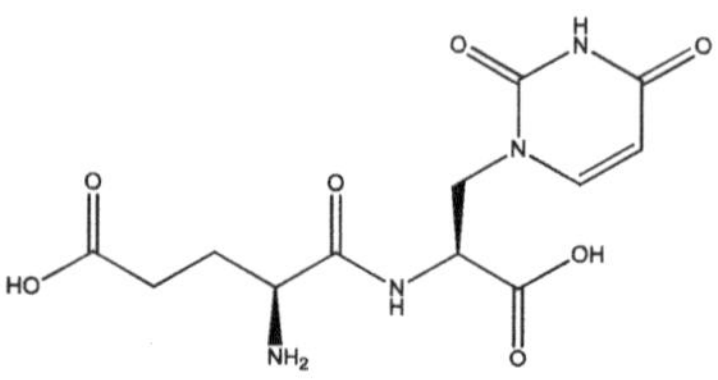

(S)-4-amino-5-(((S)-1-carboxy-2-(2,4-dioxo-3,4-dihydropyrimidin-1(2H)-yl)ethyl)amino)-5-oxopentanoic acid **(11)**

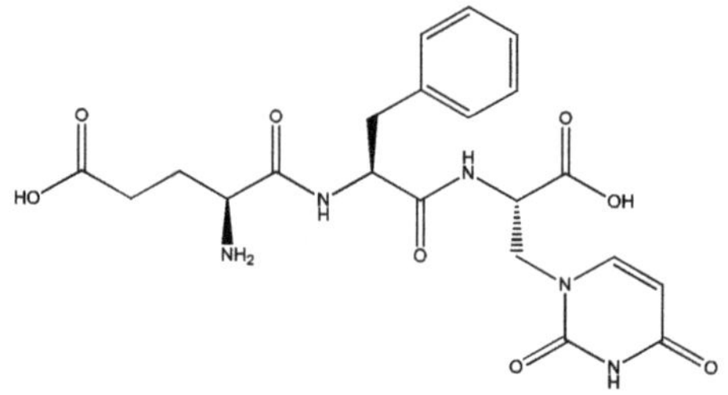

(S)-4-amino-5-(((S)-1-(((S)-1-carboxy-2-(2,4-dioxo-3,4-dihydropyrimidin-1(2H)-yl)ethyl)amino)
-1-oxo-3-phenylpropan-2-yl)amino)-5-oxopentanoic acid **(12)**

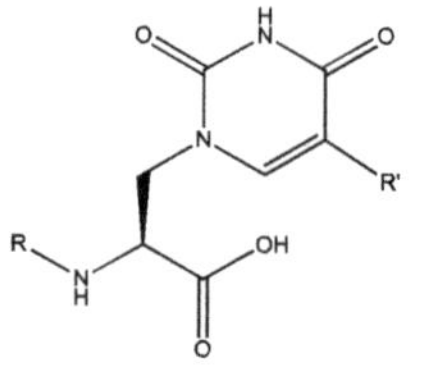

R	R'	
MMT	CH3	**(13)** MMTAlaT
Boc	H	**(14)**

H-(*Ala*A-AlaA-*Ala*T-AlaA-*Ala*T-AlaT)-Lys-NH2 **(15)**
H-(AlaA-*Ala*A-AlaT-*Ala*A-AlaT-*Ala*T)-Lys-NH2 **(16)**

Figure 3. Examples of peptides containing willardiine and the derivative **6** herein referred to as AlaT, whose structure as N-protected monomer for solid phase oligomerization is also reported (**13, 14**). A and T stand for adenine and thymine nucleobases, respectively. Italics refer to D-configuration of the amino acid residues in the sequences **15** and **16**.

Since proteins which bind nucleic acids, employing different strategies for the recognition of the structural elements in their nucleic acid targets, are able to regulate their structure and functions, their preparation with nucleoamino acid incorporations into the protein structure, aiming at reinforcing their nucleic acid-binding ability, was described in the literature [96]. The introduction of nucleoamino acid units with strategies using the Boc-protected AlaT **14** (Figure 3) into the structures of dihydrofolate reductase and the Klenow fragment of DNA polymerase I was accomplished with acceptable efficiencies [96]. In another work, nucleopeptide/nucleopeptide pairing properties and different stacking modes were studied using peptides containing AlaT moieties like those corresponding to the sequences **15** and **16** (Figure 3) [97]. Overall, the different types of nucleopeptides studied, including both L- and D-nucleoamino acid residues, proved to be valuable model systems for the investigation of processes based on nucleobase-stacking like interactions with the base stack, intercalation and electron transfer, and the findings of this study indicated that the extension of a side chain linker by a methylene group in the L-willardiine structure does affect the functional unit regarding facial orientation and also the order of donor/acceptor positions [97]. AlaT units were used also in the synthesis of short, β-turn- or helix-forming, nucleopeptides containing 2-aminoisobutyric acid (Aib). Ref. [98] More in detail, AlaT units were introduced into homo-peptide stretches. In this regard, an hexamer sequence was designed and realized in order to allow the alignment of two DNA bases on the same face of the helical structure [98]. The most interesting features of these AlaT-containing nucleopeptides emerged from this research work included a particularly rigid peptide backbone that formed helices or β-turn structures, in dependence of the main-chain length, a backbone-to-side chain H-bond limiting the nucleobase mobility and a high tendency to form dimers that were mediated by the nucleobases [98]. The synthesis of the AlaT monomer for the peptide oligomerization was accomplished starting from the L-serine β-lactone, while the peptide was synthesised in solution due to the scarce propensity of Aib to form peptide bonds, especially in the case of two or more residues of this hindered amino acid consecutively present in the peptide sequence. NMR studied performed in $CDCl_3$ on this type of nucleopeptides revealed their tendency to adopt folded conformations [98]. By introducing two AlaT nucleoamino acids and 0, 1 or 4 Aib units in alanyl-nucleoheptapeptides, and studying the corresponding conformational properties [99], it was possible to conclude that a single Aib amino acid in the sequence is sufficient to promote the adoption of a helical structure in the AlaT nucleopeptides and is enough to markedly reduce their vulnerability towards enzymatic hydrolysis as ascertained in assays conducted in murine serum [99]. On the other hand, introducing four Aib residues, out of seven residues in the nucleopeptide sequence, led to a rigid helical alanyl-nucleopeptide that was almost untouched by serum enzymes and did not show any appreciable cytotoxicity. The reported behaviour probably depends on the fact that the hydrolytic enzymes are more efficient in cleaving peptide bonds among proteinogenic amino acids, rather than amide bonds involving non-coded residues [99]. AlaT was one of the nucleobase-containing residues inserted into the peptide sequences described in a recent work aiming at establishing efficient methods for the chemical synthesis and de novo sequencing of nucleopeptides able to recognize oligonucleotide targets with high affinity [92]. Specifically, combinatorial libraries with up to 100 million biohybrid compounds were prepared and tested against RNA targets, demonstrating that the biohybrid materials bearing nucleobase moieties were endowed with higher bulk affinities for the oligonucleotide target than peptides based exclusively on canonical amino acids. By affinity selection mass spectrometry, it was also possible to discover particular nucleopeptide variants with a high affinity for the specific oligonucleotide targets, that in this studies corresponded to pre-microRNA hairpins [92].

The effect of double strand formation in alanyl-nucleopeptide/protein chimeras on the induction of a conformational switch in proteins and peptides was explored using peptides containing the guanine and cytosine analogues of **1** [100]. The importance of this study stems from the fact that the biological properties of proteins are correlated with their conformation and the three-dimensional arrangement of the secondary structure elements present in them. Moreover, protein-protein binding and conformational changes, such as those caused by molecular switches, may determine consequences in terms of biological function of the protein. Specifically, one type of molecular switches that was inserted in the peptide sequences of this work corresponded to the alanyl-nucleopeptides based on analogues of **1**, and more specifically obtained by the oligomerization of alanine units bearing the guanine and cytosine nucleobases attached to the β-position of the amino acid residues. In this context, the study of several complementary nucleopeptides with different mismatches, residue numbers, and under various buffer conditions was performed by using NMR and UV spectroscopy in order to investigate their interaction abilities and consequently the stability of their complexes [100]. Moreover, willardiine analogues were incorporated in the sequences of β-sheet-forming peptides. From a synthetic perspective, the same work aimed at developing N-methylated nucleoamino acids to be incorporated in the nucleopeptides, by using a convenient route for the synthesis of these N-methylated derivatives starting from N-methyl-serinelactone [100], while the main aim from a biomolecular and applicative viewpoint was replacing the C-terminal helix structure of humane interleukine 8 (IL-8) (residues 56–77) with a nucleopeptide moiety in order to switch from a helical-structured C terminus to another secondary structure determined by the interactions of the alanyl-nucleopeptides based on **1** analogues. However, due to the difficulties found in the incorporation of the nucleoamino acids in the peptides and proteins, probably because of the effects of the nucleobases, an enzymatic ligation was ultimately chosen for the alanyl-nucleopeptide/protein bioconjugation [100]. As for the stability of the nucleopeptide/nucleopeptide duplexes, it was evaluated by varying the number of residues, as well as the sequences of the nucleopeptides. Moreover, AlaT was introduced with the aim of avoiding aggregation problems, together with L- lysine and L-glutamine as mismatch amino acids, that are notoriously able to increase the solubility of the modified nucleopeptides. Subsequently, structural insights on the nucleopeptide constructs, as well as the exploration of their conformational changes were achieved by temperature-dependent NMR and UV spectroscopy-based experiments, finding that a single mismatch was enough to improve or decrease the self-pairing tendency. As expected, a higher aggregation propensity was revealed in pure G/C-nucleopeptides than in the modified oligomers as shown by NMR spectroscopy. In addition, single or multiple insertions of N-methylated nucleoamino acid residues in the center of the alanyl-nucleopeptide sequences led to structures with lower aggregation tendencies [100], while charged N-alkyl chains were introduced in the complementary nucleopeptides to enable NMR-spectroscopy monitoring, blocking the aggregation of the molecules [100]. In order to investigate the effects of nucleopeptide oligomers as molecular switches and in consideration of the fact that two complementary nucleopeptide sequences are able to form a stable double strand with linear topology endowed with a β-strand conformation thanks to the specific complementary DNA base recognition [100], an alanyl-nucleopeptide segment with six residues was incorporated in both a peptide and a protein in order to stabilize a β-sheet conformation in the resulting conjugates. Once inserted in the peptide and protein structures the alanyl-nucleopeptides led to the formation of a double helix that has proven to be a switchable stabilizing or destabilizing element of the resulting peptide or protein structure. Interestingly, when incorporated in β-sheet forming peptides, as well as in proteins, nucleopeptides led to a portion of a β-sheet, as observed in the case of nucleopeptide double strands, or a disturbed β-sheet, as revealed in the case of nucleopeptide single strand [100]. An N-acetylated peptide/nucleopeptide conjugate was firstly synthesised and used in binding experiments with a complementary nucleopeptide, but the expected pairing was not revealed probably because it was prevented by both sterical and electronical factors that involved both the nucleoamino acids and the amino acid residues present in the peptide/nucleopeptide conjugate. With this in mind, a shorter and more flexible peptide that carried a less rigid β-turn element was conjugated to the alanyl-nucleopeptide, and this made it possible to observe a clear nucleopeptide/nucleopeptide pairing in the binding experiment involving the new conjugate and the complementary nucleopeptide as hypothesised [100]. Furthermore, the substitution of the the C-terminal helix part of IL-8 with a peptide sequence that contained an alanyl-nucleopeptide moiety was achieved with the aim to switch from a structured C terminus to another stable secondary structure determined by the nucleopeptide molecular interactions. More in detail, the fragment of human IL-8 (residues 1–55) was expressed in *E. coli* by the 'Intein Mediated Purification with an Affinity Chitin-binding Tag' (IMPACT) purification system. Subsequently, a first approach using a native chemical ligation was employed leading to a protein/nucleopeptide conjugate that allowed to switch the protein conformation and to modify the protein binding properties by the nucleobase interactions provided by the alanyl-nucleopeptide moiety [100]. A second approach used an enzymatic ligation for the protein/nucleopeptide conjugation. To this scope, it was necessary to synthesize alanyl-nucleopeptides with different number of residues and N-terminal nucleoamino acid units [100]. The ligation step was perfomed on different sequences using the protease clostripain (from *Clostridium histolyticum*) as enzyme. Overall, the enzymatic ligation proved to be successful in the incorporation of the alanyl-nucleopeptides into the desired protein or peptide [100].

To understand the complex relationship between the various interactions involving the nucleic acid pairing nucleobases on which the functions of DNA and RNA mainly depend, model systems are needed in which the interstrand pairing is less restricted for effect of the backbone than in natural oligonucleotides. Examples of systems of this type were offered by alanyl-nucleopeptides whose peptide backbone replaced the sugar phosphate repeats of RNA or DNA. In this context, the group led by prof. Diederichsen devoted enormous efforts to the synthesis of a large number of such model systems based on an alanine-containing backbone to which both canonical and non-canonical bases were anchored and whose monomers, thus, can be seen as willardiine analogues [101]. These nucleopeptide systems were found to aggregate into structures that showed several binding motifs never observed in the case of RNA or DNA. In particular, Diederichsen et al. aimed at studying these uncommon binding motifs to get an insight into the interplay of the interactions between the DNA bases. However, the solubility of the alanyl-nucleopeptide sequences was often too low making it difficult to experimentally elucidate the geometrical arrangements with techniques like NMR or X-ray, and on the other hand UV spectroscopy could provide information solely on the overall stability of the various aggregates through measurements of melting temperatures [101]. Therefore, in the attempt to corroborate the knowledge about the geometrical structure and the bonding motifs of nucleobases and to study the effects which governed the trends in the stabilities of the alanyl-nucleopeptide systems, new insights on the interplay of their various interactions were obtained by theoretical approaches [101]. Specifically, as a simplification this was achieved without evaluating the absolute stabilities, even because many contributions such as dynamic and entropic effects were very similar for close systems and thus, were considered less important in the described approach [101]. Since the noncanonical nucleobases were not implemented in the standard version of the Amber4.1 force field, for the investigation of all experimentally tested willardiine analogues-based nucleopeptides, it was essential to parameterize them, which was performed by adding the missing parameters to the Amber Force Field. The construction of all possible pairing modes for the nucleopeptide model dimer was accomplished at the beginning of the theoretical investigation. However, not all the pairing modes were realizable because of the restriction of the nucleopeptide backbone, as well as the geometrical arrangement of the dimer. For some pairing modes it was possible to obtain a construction, but the geometrical restrictions of the nucleopeptide backbone led to a particularly high strain in the system that made them fall apart in a first geometry optimization step. On the other hand, molecular dynamics (MD) simulations were run on those systems that revealed to be stable [101]. By starting several geometry optimizations at different points of the MD simulation it was possible to get useful information about their geometrical arrangements at zero temperature (T = 0 K) and, in particular, the obtained geometries resulted identical. Moreover, by using a two-steps method in which the influence of the backbone and the effects linked to the DNA bases were evaluated separately, it was possible to describe the interactions that took place within a nucleopeptide dimer at T = 0 K [101]. Overall, the described theoretical study led to an efficient protocol that worked for the description of the nucleopeptides based on willardiine analogues, especially in consideration of the good correlation found between the computed stabilization energies and the measured melting temperature (T_m) values. In particular, a very good correlation between the experimental T_ms and the calculated stabilization energies was observed for the respective most stable pairing modes of the dimer formed by the various nucleopeptide oligomers. This suggested that the T = 0 K model depicts accurately the main effects which led to the trends observed in the investigated systems and, thus, that the enthalpic and entropic effects were not important for the relative stabilities of the alanyl-nucleopeptide dimer but only in the case of their absolute stabilities [101].

3.1. Charge Transfer and Transport in Nucleopeptides

The research field that studies the charge transfer and transport in nucleic acids occupies an important role in the design of DNA-based devices for molecular electronics, as well as in the investigation of the oxidative damage to double helical oligonucleotides [102]. Interestingly, a study [103] was conducted about the femtosecond time resolved electron transfer dynamics of double-stranded alanyl-nucleopeptides in which both peptide strands contained willardiine analogues and hosted a molecule of 9-amino-6-chloro-2-methoxy-acridine in its protonated state intercalated in the interior of the double strand. In this work several alanyl-nucleopeptides were designed and synthesised in order to realize rigid and linear pairing complexes [103]. From a structural viewpoint, the main results of this study were in agreement with the notion of alanyl-nucleopeptide stacking distances in the base staple similar to the one in the B-DNA model, thus leading to structural evidences for base stacking in double-stranded alanyl-nucleopeptides [103].

3.2. Prebiotic Role of Nucleopeptides

The scientific research on the origins of life is a very attractive field and in this context a great relevance is given to prebiotic chemistry according to which some of the chemical precursors of proteins and nucleic acids might be produced under prebiotic conditions [104,105]. Remarkably, nucleopeptides were proposed as potential primordial nucleic material in a scenario preceding the so-called 'RNA world', due to their characteristics of self-replicating molecules. On the other hand, the interaction between RNAs and nucleopeptides could have played a fundamental role in the transition from the hypothesised nucleopeptide-driven world to the current DNA, RNA and protein-based genetic system [106]. In this regard, chimeric nucleobase-peptide derivatives were used as models of nucleopeptides to be studied to prove their ability to replicate non-enzymatically, which is a pivotal mechanism on which chemical evolution is based [106]. More in particular, it was shown that the replication of complementary nucleobase-bearing peptides followed various mechanisms with a net selection of a single structure over the several other possible ones sustaining, thus, the hypothesis that similar processes might be at the origin of the first functional nucleopeptide assemblies, which subsequently determined the occurrence of biological structures like ribosomes [106]. Moreover, the processes of selection and self-organization were demonstrated to occur in mixtures of complementary nucleopeptides consisting of a limited number of residues, and specifically in nucleopeptides build-up of eight amino acids conjugated at their N-termini to thymine and adenine bases through carboxymethylene bridges. Both theoretical simulations and experimental studies were conducted on these molecules, with their main results suggesting that the template-directed replication processes of cross-catalytic and autocatalytic type took place within these networks leading to the formation of the products [106].

Alanyl-nucleopeptides based on willardiine might have been involved in a prebiotic scenario. In fact, this nucleoamino acid is not only a metabolic product of plants but also a potential prebiotic product. In particular the prebiotic synthesis of willardiine was demonstrated to occur via pyridoxal-5′-phosphate (PLP)-catalyzed derivatization of O-acyl L-serine or L-serine with the appropriate base that in case of willardiine is uracil [107]. Notwithstanding the hypothesised ability of willardiine-based nucleopeptides with an appropriate arrangement of the bases to recognize a complementary sequence of the natural nucleic acid, which is considered a requirement in the transition from the prebiotic nucleopeptide world to the current based on DNA, RNA and proteins, the nucleopeptides synthesized initially to this scope failed to recognise effectively their complementary oligonucleotide strands, probably due to the use of racemic monomers. Nonetheless, several synthetic efforts including those aimed at controlling the chirality of the backbone led to nucleopeptides that could bind the complementary natural nucleic strands [107].

4. Conclusions

Following the enormous efforts devoted to modifying the structure of uracylalanines to obtain new glutamate receptor binders, a large number of willardiine analogues is now available. Some of them act as agonists of ionotropic glutamate receptors agonizing especially the non-NMDA glutamate receptors. They showed different binding affinities for the non-NMDA receptors and had a key role in the structural studies of AMPA and kainate receptors, but also in the investigation of the specific effects caused by the receptor activation in the different brain regions. Many neurological disorders are determined by alterations in glutamate signaling, and thus, ligands for AMPA and kainate receptors like the willardiine analogues deserve attention as potential therapeutics. Some halogen-based willardiine analogues show binding affinities and stability significantly higher than the natural willardiine, as emerged in studies with the AMPA receptor. On the other side, synthetic carboxylbenzyl-substituted willardiine analogues have proved to be AMPA and kainate glutamate receptor antagonists, with therapeutic potential for a variety of neurological diseases caused by the aberrant activation of kainate or AMPA receptors.

Not less importantly, the willardiine and its thymine-based analogue AlaT have a key role in peptide chemistry. In fact, besides the natural γ-glutamylwillardiine and γ-glutamylphenylalanylwillardiine, isolated from vegetal sources, there are different examples in which the nucleoamino acid and its analogue AlaT were employed for nucleopeptide development. The applications are various including nucleopeptide/DNA chimera fabrication for DNA diagnostics, nucleoamino acid-derivatization of proteins for facilitating protein-nucleic acid interaction, and nucleopeptide-nucleopeptide interaction as model systems in various structural studies or for nanotechnological strategies. All the above aspects on chemistry and biotechnological applications of both willardine analogues and the corresponding nucleopeptides encourage clearly future efforts for the development of novel willardiine-inspired drugs able to interact with receptor family members in CNS function with greater selectivity for individual receptor subunits and as potentially novel therapeutics.

Author Contributions: All authors have contributed equally to this work. All authors have read and agreed to the published version of the manuscript.

Funding: This research received no external funding.

Institutional Review Board Statement: Not applicable.

Informed Consent Statement: Not applicable.

Data Availability Statement: Data sharing not applicable.

Acknowledgments: The authors thank Antonella Gargiulo (IBB-CNR) for helping with literature checking.

Conflicts of Interest: The authors declare no conflict of interest.

References

1. Gmelin, R. Die freien Aminosäuren der Samen von Acacia Willardiana (Mimosaceae). Isolierung von Willardiin, Einer Neuen Pflanzlichen Aminosäure, Vermutlich L-Uracil-[β-(α-amino-propionsäure)]-(3). *Hoppe-Seyler's Zeitschrift für Physiol. Chem.* **1959**, *316*, 164–169. [CrossRef] [PubMed]
2. Lorenz, K.B.; Diederichsen, U. Nucleo amino acids as arginine mimetics in cyclic peptides. *Lett. Pept. Sci.* **2003**, *10*, 111–117. [CrossRef]
3. Cheikh, A.B.; Orgel, L.E. Polymerization of amino acids containing nucleotide bases. *J. Mol. Evol.* **1990**, *30*, 315–321. [CrossRef]
4. Musumeci, D.; Ullah, S.; Ikram, A.; Roviello, G.N. Novel insights on nucleopeptide binding: A spectroscopic and In Silico investigation on the interaction of a thymine-bearing tetrapeptide with a homoadenine DNA. *J. Mol. Liq.* **2022**, *347*, 117975. [CrossRef]
5. Negi, V.S.; Pal, A.; Borthakur, D. Biochemistry of plants N–heterocyclic non-protein amino acids. *Amino Acids* **2021**, *53*, 801–812. [CrossRef] [PubMed]
6. Stensbol, T.; Madsen, U.; Krogsgaard-Larsen, P. The AMPA receptor binding site: Focus on agonists and competitive antagonists. *Curr. Pharm. Des.* **2002**, *8*, 857–872. [CrossRef]
7. Kew, J.N.; Kemp, J.A. Ionotropic and metabotropic glutamate receptor structure and pharmacology. *Psychopharmacology* **2005**, *179*, 4–29. [CrossRef]
8. Yelshanskaya, M.; Sobolevsky, A. Structural Insights into Function of Ionotropic Glutamate Receptors. *Biochem. Suppl. Ser. A Membr. Cell Biol.* **2022**, *16*, 190–206. [CrossRef]
9. Bowie, D. *The Many Faces of the AMPA-Type Ionotropic Glutamate Receptor*; Elsevier: Amsterdam, The Netherlands, 2022; p. 108975.
10. Pinzón-Parra, C.A.; Coatl-Cuaya, H.; Díaz, A.; Guevara, J.; Rodríguez-Moreno, A.; Flores, G. Long-term effect of neonatal antagonism of ionotropic glutamate receptors on dendritic spines and cognitive function in rats. *J. Chem. Neuroanat.* **2022**, *119*, 102054. [CrossRef]

11. Hu, T.-M.; Wu, C.-L.; Hsu, S.-H.; Tsai, H.-Y.; Cheng, F.-Y.; Cheng, M.-C. Ultrarare Loss-of-Function Mutations in the Genes Encoding the Ionotropic Glutamate Receptors of Kainate Subtypes Associated with Schizophrenia Disrupt the Interaction with PSD95. *J. Pers. Med.* **2022**, *12*, 783. [CrossRef]
12. Bowie, D. Ionotropic glutamate receptors & CNS disorders. *CNS Neurol. Disord. Drug Targets* **2008**, *7*, 129–143.
13. Negrete-Díaz, J.V.; Falcón-Moya, R.; Rodríguez-Moreno, A. Kainate receptors: From synaptic activity to disease. *FEBS J.* **2021**, *289*, 5074–5088. [CrossRef]
14. Postila, P.A.; Ylilauri, M.; Pentikäinen, O.T. Full and partial agonism of ionotropic glutamate receptors indicated by molecular dynamics simulations. *J. Chem. Inf. Model.* **2011**, *51*, 1037–1047. [CrossRef]
15. Henley, J.M.; Wilkinson, K.A. Synaptic AMPA receptor composition in development, plasticity and disease. *Nat. Rev. Neurosci.* **2016**, *17*, 337–350. [CrossRef]
16. Sihra, T.S.; Rodríguez-Moreno, A. Presynaptic kainate receptor-mediated bidirectional modulatory actions: Mechanisms. *Neurochem. Int.* **2013**, *62*, 982–987. [CrossRef]
17. Rodríguez-Moreno, A.; Sihra, T.S. Metabotropic actions of kainate receptors in the control of glutamate release in the hippocampus. *Adv. Exp. Med. Biol.* **2011**, *717*, 39–48. [CrossRef]
18. Sihra, T.S.; Rodríguez-Moreno, A. Metabotropic actions of kainate receptors in the control of GABA release. *Adv. Exp. Med. Biol.* **2011**, *717*, 1–10. [CrossRef]
19. Neto, J.X.L.; Fulco, U.L.; Albuquerque, E.L.; Corso, G.; Bezerra, E.M.; Caetano, E.W.; Da Costa, R.F.; Freire, V.N. A quantum biochemistry investigation of willardiine partial agonism in AMPA receptors. *Phys. Chem. Chem. Phys.* **2015**, *17*, 13092–13103. [CrossRef]
20. Pasternack, A.; Coleman, S.K.; Jouppila, A.; Mottershead, D.G.; Lindfors, M.; Pasternack, M.; Keinänen, K. α-Amino-3-hydroxy-5-methyl-4-isoxazolepropionic acid (AMPA) receptor channels lacking the N-terminal domain. *J. Biol. Chem.* **2002**, *277*, 49662–49667. [CrossRef]
21. Kato, A.S.; Gill, M.B.; Yu, H.; Nisenbaum, E.S.; Bredt, D.S. TARPs differentially decorate AMPA receptors to specify neuropharmacology. *Trends Neurosci.* **2010**, *33*, 241–248. [CrossRef]
22. Traynelis, S.F.; Wollmuth, L.P.; McBain, C.J.; Menniti, F.S.; Vance, K.M.; Ogden, K.K.; Hansen, K.B.; Yuan, H.; Myers, S.J.; Dingledine, R. Glutamate receptor ion channels: Structure, regulation, and function. *Pharmacol. Rev.* **2010**, *62*, 405–496. [CrossRef]
23. More, J.C.; Troop, H.M.; Dolman, N.P.; Jane, D.E. Structural requirements for novel willardiine derivatives acting as AMPA and kainate receptor antagonists. *Br. J. Pharmacol.* **2003**, *138*, 1093–1100. [CrossRef]
24. Immel, J.R.; Bloom, S. carba-Nucleopeptides (cNPs): A Biopharmaceutical Modality Formed through Aqueous Rhodamine B Photoredox Catalysis. *Angew. Chem. Int. Ed.* **2022**, *61*, e202205606. [CrossRef]
25. Giraud, T.; Hoschtettler, P.; Pickaert, G.; Averlant-Petit, M.-C.; Stefan, L. Emerging low-molecular weight nucleopeptide-based hydrogels: State of the art, applications, challenges and perspectives. *Nanoscale* **2022**, *14*, 4908–4921. [CrossRef]
26. Huganir, R.L.; Nicoll, R.A. AMPARs and synaptic plasticity: The last 25 years. *Neuron* **2013**, *80*, 704–717. [CrossRef]
27. Kessels, H.W.; Malinow, R. Synaptic AMPA receptor plasticity and behavior. *Neuron* **2009**, *61*, 340–350. [CrossRef]
28. Cheng, G.-R.; Li, X.-Y.; Xiang, Y.-D.; Liu, D.; McClintock, S.M.; Zeng, Y. The implication of AMPA receptor in synaptic plasticity impairment and intellectual disability in fragile X syndrome. *Physiol. Res.* **2017**, *66*, 715–727. [CrossRef]
29. Lee, K.; Goodman, L.; Fourie, C.; Schenk, S.; Leitch, B.; Montgomery, J.M. AMPA receptors as therapeutic targets for neurological disorders. *Adv. Protein Chem. Struct. Biol.* **2016**, *103*, 203–261.
30. Lees, A.; Fahn, S.; Eggert, K.M.; Jankovic, J.; Lang, A.; Micheli, F.; Maral Mouradian, M.; Oertel, W.H.; Olanow, C.W.; Poewe, W. Perampanel, an AMPA antagonist, found to have no benefit in reducing "off" time in Parkinson's disease. *Mov. Disord.* **2012**, *27*, 284–288. [CrossRef]
31. Walker, F.O. Huntington's disease. *Lancet* **2007**, *369*, 218–228. [CrossRef]
32. Salpietro, V.; Dixon, C.L.; Guo, H.; Bello, O.D.; Vandrovcova, J.; Efthymiou, S.; Maroofian, R.; Heimer, G.; Burglen, L.; Valence, S. AMPA receptor GluA2 subunit defects are a cause of neurodevelopmental disorders. *Nat. Commun.* **2019**, *10*, 3094. [CrossRef] [PubMed]
33. Kim, J.-W.; Park, K.; Kang, R.J.; Gonzales, E.L.T.; Kim, D.G.; Oh, H.A.; Seung, H.; Ko, M.J.; Kwon, K.J.; Kim, K.C. Pharmacological modulation of AMPA receptor rescues social impairments in animal models of autism. *Neuropsychopharmacology* **2019**, *44*, 314–323. [CrossRef] [PubMed]
34. Jaso, B.A.; Niciu, M.J.; Iadarola, N.D.; Lally, N.; Richards, E.M.; Park, M.; Ballard, E.D.; Nugent, A.C.; Machado-Vieira, R.; Zarate, C.A. Therapeutic modulation of glutamate receptors in major depressive disorder. *Curr. Neuropharmacol.* **2017**, *15*, 57–70. [CrossRef] [PubMed]
35. Shepherd, J.D. Memory, plasticity and sleep-A role for calcium permeable AMPA receptors? *Front. Mol. Neurosci.* **2012**, *5*, 49. [CrossRef] [PubMed]
36. Shaltiel, G.; Maeng, S.; Malkesman, O.; Pearson, B.; Schloesser, R.; Tragon, T.; Rogawski, M.; Gasior, M.; Luckenbaugh, D.; Chen, G. Evidence for the involvement of the kainate receptor subunit GluR6 (GRIK2) in mediating behavioral displays related to behavioral symptoms of mania. *Mol. Psychiatry* **2008**, *13*, 858–872. [CrossRef]
37. Schiffer, H.; Heinemann, S. Association of the human kainate receptor GluR7 gene (GRIK3) with recurrent major depressive disorder. *Am. J. Med. Genet. Part B Neuropsychiatr. Genet.* **2007**, *144*, 20–26. [CrossRef]

38. Nair, J.D.; Wilkinson, K.A.; Henley, J.M.; Mellor, J.R. Kainate receptors and synaptic plasticity. *Neuropharmacology* **2021**, *196*, 108540. [CrossRef]
39. Li, H.; Li, J.; Guan, Y.; Wang, Y. The emerging role of kainate receptor functional dysregulation in pain. *Mol. Pain* **2021**, *17*, 1744806921990944. [CrossRef]
40. Henley, J.M.; Nair, J.D.; Seager, R.; Yucel, B.P.; Woodhall, G.; Henley, B.S.; Talandyte, K.; Needs, H.I.; Wilkinson, K.A. Kainate and AMPA receptors in epilepsy: Cell biology, signalling pathways and possible crosstalk. *Neuropharmacology* **2021**, *195*, 108569. [CrossRef]
41. Crépel, V.; Mulle, C. Physiopathology of kainate receptors in epilepsy. *Curr. Opin. Pharmacol.* **2015**, *20*, 83–88. [CrossRef]
42. Patneau, D.K.; Mayer, M.L.; Jane, D.E.; Watkins, J.C. Activation and desensitization of AMPA/kainate receptors by novel derivatives of willardiine. *J. Neurosci.* **1992**, *12*, 595–606. [CrossRef] [PubMed]
43. Hawkins, L.; Beaver, K.; Jane, D.; Taylor, P.; Sunter, D.; Roberts, P. Characterization of the pharmacology and regional distribution of (S)-[3H]-5-fluorowillardiine binding in rat brain. *Br. J. Pharmacol.* **1995**, *116*, 2033–2039. [CrossRef] [PubMed]
44. Lunn, M.; Ganakas, A.; Mercer, L.; Lawrence, A.; Beart, P. Localisation and properties of AMPA-insensitive kainate sites: Receptor autoradiography and gene expression in rat brain. *Neurosci. Lett.* **1996**, *204*, 121–124. [CrossRef]
45. Larm, J.A.; Cheung, N.S.; Beart, P.M. (S)-5-fluorowillardiine-mediated neurotoxicity in cultured murine cortical neurones occurs via AMPA and kainate receptors. *Eur. J. Pharmacol.* **1996**, *314*, 249–254. [CrossRef]
46. JENSEN, R.J. Responses of directionally selective retinal ganglion cells to activation of AMPA glutamate receptors. *Vis. Neurosci.* **1999**, *16*, 205–219. [CrossRef]
47. Olivera, S.; Rodriguez-Ithurralde, D.; Henley, J.M. Regional localization and developmental profile of acetylcholinesterase-evoked increases in [3H]-5-fluorowillardiine binding to AMPA receptors in rat brain. *Br. J. Pharmacol.* **2001**, *133*, 1055–1062. [CrossRef]
48. Kessler, M.; Arai, A.C. Use of [3H] fluorowillardiine to study properties of AMPA receptor allosteric modulators. *Brain Res.* **2006**, *1076*, 25–41. [CrossRef]
49. Dolman, N.P.; Troop, H.M.; More, J.C.; Alt, A.; Knauss, J.L.; Nistico, R.; Jack, S.; Morley, R.M.; Bortolotto, Z.A.; Roberts, P.J. Synthesis and pharmacology of willardiine derivatives acting as antagonists of kainate receptors. *J. Med. Chem.* **2005**, *48*, 7867–7881. [CrossRef]
50. Martinez, M.; Ahmed, A.H.; Loh, A.P.; Oswald, R.E. Thermodynamics and Mechanism of the Interaction of Willardiine Partial Agonists with a Glutamate Receptor: Implications for Drug Development. *Biochemistry* **2014**, *53*, 3790–3795. [CrossRef]
51. Ahmed, A.H.; Thompson, M.D.; Fenwick, M.K.; Romero, B.; Loh, A.P.; Jane, D.E.; Sondermann, H.; Oswald, R.E. Mechanisms of antagonism of the GluR2 AMPA receptor: Structure and dynamics of the complex of two willardiine antagonists with the glutamate binding domain. *Biochemistry* **2009**, *48*, 3894–3903. [CrossRef] [PubMed]
52. GN Roviello, G Oliviero, A Di Napoli, N Borbone Synthesis, self-assembly-behavior and biomolecular recognition properties of thyminyl dipeptides. *Arab. J. Chem.* **2020**, *13*, 1966–1974. [CrossRef]
53. Roviello, G.N.; Vicidomini, C.; Costanzo, V.; Roviello, V. Nucleic acid binding and other biomedical properties of artificial oligolysines. *Int. J. Nanomed.* **2016**, *11*, 5897–5904. [CrossRef]
54. Roviello, G.N. Novel insights into nucleoamino acids: Biomolecular recognition and aggregation studies of a thymine-conjugated l-phenyl alanine. *Amino Acids* **2018**, *50*, 933–941. [CrossRef]
55. Musumeci, D.; Mokhir, A.; Roviello, G.N. Synthesis and nucleic acid binding evaluation of a thyminyl L-diaminobutanoic acid-based nucleopeptide. *Bioorg. Chem.* **2020**, *100*, 103862. [CrossRef]
56. Roviello, G.N.; Benedetti, E.; Pedone, C.; Bucci, E.M. Nucleobase-containing peptides: An overview of their characteristic features and applications. *Amino Acids* **2010**, *39*, 45–57. [CrossRef] [PubMed]
57. Roviello, G.N.; Musumeci, D. Synthetic approaches to nucleopeptides containing all four nucleobases, and nucleic acid-binding studies on a mixed-sequence nucleo-oligolysine. *RSC Adv.* **2016**, *6*, 63578–63585. [CrossRef]
58. Roviello, G.N.; Musumeci, D.; Moccia, M.; Castiglione, M.; Sapio, R.; Valente, M.; Bucci, E.M.; Perretta, G.; Pedone, C. dabPNA: Design, synthesis, and DNA binding studies. *Nucleosides Nucleotides Nucleic Acids* **2007**, *26*, 1307–1310. [CrossRef]
59. Terracciano, M.; De Stefano, L.; Borbone, N.; Politi, J.; Oliviero, G.; Nici, F.; Casalino, M.; Piccialli, G.; Dardano, P.; Varra, M.; et al. Solid phase synthesis of a thrombin binding aptamer on macroporous silica for label free optical quantification of thrombin. *RSC Adv.* **2016**, *6*, 86762–86769. [CrossRef]
60. Oliviero, G.; Amato, J.; Borbone, N.; D'Errico, S.; Piccialli, G.; Bucci, E.; Piccialli, V.; Mayol, L. Synthesis of 4-N-alkyl and ribose-modified AICAR analogues on solid support. *Tetrahedron* **2008**, *64*, 6475–6481. [CrossRef]
61. Oliviero, G.; Amato, J.; Borbone, N.; D'Errico, S.; Piccialli, G.; Mayol, L. Synthesis of N-1 and ribose modified inosine analogues on solid support. *Tetrahedron Lett.* **2007**, *48*, 397–400. [CrossRef]
62. Oliviero, G.; D'Errico, S.; Borbone, N.; Amato, J.; Piccialli, V.; Varra, M.; Piccialli, G.; Mayol, L. A solid-phase approach to the synthesis of N-1-alkyl analogues of cyclic inosine-diphosphate-ribose (cIDPR). *Tetrahedron* **2010**, *66*, 1931–1936. [CrossRef]
63. D'Errico, S.; Oliviero, G.; Amato, J.; Borbone, N.; Cerullo, V.; Hemminki, A.; Piccialli, V.; Zaccaria, S.; Mayol, L.; Piccialli, G. Synthesis and biological evaluation of unprecedented ring-expanded nucleosides (RENs) containing the imidazo[4,5-d][1,2,6]oxadiazepine ring system. *Chem. Commun.* **2012**, *48*, 9310. [CrossRef]
64. Oliviero, G.; D'Errico, S.; Borbone, N.; Amato, J.; Piccialli, V.; Piccialli, G.; Mayol, L. Facile Solid-Phase Synthesis of AICAR 5'-Monophosphate (ZMP) and Its 4-N-Alkyl Derivatives. *Eur. J. Org. Chem.* **2010**, *2010*, 1517–1524. [CrossRef]

65. D'Errico, S.; Oliviero, G.; Borbone, N.; Amato, J.; D'Alonzo, D.; Piccialli, V.; Mayol, L.; Piccialli, G. A Facile Synthesis of 5′-Fluoro-5′-deoxyacadesine (5′-F-AICAR): A Novel Non-phosphorylable AICAR Analogue. *Molecules* **2012**, *17*, 13036–13044. [CrossRef] [PubMed]
66. Galeone, A.; Mayol, L.; Oliviero, G.; Piccialli, G.; Varra, M. Synthesis of a novel N-1 carbocyclic, N-9 butyl analogue of cyclic ADP ribose (cADPR). *Tetrahedron* **2002**, *58*, 363–368. [CrossRef]
67. Roviello, G.N.; Gaetano, S.D.; Capasso, D.; Cesarani, A.; Bucci, E.M.; Pedone, C. Synthesis, spectroscopic studies and biological activity of a novel nucleopeptide with Moloney murine leukemia virus reverse transcriptase inhibitory activity. *Amino Acids* **2009**, *38*, 1489–1496. [CrossRef] [PubMed]
68. Roviello, G.N.; Di Gaetano, S.; Capasso, D.; Franco, S.; Crescenzo, C.; Bucci, E.M.; Pedone, C. RNA-Binding and Viral Reverse Transcriptase Inhibitory Activity of a Novel Cationic Diamino Acid-Based Peptide. *J. Med. Chem.* **2011**, *54*, 2095–2101. [CrossRef] [PubMed]
69. Seley-Radtke, K.L.; Yates, M.K. The evolution of nucleoside analogue antivirals: A review for chemists and non-chemists. Part 1: Early structural modifications to the nucleoside scaffold. *Antivir. Res.* **2018**, *154*, 66–86. [CrossRef] [PubMed]
70. Watanabe, S.; Tomizaki, K.-y.; Takahashi, T.; Usui, K.; Kajikawa, K.; Mihara, H. Interactions between peptides containing nucleobase amino acids and T7 phages displayingS. cerevisiae proteins. *Biopolymers* **2007**, *88*, 131–140. [CrossRef]
71. Uozumi, R.; Takahashi, T.; Yamazaki, T.; Granholm, V.; Mihara, H. Design and conformational analysis of natively folded β-hairpin peptides stabilized by nucleobase interactions. *Biopolymers* **2010**, *94*, 830–842. [CrossRef] [PubMed]
72. Roviello, V.; Musumeci, D.; Mokhir, A.; Roviello, G.N. Evidence of protein binding by a nucleopeptide based on a thymine-decorated L-diaminopropanoic acid through CD and in silico studies. *Curr. Med. Chem.* **2021**, *28*, 5004–5015. [CrossRef]
73. Musumeci, D.; Roviello, V.; Roviello, G.N. DNA- and RNA-binding ability of oligoDapT, a nucleobase-decorated peptide, for biomedical applications. *Int. J. Nanomed.* **2018**, *13*, 2613–2629. [CrossRef]
74. Roviello, G.N.; Vicidomini, C.; Di Gaetano, S.; Capasso, D.; Musumeci, D.; Roviello, V. Solid phase synthesis and RNA-binding activity of an arginine-containing nucleopeptide. *RSC Adv.* **2016**, *6*, 14140–14148. [CrossRef]
75. Geotti-Bianchini, P.; Beyrath, J.; Chaloin, O.; Formaggio, F.; Bianco, A. Design and synthesis of intrinsically cell-penetrating nucleopeptides. *Org. Biomol. Chem.* **2008**, *6*, 3661. [CrossRef]
76. Roviello, G.; Musumeci, D.; Castiglione, M.; Bucci, E.M.; Pedone, C.; Benedetti, E. Solid phase synthesis and RNA-binding studies of a serum-resistant nucleo-epsilon-peptide. *J. Pept. Sci.* **2009**, *15*, 155–160. [CrossRef]
77. De Napoli, L.; Messere, A.; Montesarchio, D.; Piccialli, G.; Benedetti, E.; Bucci, E.; Rossi, F. A new solid-phase synthesis of oligonucleotides 3′-conjugated with peptides. *Bioorg. Med. Chem.* **1999**, *7*, 395–400. [CrossRef]
78. Roviello, G.N.; Roviello, V.; Musumeci, D.; Bucci, E.M.; Pedone, C. Dakin–West reaction on 1-thyminyl acetic acid for the synthesis of 1,3-bis(1-thyminyl)-2-propanone, a heteroaromatic compound with nucleopeptide-binding properties. *Amino Acids* **2012**, *43*, 1615–1623. [CrossRef] [PubMed]
79. Bucci, R.; Bossi, A.; Erba, E.; Vaghi, F.; Saha, A.; Yuran, S.; Maggioni, D.; Gelmi, M.L.; Reches, M.; Pellegrino, S. Nucleobase morpholino β amino acids as molecular chimeras for the preparation of photoluminescent materials from ribonucleosides. *Sci. Rep.* **2020**, *10*, 19331. [CrossRef]
80. Wang, H.; Feng, Z.; Qin, Y.; Wang, J.; Xu, B. Nucleopeptide Assemblies Selectively Sequester ATP in Cancer Cells to Increase the Efficacy of Doxorubicin. *Angew. Chem. Int. Ed. Engl.* **2018**, *57*, 4931–4935. [CrossRef]
81. Höger, G.A.; Wiegand, M.; Worbs, B.; Diederichsen, U. Membrane-Associated Nucleobase-Functionalized β-Peptides (β-PNAs) Affecting Membrane Support and Lipid Composition. *ChemBioChem* **2020**, *21*, 2599–2603. [CrossRef]
82. Dolman, N.P.; More, J.C.; Alt, A.; Knauss, J.L.; Troop, H.M.; Bleakman, D.; Collingridge, G.L.; Jane, D.E. Structure-activity relationship studies on N3-substituted willardiine derivatives acting as AMPA or kainate receptor antagonists. *J. Med. Chem.* **2006**, *49*, 2579–2592. [CrossRef] [PubMed]
83. Mik, V.; Mickova, Z.; Dolezal, K.; Frebort, I.; Pospisil, T. Activity of (+)-Discadenine as a Plant Cytokinin. *J. Nat. Prod.* **2017**, *80*, 2136–2140. [CrossRef] [PubMed]
84. Rozan, P.; Kuo, Y.H.; Lambein, F. Amino acids in seeds and seedlings of the genus Lens. *Phytochemistry* **2001**, *58*, 281–289. [CrossRef]
85. Ignatowska, J.; Mironiuk-Puchalska, E.; Grześkowiak, P.; Wińska, P.; Wielechowska, M.; Bretner, M.; Karatsai, O.; Jolanta Rędowicz, M.; Koszytkowska-Stawińska, M. New insight into nucleo α-amino acids—Synthesis and SAR studies on cytotoxic activity of β-pyrimidine alanines. *Bioorg. Chem.* **2020**, *100*, 103864. [CrossRef]
86. Walsh, C.T.; Zhang, W. Chemical Logic and Enzymatic Machinery for Biological Assembly of Peptidyl Nucleoside Antibiotics. *ACS Chem. Biol.* **2011**, *6*, 1000–1007. [CrossRef]
87. Filippov, D.; Kuyl-Yeheskiely, E.; Van der Marel, G.A.; Tesser, G.I.; Van Boom, J.H. Synthesis of a nucleopeptide fragment from poliovirus genome. *Tetrahedron Lett.* **1998**, *39*, 3597–3600. [CrossRef]
88. Roviello, G.N.; Roviello, V.; Autiero, I.; Saviano, M. Solid phase synthesis of TyrT, a thymine-tyrosine conjugate with poly(A) RNA-binding ability. *RSC Adv.* **2016**, *6*, 27607–27613. [CrossRef]
89. Kramer, R.A.; Bleicher, K.H.; Wennemers, H. Design and Synthesis of Nucleoproline Amino Acids for the Straightforward Preparation of Chiral and Conformationally Constrained Nucleopeptides. *Helv. Chim. Acta* **2012**, *95*, 2621–2634. [CrossRef]

90. Roviello, G.N.; Crescenzo, C.; Capasso, D.; Di Gaetano, S.; Franco, S.; Bucci, E.M.; Pedone, C. Synthesis of a novel Fmoc-protected nucleoaminoacid for the solid phase assembly of 4-piperidyl glycine/L-arginine-containing nucleopeptides and preliminary RNA: Interaction studies. *Amino Acids* **2010**, *39*, 795–800. [CrossRef]
91. Xie, J.; Noel, O. Synthesis of Nucleo Aminooxy Acid Derivatives. *Synthesis* **2012**, *45*, 134–140. [CrossRef]
92. Pomplun, S.; Gates, Z.P.; Zhang, G.; Quartararo, A.J.; Pentelute, B.L. Discovery of nucleic acid binding molecules from combinatorial biohybrid nucleobase peptide libraries. *J. Am. Chem. Soc.* **2020**, *142*, 19642–19651. [CrossRef] [PubMed]
93. Kristensen, I.; Larsen, P.O. γ-Glutamylwillardiine and γ-glutamylphenylalanylwillardiine from seeds of Fagus silvatica. *Phytochemistry* **1974**, *13*, 2799–2802. [CrossRef]
94. Sisido, M.; Kuwahara, M. Novel Peptide Nucleic Acids with Improved Solubility and DNA-Binding Ability. In *Self-Assembling Peptide Systems in Biology, Medicine and Engineering*; Springer: Berlin/Heidelberg, Germany, 2002; pp. 295–309.
95. Roviello, G.N.; Gröschel, S.; Pedone, C.; Diederichsen, U. Synthesis of novel MMT/acyl-protected nucleo alanine monomers for the preparation of DNA/alanyl-PNA chimeras. *Amino Acids* **2010**, *38*, 1301–1309. [CrossRef] [PubMed]
96. Talukder, P.; Dedkova, L.M.; Ellington, A.D.; Yakovchuk, P.; Lim, J.; Anslyn, E.V.; Hecht, S.M. Synthesis of alanyl nucleobase amino acids and their incorporation into proteins. *Bioorg. Med. Chem.* **2016**, *24*, 4177–4187. [CrossRef]
97. Diederichsen, U.; Weicherding, D.; Diezemann, N. Side chain homologation of alanyl peptide nucleic acids: Pairingselectivity and stacking. *Org. Biomol. Chem.* **2005**, *3*, 1058–1066. [CrossRef] [PubMed]
98. Geotti-Bianchini, P.; Crisma, M.; Peggion, C.; Bianco, A.; Formaggio, F. Conformationally controlled, thymine-based α-nucleopeptides. *Chem. Commun.* **2009**, *22*, 3178–3180. [CrossRef]
99. Geotti-Bianchini, P.; Moretto, A.; Peggion, C.; Beyrath, J.; Bianco, A.; Formaggio, F. Replacement of Ala by Aib improves structuration and biological stability in thymine-based α-nucleopeptides. *Org. Biomol. Chem.* **2010**, *8*, 1315–1321. [CrossRef]
100. Ranevski, R. Synthese und Untersuchung von Alanyl-PNA Oligomeren und deren Einfluß auf β-Faltblatt Strukturen. Ph.D. Thesis, Georg-August-Universitat, Göttingen, Germany, 2006.
101. Sturm, C. Theoretical Investigation of the Geometrical Arrangements of Alpha-Alanyl-Peptide Nucleic Acid Hexamer Dimers and the Underlying Interstrand Binding Motifs. Ph.D. Thesis, Universität Würzburg, Würzburg, Germany, 2006.
102. Lewis, F.D.; Liu, X.; Liu, J.; Miller, S.E.; Hayes, R.T.; Wasielewski, M.R. Direct measurement of hole transport dynamics in DNA. *Nature* **2000**, *406*, 51–53. [CrossRef]
103. Weicherding, D.; Davis, W.; Hess, S.; Von Feilitzsch, T.; Michel-Beyerle, M.; Diederichsen, U. Femtosecond time-resolved guanine oxidation in acridine modified alanyl peptide nucleic acids. *Bioorg. Med. Chem. Lett.* **2004**, *14*, 1629–1632. [CrossRef]
104. Cleaves, H.J. Prebiotic chemistry: What we know, what we don't. *Evol. Educ. Outreach* **2012**, *5*, 342–360. [CrossRef]
105. Leslie, E.O. Prebiotic chemistry and the origin of the RNA world. *Crit. Rev. Biochem. Mol. Biol.* **2004**, *39*, 99–123. [CrossRef] [PubMed]
106. Scognamiglio, P.L.; Platella, C.; Napolitano, E.; Musumeci, D.; Roviello, G.N. From prebiotic chemistry to supramolecular biomedical materials: Exploring the properties of self-assembling nucleobase-containing peptides. *Molecules* **2021**, *26*, 3558. [CrossRef] [PubMed]
107. Frenkel-Pinter, M.; Samanta, M.; Ashkenasy, G.; Leman, L.J. Prebiotic peptides: Molecular hubs in the origin of life. *Chem. Rev.* **2020**, *120*, 4707–4765. [CrossRef] [PubMed]

Review

Peptide-Based Vaccine against Breast Cancer: Recent Advances and Prospects

Muhammad Luqman Nordin [1,2], Ahmad Khusairi Azemi [3], Abu Hassan Nordin [4], Walid Nabgan [5], Pei Yuen Ng [6], Khatijah Yusoff [7], Nadiah Abu [8], Kue Peng Lim [9], Zainul Amiruddin Zakaria [10], Noraznawati Ismail [3,*] and Fazren Azmi [1,*]

1. Centre for Drug Delivery Technology, Faculty of Pharmacy, Universiti Kebangsaan Malaysia (UKM) Kuala Lumpur Campus, Jalan Raja Muda Abdul Aziz, Kuala Lumpur 50300, Malaysia; luqman.n@umk.edu.my
2. Department of Veterinary Clinical Studies, Faculty of Veterinary Medicine, Universiti Malaysia Kelantan (UMK), Pengkalan Chepa, Kota Bharu 16100, Kelantan, Malaysia
3. Institute of Marine Biotechnology, Universiti Malaysia Terengganu, Kuala Terengganu 21030, Malaysia; madkucai89@gmail.com
4. Faculty of Applied Sciences, Universiti Teknologi MARA (UiTM), Arau 02600, Malaysia; abuhassannordin@gmail.com
5. Departament d'Enginyeria Química, Universitat Rovira I Virgili, Av. Països Catalans 26, 43007 Tarragona, Spain; wnabgan@gmail.com
6. Drug and Herbal Research Centre, Faculty of Pharmacy, Universiti Kebangsaan Malaysia (UKM), Jalan Raja Muda Abdul Aziz, Kuala Lumpur 50300, Malaysia; pyng@ukm.edu.my
7. National Institutes of Biotechnology, Malaysia Genome and Vaccine Institute, Jalan Bangi, Kajang 43000, Malaysia; kyusoff@upm.edu.my
8. UKM Medical Molecular Biology Institute (UMBI), UKM Medical Centre, Jalan Ya'acob Latiff, Bandar Tun Razak, Cheras, Kuala Lumpur 56000, Malaysia; nadiah.abu@ppukm.ukm.edu.my
9. Cancer Immunology & Immunotherapy Unit, Cancer Research Malaysia, No. 1 Jalan SS12/1A, Subang Jaya 47500, Malaysia; kuepeng.lim@cancerresearch.my
10. Borneo Research on Algesia, Inflammation and Neurodegeneration (BRAIN) Group, Faculty of Medicine and Health Sciences, Universiti Malaysia Sabah, Jalan UMS, Kota Kinabalu 88400, Malaysia; zaz@ums.edu.my
* Correspondence: noraznawati@umt.edu.my (N.I.); fazren.azmi@ukm.edu.my (F.A.); Tel.: +60-6683240 (N.I.)

Citation: Nordin, M.L.; Azemi, A.K.; Nordin, A.H.; Nabgan, W.; Ng, P.Y.; Yusoff, K.; Abu, N.; Lim, K.P.; Zakaria, Z.A.; Ismail, N.; et al. Peptide-Based Vaccine against Breast Cancer: Recent Advances and Prospects. *Pharmaceuticals* **2023**, *16*, 923. https://doi.org/10.3390/ph16070923

Academic Editor: Rakesh Tiwari

Received: 16 April 2023
Revised: 7 June 2023
Accepted: 15 June 2023
Published: 25 June 2023

Copyright: © 2023 by the authors. Licensee MDPI, Basel, Switzerland. This article is an open access article distributed under the terms and conditions of the Creative Commons Attribution (CC BY) license (https://creativecommons.org/licenses/by/4.0/).

Abstract: Breast cancer is considered the second-leading cancer after lung cancer and is the most prevalent cancer among women globally. Currently, cancer immunotherapy via vaccine has gained great attention due to specific and targeted immune cell activity that creates a potent immune response, thus providing long-lasting protection against the disease. Despite peptides being very susceptible to enzymatic degradation and poor immunogenicity, they can be easily customized with selected epitopes to induce a specific immune response and particulate with carriers to improve their delivery and thus overcome their weaknesses. With advances in nanotechnology, the peptide-based vaccine could incorporate other components, thereby modulating the immune system response against breast cancer. Considering that peptide-based vaccines seem to show remarkably promising outcomes against cancer, this review focuses on and provides a specific view of peptide-based vaccines used against breast cancer. Here, we discuss the benefits associated with a peptide-based vaccine, which can be a mainstay in the prevention and recurrence of breast cancer. Additionally, we also report the results of recent trials as well as plausible prospects for nanotechnology against breast cancer.

Keywords: breast cancer; immunotherapy; peptide-based vaccine; nanoparticle; metastasis

1. Introduction

Among women worldwide, breast cancer is considered the most frequently occurring cancer. According to the World Health Organization (WHO), 2.1 million females have been diagnosed with breast cancer every year, which is responsible for approximately 15% of all cancer deaths [1]. Breast cancer is a cancer that develops from the epithelial cells of the mammary gland, duct, or lobules. Breast cancer occurrence also exists in males; however, it is relatively rare (around 1%) [2–4]. Although pathophysiological knowledge of breast cancer is minimal, certain established risk factors, such as genetic predisposition (family history), diet, and an unhealthy lifestyle, are unquestionably linked to the development of breast cancer [5–8]. To highlight, both genetic and environmental factors influence the diversity of breast cancer etiology [9,10].

It is widely accepted that BRCA 1 and BRCA 2 genes are responsible for repairing DNA dysregulation, alteration, and damage. These genes are predicted to be accountable for approximately 40–85% of the risk of hereditary breast cancer when they are mutated. BRCA 1 and BRCA 2 are located on chromosomes 17q and 13q12–13, enabling the inference of more distally mutated loci associated with mutations to affect their functional enhancers and promoters' actions. Besides BRCA germline families, mutations in p53, PTEN, CHEK2, ATM, PALB2, RAD51C, and RAD51D have also been associated with breast cancer [11,12]. Additionally, mutations in BRCA genes can lead to the acquisition of a multi-drug resistant phenotype, subsequently contributing to a major limitation in clinical treatment for breast cancer [13–15].

The condition can worsen if the genes are inherited from one generation to another and become inheritable mutations. It can happen when epigenetically mutated cells accumulate and then further create a microenvironment that improves drug efflux, drug evasion, the anti-apoptotic pathway, and other escape mechanisms from the immune system. Epigenetic mutations, resulting mostly in DNA methylation patterns, histone acetylation, and phosphorylation, are known to have a profound effect on gene expression, resulting in the activation of tumor suppressor genes and leading to the emergence of cancer drug resistance [16]. When malignant cells continue to metastasize, they overly express immune checkpoint inhibition (CPI) signals, resulting in stimulation of inhibitory co-stimulatory molecules (PD1/PD-L1/LAG-3, CTLA4) and anti-apoptotic signaling pathways, causing tumor cells to deactivate immune activation and immune detection; hence, the tumor cells escape and progress to cancerous form [17]. Interactions of cancer in the tumor microenvironment can activate cellular components in the environment, including tumor-associated macrophages (TAM), cancer-associated fibroblasts (CAF), and mesenchymal cells, to protect cancer cells from being susceptible to drugs and promote drug resistance [18]. Some of the cancer biomarkers can also hinder tumor antigen expression, leading to the failure of the intended drug to penetrate cancer due to the unfavorable and diverse mutational landscape possessed by the cancer's microenvironment, which additionally creates immunometabolism barriers [19].

Myeloid-derived suppressor cells (MDSCs) are one form of immune cell that plays a major role in tumor immunosuppression. These cells consist of immature monocytes and granulocytes released from the bone marrow into the blood during disease conditions, including cancer. Tumor-associated macrophages (TAMs) are another type of cell that functions similarly to MDSCs. The ability of MDSCs and TAMs to suppress the antitumor response is the subject of many recent studies [20,21]. MDSCs could suppress not only natural killer (NK) cells and dendritic cells (DCs) but also T cells. T cells were suppressed through the production of inducible nitric oxide (NO), nitric oxide species (iNOS), reactive oxygen species (ROS), arginine, and cysteine deprivation. Meanwhile, MDSC is able to synergize T regulatory cells (Treg) and TAMs and cause downregulation of IL-12 production by TAMs, which is an important cytokine involved in T cell and NK cell activation through membrane-bound TGF-β [22–24]. Figure 1 demonstrates some of the mechanisms of cancer cell evasion via hijacking the immune system.

The presence of tumor immunogenicity in the breast cancer microenvironment has necessitated the use of immunotherapy as a potential cancer treatment [25,26]. Immunotherapy can target specific cells that are involved in hijacking the immune cells; thus, it seems to be a good idea for the therapy's success. Immunotherapy in the form of a vaccine functions by utilizing the patient's immune system to identify and eradicate cancerous cells. Cancer cells produce chemokines, cytokines, and prostaglandins that attract diverse infiltrating immune cells consisting mainly of macrophages, neutrophils, and lymphocytes [27]. These infiltrating immune cells can stimulate tumor necrosis factor (TNF), IFN, matrix metalloproteinases, natural killer (NK) cells, and T cells, leading to the destruction of cancer cells. Most targeted therapies in recent years for cancer immunotherapy involve utilizing and targeting enough tumor-infiltrating T lymphocytes (TILs) and cytotoxic T cells (CTLs), which may correlate with the presence of antigen loads and suppress immune inhibitory signals responsible for local immunosuppression of the tumor microenvironment [28,29]. The discovery of breast cancer tumor-associated antigens (TAAs) or tumor-specific antigens (TSAs), which are expressed in breast cancer cells, has made it possible to develop a vaccine against breast cancer. Therefore, an understanding of the immune cell population may have significant consequences for the prevention of breast cancer, enhanced risk management strategies, and the control of breast carcinogenesis.

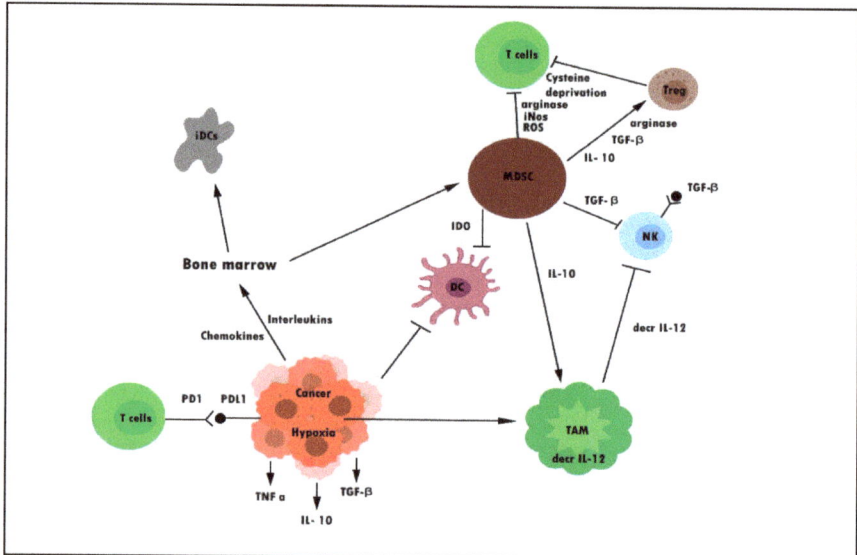

Figure 1. Mechanisms of cancer escaping pathways from the immune system. DC denotes dendritic cell, iDC denotes immature dendritic cell, MDSC denotes myeloid-derived suppressor cell, decr IL-12 denotes a decrease in interleukin-12, TAM denotes tumor-associated macrophage, and Treg denotes regulatory T cell.

In this review, we provide an overview of the recent advancements in the development of a peptide-based vaccine against breast cancer, with emphasis given to antigen selection and vaccine design. Additionally, breast cancer immunology and considerations for future directions for rational breast cancer vaccine design will be discussed.

2. Cancer Vaccines

The fundamental understanding of tumor immunology and its plausible mechanism of action has opened the route to employing the body's immunity against cancer [30]. Immunotherapy in the form of a vaccine has great potential for breast cancer treatment over chemotherapy and endocrine therapies due to several issues, including relapse and drug resistance. Recent reports demonstrate that about 80% of treatment failures are due to metastases and drug resistance from several mechanisms of action, such as genetic mutation [13,14]. The management of advanced malignant breast cancer, with median overall survival ranging from 4 to 5 years for luminal-like tumors to 1 year for triple-negative disease, remains minimal and is considered short [31]. Until now, many scientists have tried to discover how to overcome therapeutic resistance because more than one mechanism may be responsible for oncogenesis. Even though there are several immunotherapy forms, including adoptive cell transfer, checkpoint blockage, and antibody-based drugs, vaccines are seemingly more tempting due to their wide safety profile and lifelong protection [32–34]. However, until now, no breast cancer vaccine has been authorized by the U.S. Food and Drug Administration (FDA) for either therapeutic or prophylactic purposes. The immune system in humans is incredibly complex. Even though, until now, no breast cancer vaccine has been authorized, many are still in clinical trials. It is only a matter of time. The tumor microenvironment (TME) in tumors remains one of the major hurdles in developing a therapy against breast cancer. Breast cancer is generally infiltrated by immune cells triggered by the cancer cells, which can create an immunosuppressive microenvironment that encourages tumor growth by inhibiting immune cells [35]. Besides TME, the following factors are thought to be responsible for the challenges of breast cancer vaccine development: (i) the stage of breast cancer; (ii) the choice of TAAs to target; and (iii) the vaccine's low immunogenicity as a result of the antigen selected or as a result of the vaccine delivery platform used [36].

There is, however, increasing attention in clinical research that evaluates vaccines derived from the peptide. The rationale for this interest is based on the aberrant expression of proteins or antigens by breast cancer. With the discovery of breast cancer antigens, the peptide-based vaccine is becoming a potential alternative to conventional therapies, which are known to have serious drawbacks. The vaccines would modulate the immune system of the body to specifically attack cancer cells based on the recognition of tumor associate antigens (TAAs) or tumor-specific antigens (TSAs) on the surface of cancer. Interestingly, recognition of these antigens eventually allows the immune system to recognize these antigens, to have long-lasting immunity, and to solve the relapse issue after completion of the treatment. A robust, fundamental, and precise comprehension mechanism for the action of peptide vaccines is required to establish potent and effective cancer vaccines.

3. Peptide-Based Vaccine and Key Regulator in Breast Cancer Immunogenicity

A peptide-based cancer vaccine is a short chain of amino acids that contain epitopes that are reactive to T cells. The major objective of peptide-based cancer vaccines is to induce the necessary host immune response to recognize and eliminate targeted cancer cells based on a defined set of TAAs and TSAs. The peptide-based vaccine follows the principle of immunotherapy, which modulates the immune system of the body to specifically attack cancer cells based on the recognition of aberrant expression of tumor antigens or proteins in the cancer cell. Interestingly, recognition of these antigens eventually allows the immune system to recognize these antigens, to have long-lasting immunity, and to solve the relapse issue after completion of the treatment.

Administration of a peptide vaccine functionalized CD8+ cytotoxic T lymphocytes (CTL) cells to attack tumor cells through the release of granozymes, granulysin, perforin, and Fas ligand (FasL) through Fas death receptor binding to cancer cells for apoptosis to occur (Figure 2). Cytokine release helps with lymphocyte migration, B cell development, T cell activation, and expansion. After activation, CD4+ T cells further differentiate to develop dominant anticancer pathways and responses. In order to modulate tumor-specific immune responses and inhibit proliferation in the body, significant interventions have been rendered to identify tumor-expressed antigen cells or those recognized as tumor-associated antigens (TAAs) utilizing T cells [37,38]. Identified as an example of HER2 antigen in breast cancer and transformed into a vaccine component capable of triggering a specific and systemic immune response that may contribute to the suppression, removal, and destruction of cancer in body tissues [39]. When TAAs are found in the body from cancer cells, specific fragments of cancer proteins are expressed on the cell surface and then attached to the MHC 1 complex [40]. It would ultimately be recognized by NK cells and CD8+ cells. The dissimilarity between endogenous and exogenous peptides is crucial for a functional immune response.

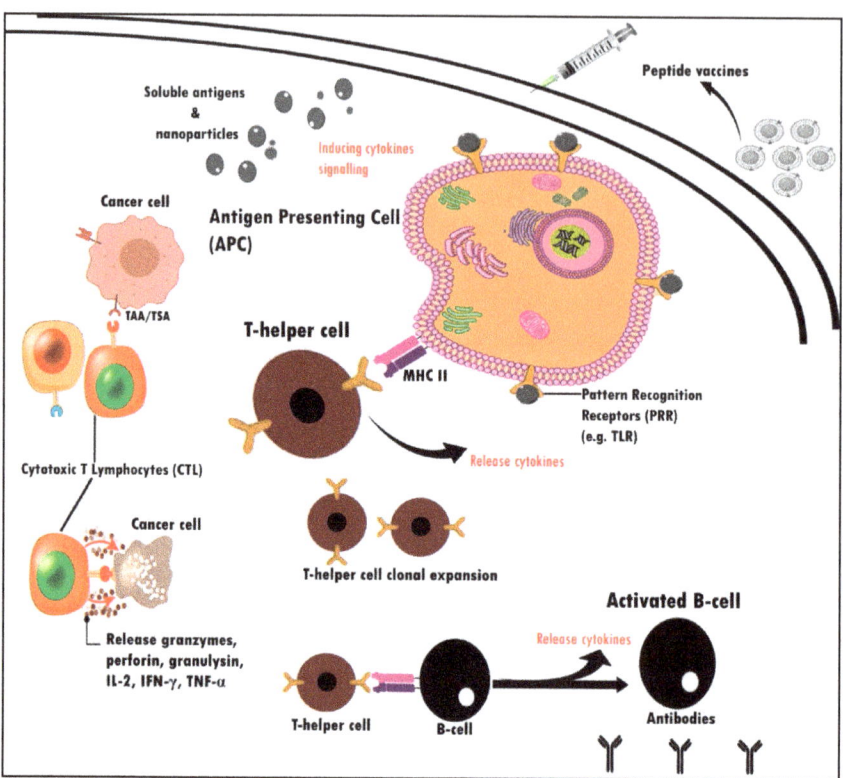

Figure 2. Schematic illustration mechanism of action for peptide vaccines. These immunological events are essential for enhancing the cell-mediated and humoral immune response against peptide vaccines. PRR is expressed by APCs and acts by recognizing and binding with antigens.

T-cell responses are specific and triggered after peptides have been taken up and processed by antigen presentation cells (APCs), which then transit to lymph nodes and expose the antigen on their cell surface as foreign molecules through MHC-I and II [41,42]. Prolonged activation of T cells leads to further differentiation into CD8+ cytotoxic and CD4+ helper T cells through the CD40-CD40L pathway. Studies have shown that CD40-CD40L associations are capable of inducing humoral and cellular thymus-dependent (TD) reactions [43]. When APCs are triggered through CD40-CD40L, it is found that ligation of CD40-CD40L to APCs, particularly dendritic cells, is capable of generating cytotoxic CD8+ T cells [43]. This may grant CD8+ T cells the potential to modulate antigen-specific immune responses similar to the CD4+ T cell response, which is often correlated with CD40L expression [44,45]. The use of TAA-related peptides such as GP2 tends to be an efficient way to stimulate CD8+ and CD4+ production against cancer cells.

Peptide-based vaccines offer several advantages over other types of cancer vaccines. Peptide-based vaccines can be easily customized with minimal epitopes while still being able to induce desired immune responses safely. Peptides are manufactured almost entirely using synthetic chemical approaches. Therefore, peptide antigens can be completely and specifically identified as chemical entities. Hence, all issues associated with the biological contamination of antigens are effectively eliminated.

Despite its benefits, peptide-based vaccination also exhibits several drawbacks, including a limited half-life, an insufficient immunogenic response, being easily degradable, and low bioavailability. However, due to the heterogeneity between solid tumors and the external microenvironment, the efficacy of the immune response in solid tumors is not as anticipated [46]. Therefore, by modifying the delivery system of peptides, such as nanocarriers that have a protective layer and are bound to the TAAs, it is then possible to prevent the degradation of these proteases and improve the association between peptide vaccines and cancer. Recent studies and many clinical studies have uncovered the potential use of peptide-based vaccines as immunotherapeutic agents that may weaken or break the immune tolerance of cancer patients. However, several modifications need to be made to the peptide vaccine to reach ideal potency.

The clinical efficacy of peptides can be easily enhanced by covalently conjugating or linking chemically with specific immunostimulatory molecules at specific positions within the peptide sequence. The use of only minimal antigens is capable of triggering humoral and cell-mediated immunities. Peptides applied in breast cancer vaccines (Table 1) and selected breast cancer peptides in clinical trials (Table 2) provide convincing evidence that peptide-based vaccines are a viable strategy for the treatment of breast cancer. A peptide-based vaccine depends on mobilizing cytotoxic CD8+ T cells and NK cells to kill the cancerous cells. To stimulate a tumor-specific immune response, TAAs or TSAs must be presented to the APC and make the immune system of the host recognize them. Several TAAs are mainly identified as immune targets for a vaccine against breast cancer. This includes HER2, MUC-1, EphA, Survivin, SART3, CEA, p53, and WT1 [47–49]. These antigens have provided convincing evidence as immune targets in preclinical and clinical studies and warrant further research.

Present clinical studies have demonstrated the therapeutic value of peptide-based vaccines to reduce cancer recurrence and enhance overall patient survival [50,51]. NeuVax™ (NCT01479244) is the most mature level of production for a peptide-derived breast cancer vaccine. It was in Phase III clinical trials and was initiated by the US National Cancer Institute in 2011 [52]. Targeting precise TAAs is vital to induce successful T-cell differentiation and alarm signals for tumor destruction mechanisms. In order to obtain a TAA-specific T cell response and upregulate stimulatory signals, APC needs to be provided with enough TAAs and be in a mature state. Otherwise, antigens may not promote oncogenesis and trigger T helper-cell clonal expansion. This is essential because CTLs are a significant cell type responsible for killing cancer cells. Various peptide breast cancer antigens that may trigger an immune response in patients have been identified and used as targets for breast cancer vaccines. However, the reduced immunogenicity of these peptide cancer antigens and cancer immune evasion mechanisms makes the development of breast cancer vaccines challenging. This condition necessitated the need to build an efficient vaccine delivery system with powerful immunostimulatory properties to promote APC activation, thus eliciting a strong T cell response and weakening and breaking the immunotolerance of cancer antigens in the tumor microenvironment.

Noteworthy, breast cancer is known to be a complicated, non-infectious, and immunogenic disease with the ability to alter the tumor microenvironment, making it resistant to treatment. In this case, manipulating the body's immunological reaction to cancer will provide some merit for potential cancer vaccine research. Peptide detection by Human Leukocyte Antigens (HLA), which are gene coding for the Major Histocompatibility Complex (MHC), has also become a challenge because each HLA has its own subtypes, and each subtype can be specific to TAA proteins. For instance, GP2 is a Class I peptide MHC that is restricted to HLA-A2 and/or HLA-A3 [51]. Nevertheless, the function of HLA in determining the prognosis and effectiveness of peptide vaccines is still uncertain [53–55]. Research published by Jackson et al. [53] showed that HLA-A2 expression did not substantially correlate with the prognosis of women with breast cancer. This finding is essential for the selection of candidates for the HLA-A2 breast cancer vaccine. A peptide can target important ligands in breast cancer regulatory pathways, including chemokines, Y-box binding protein-1 (YB-1), Sin3, TNF 1-related ligand-inducing apoptosis (TRAIL), and FasL [56].

4. Identified Tumor-Associated Antigens in Peptides Vaccine Development for Breast Cancer

4.1. HER2

HER2, also known as ERBB2, NEU, and CD34, is a human epidermal growth factor receptor 2 and a component of transmembrane glycoprotein that is overexpressed in approximately 20–30% of primary breast carcinomas [47,57] for tyrosine kinase activity. The HER2/neu cell surface receptor is the most frequently targeted TAA; thus, the HER2-derived peptide vaccine has shown excellent potential in developing breast cancer vaccines. Upon dimerization of the antigen receptor, the numerous intracellular signaling pathways are activated by transphosphorylation, which mediates cell proliferation and differentiation. However, when inappropriate activation happens, it contributes to the production of many malignancies [58]. Slamon et al. [59] first discovered the function of HER2 as a marker with a prognosis value for treating breast cancer in 1987. It has been confirmed and proved by several scholars [60–62]. Studies conducted by Clynes and colleagues reported monoclonal antibodies targeting HER2 to provide clinical benefits against HER2 overexpressing breast carcinomas [63,64]. To date, the application of HER2 as a therapeutic marker and predictor in invasive breast carcinomas has been commonly utilized and continues to develop.

To sum up, HER2/neu is a well-known therapeutic target that is a hallmark of HER2-positive breast cancer. With HER2-targeted vaccinations, targeting HER2 appears to be a reasonable strategy for the dysregulation of numerous signaling cascades that promote oncogenesis. It has been extensively used with a GM-CSF adjuvant. They offer little or no risk and the chance of producing a memory antibody against the same disease. Ex vivo expansion of cellular immunity, including activation of CD8+ CTL against breast carcinoma, will be enabled by the production of anti-HER2 immunization. Vaccinated patients showed high levels of CD8+ and mediated delayed-type hypersensitivity reactions [64]. Established by Mansourian et al. [65], p5 peptide encapsulated with liposomes co-administrated with CpG-ODN has been shown to decrease tumor size and, at the same time, improve animal survival period in mice of the breast cancer model. This was confirmed by Farzad et al.'s study [57], which displayed another peptide, the P435 HER2-derived peptide, conjugated to liposomes capable of inducing CTL responses, therefore improving prognosis in the TUBO murine breast cancer model. Another study revealed that Nelipepimut-S (E75) was a nine-amino acid peptide extracted from the HER2 protein capable of increasing the patient's survival rate. Research from recent clinical trials has demonstrated positive effects of HER2-specific vaccinations that, when paired with chemo-drugs, could synergistically inhibit the recurrence of breast cancer, creating robust immunity and sustaining elevated CTL rates. One of the hurdles to the HER2-based vaccine is against TNBC subtypes due to the lack of HER-2 (ERBB2), progesterone, and estrogen receptors. Costa and colleagues suggested that combinations of HER2-based vaccines with pembrolizumab or nivolumab (immune checkpoint inhibitor antibodies) merit promising outcomes [66]. Combinations with other therapies might produce synergistic effects and resensitize other cancer cell death programs.

4.2. MUC-1

Transmembrane mucin-like glycoprotein Mucin 1 (MUC-1) is a type of glycoprotein comprising a single polypeptide chain with multiple oligosaccharide side chains with oxygen linkages to serine, proline, and threonine residues [67] frequently overexpressed in glandular and epithelial mammary, lung, and colon cancers [68]. MUC-1 overexpression can be used as a marker for cancer that suggests that the cancer is progressing [49].

Several studies have demonstrated that targeting MUC-1 was a successful option for the cancer vaccine because it is broadly dispersed in all tumors and cancers, including stem cell cancer [69]. Covalently bound to the TLR agonist, the completely synthetic glycosylated MUC-1 peptide vaccine exhibited good humoral and cellular immune responses [70]. The pioneered MUC-1 peptide vaccine study was performed in 1995 against patients with breast carcinoma. The majority of patients reacted to the medication, and no toxicity was found [71]. The extension of the research was carried out up until the Phase III clinical trial. A pilot Phase III analysis of 31 early Stage II breast cancer patients utilizing oxidized mannan-MUC-1 immunotherapy found that MUC1 immunotherapy is effective [72]. The MUC-1 peptide vaccine candidates have demonstrated an improved survival rate. MUC-1 is proposed as a potential biomarker to be targeted in breast cancer therapy because patients would typically overexpress the MUC-1 biomarker (approximately 90%) for immune system detection. Antibodies against MUC-1 can efficiently cause CTL and TLR. Hence, reasonable disease regulation is accomplished as patients produce strong antibody titers of MUC1 IgG. In clinical studies on women with Stage I and Stage II breast cancer, MUC-1 IgG and IgM antibodies were tested and assessed for their association with disease-specific survival [49,73]. MUC1 is a possible antigen to be utilized as a site-specific target for the deployment of therapeutic agents as a vaccine against MUC1 for breast carcinoma. Recent studies have shown that MUC1 can induce antigen-specific cellular and humoral responses not only to trigger MUC1-specific CD4+ and CD8+ T-cells but also to generate antibodies [69].

4.3. EphA

EphA is a type of transmembrane glycoprotein with tyrosine kinase (RTK) receptors on the surface that play a significant role as tumor-specific cell-surface receptors for drug-targeting sites. It is the largest group among tyrosine kinase receptor families, and among them, EphA2 is commonly overexpressed in breast cancer. The activation and overexpression of EphA2 frequently lead to its ligand-independent oncogenic and angiogenesis activation, which are triggered by dwindled contact with the ligand, ephrin-A (EphA2). Loss of the ligated EphA2 receptor decreases the intrinsic tumor-suppressive signaling pathways, accompanied by downregulation of the PI3K/Akt and the ERK pathways, thus decreasing the tumor volume and size.

As a therapeutic target, EphA2 receptors remain an essential marker. Overexpression of EphA2 receptors has been correlated with low survival in all patients with breast cancer subtypes due to the EphA2 activity that enhances tumorigenesis and the progression of metastases [74]. A monoclonal antibody (mAb EA5) has been studied to suppress EphA2 receptors in ER-positive breast cancer and to minimize cancer invasiveness [75]. The outcome was promising, and the study proceeded in the presence of tamoxifen. Furthermore, the monoclonal antibody EPhA2 can specifically target antigens and suppress the development of breast cancer cells and tumorigenesis.

YSA and SWL are peptide-based EpHAs that target EpHA2 receptors on the surface of tumor cells. Scarberry et al. [76] reported the success of using a magnetic $CoFe_2O_4$ nanoparticle-YSA peptide conjugate to extract ovarian cells from blood and fluid in mice. Even though the EpHA antigens are overexpressed in blood cancer and tumor cells, further exploration with regard to EphA as a peptide-based vaccine is very limited. Perhaps an EpHA-based vaccine does not elicit a potent immune response to eliminate various classifications of breast cancer. This may be triggered by cross-reactions between drugs and other proteins or by incomplete subcellular internalization of antibody-drug conjugates (ADCs). Thus, in order to address this issue, Salem et al. [77] proposed peptide-based targeting drugs that were less harmful but efficient and inexpensive. The aim of breast cancer therapy may be to merge EphA2 expression with carcinogenesis. Strategies focused on EphA2 targeting have been groundbreaking developments in therapeutic discovery. The targeted drug, for example, trastuzumab, is yet to be used; the concern of cardiotoxicity persists. Immunotherapy, similar to a cancer vaccination, tends to be an effective solution to treating metastatic breast carcinoma.

4.4. Survivin

Survivin, a 16.5 kDa intracellular acidic protein of 142 amino acids encoded by the BIRC5 gene, is a multifunctional protein that belongs to the smallest member of the inhibitory apoptosis protein family (IAPs). It regulates cell cycle progression through inhibition of the apoptosis pathway [78–81]. A high level of survivin expression is significantly associated with breast, urothelial, and colorectal cancer invasiveness and its low prognosis [82]. Survivin is undetectable in healthy tissue, indicating that it is exclusively presented as a biomarker when there is tumor transformation and acts as a transcriptome that is expressed in breast cancer. It seems to play a role in the antiapoptotic function of a protein, preventing the cell program from happening. A study by Ryan et al. [81] showed that a high level of survivin expression patterns is often associated with HER2/neu positive breast cancer and correlates with the prognosis.

This was confirmed by Lyu et al. [48], who reported that dysregulation of survivin was found in HER2/neu breast cancer, and survivin was identified as a desirable therapeutic target for blocking its IAP functions. A gene and immunotherapy named sepantronium bromide (YM155) have been developed to block survival. They provided a positive outcome for in vivo research by lowering the expression of survivin, raising the regression of the tumor, and prolonging the life of the mouse. However, YM155 failed to demonstrate an improvement in treatment response in metastatic non-small-cell lung cancer patients (NSCLC). This failure is probably due to the presence of multiple pathways linking survivin with other regulated proteins, making it more complicated [80,81].

Another research performed by Tanaka et al. [82] found that cytoplasm-responsive nanocarriers conjugated with a functional cell-penetrating peptide could facilitate the delivery of anti-vascular endothelial growth factor siRNA (siVEGF) complexes to tumor tissues after systemic injection and could elicit a potent anti-tumor effect. In a study from Rodel et al. [83], survivin as an antigen vaccine conferred peptide-specific CTL induction of urothelial cancers in patients without significant adverse reactions. On the other hand, the latest research indicated that the presence of the survivin antigen in breast carcinoma revealed a connection between expression and therapeutic outcomes [84].

4.5. SART3

SART3 is a tumor rejection antigen consisting of 3806 bp of nucleotides encoded by a 140-kilodalton (kDa) protein expressed in the cytosol of most of the cell proliferation and has been shown during gene transcription and mRNA synthesis of cancer cells. Similar to survivin, the SART3 oncogene is absent in normal tissues except for the fetal liver and testicles [85]. This antigen exhibits strong binding with HLA-A24-restricted CTL epitopes and may be useful for specific immunotherapy.

A Phase 1 clinical trial was recorded by Miyagi et al. [86] utilizing the SART3 peptide vaccine in colorectal cancer patients. The findings revealed a significant induction of cellular immune responses in 7 out of 11 patients. However, no explicit activation of the humoral immune response (IgG or IgE) has been recorded for peptides. SART3 led to the regulation of pro-inflammatory cytokine expression and the association of the degree of expression with malignancies and the prognosis for patients with breast cancer [87,88]. Given its positive outcome against SART antigen-expressed cancer, no clinical trials of SART-associated breast cancer vaccine goals have been reported.

4.6. CEA

CEA is a 180-kDa glycoprotein widely recognized as an oncofetal antigen that is found in numerous cancers, including colorectal, breast, gastric, pancreatic, and non-small cell lung cancers. CEA is one of the earliest tumor markers used to identify and anticipate the recurrence of tumors following surgical resection [89]. Its overexpression leads to the progression of the tumor. High secretions of CEA from cancer cells in the blood serum and over-expressed CEA on the surface of tumor cells make it accessible for use as a selective marker for cancer immunotherapy. CEA has been used as the foundation for numerous cancer vaccines, including DNA-based vaccines, dendritic cell-based vaccines, recombinant vector-based vaccines, protein-based vaccines, and anti-idiotype antibody vaccines, with the potential to induce both humoral and cell-mediated immunity that contributes to the killing of cancer cells [90]. Ojima et al. [91] demonstrated that genetically modified dendritic cells that express CEA administered simultaneously with interleukin 12 (IL-12), GM-CSF enhanced the therapeutic effects in CEA transgenic mice through the improvement of CEA-specific T-cell responses. Interestingly, the vaccination therapy eliminated colon cancer up to 2×10^3 mm^3 sizes. Furthermore, no detrimental results were found after the experiment. The research performed by Gulley et al. [92] found that 9 out of 16 patients diagnosed with recombinant CEA-MUC-1-TRICOM poxviral-based robust tumor vaccines had an increase in both CD8+ and CD4+ immune responses. To sum up, targeting CEA could be a successful vaccination technique for the clinical application of peptide vaccines to achieve a positive antitumor response.

4.7. p53

p53 is a tumor suppressor protein that plays a vital role in regulating genomic stability by controlling the cell cycle and inducing apoptosis when cell damage is beyond repair. p53 mutations occur in about 18–25% of primary breast cancer, rendering them potential biomarkers for cancer immunotherapy. Missense mutations within the p53 gene could potentially cause the accumulation of mutant proteins within the cell nucleus through posttranscriptional modification. The prognosis value of the patient appeared to be associated with the p53 level. Approximately 80% of TNBC patients have been identified with high p53 gene levels, and so far, no immunotherapeutic medication scientifically used for TNBC has been proven to develop a peptide-derived vaccine [93]. Many 2-sulfonyl pyrimidine compounds, such as PK11007 and PK11000, are successful in killing cancer cells by explicitly attacking mutant P53 thiol groups, thereby reducing oxidative stress levels (e.g., ROS) and eventually retaining a redox state [94,95].

PRIMA-1MET (APR-246) has been clinically studied in a Phase I clinical trial and is currently undergoing more clinical review. It inhibits cancer cell growth by targeting mutant p53 and inactivating it in triple-negative breast cancer (TNBC) [96]. In the Phase I/II clinical trial, ten colorectal cancer patients were vaccinated with p53-derived synthetic long peptides (SLPs). Rapid p53-specific T-cell responses were observed in blood samples obtained six months after the last vaccine [97]. Although the theory suggests that SLP would activate a high level of T cells in vaccinated patients, the clinical findings have unfortunately not been compatible. Perhaps targeting p53 alone is not enough to eliminate breast cancer cells. Thus, multiple peptides that target multiple antigens while stimulating multi-antigenic immune responses tend to be the right approach to improving the immunogenicity and clinical efficacy of p53-directed immunotherapies [98,99].

4.8. WT1

Wilms' tumor 1 (WT1) is a gene located at chromosome 11p13, initially discovered in childhood kidney cancers and overexpressed in other hematological malignancies (leukemia) and solid cancers, including breast cancer, ovarian cancer, pancreatic cancer, renal cancer, endometrial carcinoma, and glioblastoma [100–102]. Additionally, WT1 expression has been identified and used as a potent transcriptional regulator and marker for myelodysplastic syndromes (MDS), acute myeloid leukemia (AML), and solid tumors, including breast carcinoma. This information indicates that WT1 can be a targeted antigen for cancer immunotherapy, and reducing its level shows inhibition of tumor progression [97]. Coosemans et al. [102] proved that 36 patients with endometrial cancers showed overexpression of WT1 in endometrial cells. This research is in conjunction with other reliable cancer reports [103–105]. Due to the fact that immunization principles for inducing an immune response are more or less identical in different peptide protocols, WT1 immunotherapy, fortunately, may provide diagnostic tools and prognostic markers for all solid cancers.

5. Strategies to Improve the Immunogenicity of Peptide-Based Breast Cancer Vaccines

Since peptides are not so immunogenic, few strategies for improving the effectiveness of peptide-based vaccines have been developed. The administration of immunostimulant agents (adjuvants), either mixed or chemically conjugated to the peptides, helps improve the body's immunity against peptide antigens derived from the tumor. A rationale for incorporating immunostimulatory within the vaccine designs is to develop the synchronous activation of APCs (especially dendritic and macrophage cells) and foster T-cell responses without jeopardizing the quality and safety of the vaccine formulation [106]. Various types of vaccine-adjuvant formulations used in clinical trials are shown in Table 2.

5.1. Multi-Epitope Peptide Vaccine Antigens

A key step in designing peptide-based vaccines is the choice of an epitope that is capable of stimulating robust, longer-lasting, targeted both cellular and humoral (or either one) immunity against the desired pathogen. Therefore, it is important to initially recognize appropriate peptide epitopes on the protein of interest. The selection of the epitope must also take into account the plausible hypersensitivity reaction that arises in correlation with certain antigens. Peptide vaccine development can be presented to the immune system in multi-antigenic forms, adopting the nature of pathogen properties. Ghaffari-Nazari et al. [107] reported that peptides containing cytotoxic T lymphocyte (CTL) epitopes elicit robust protective immunity against tumors. This was supported by Zamani and others, who demonstrated that CTL and T-cell epitopes derived from TAAs simultaneously stimulate the CD8+ cytotoxic and CD4+ T-cell responses. The pan-DR-biding epitope (PADRE) is an example of a T helper epitope that activates CD4+ T cells and has proven safe and well tolerated [104].

Studies conducted by Wu et al. [108] in mice using an E7 peptide-based vaccine and PADRE as an adjuvant showed stronger CTL responses compared to those without the T helper epitope. B-cell and CTL epitopes can be incorporated to generate specific and robust immune responses involving humoral and cellular immune responses. Each of the epitopes can stimulate immune responses. Therefore, combining multiple epitopes in a single system could provide synergism, increase IFN- production, and increase peptide affinity to MHC molecules, thus enhancing CD4+ and CD8+ T cell generation [109–111] because each epitope has its ability to trigger an immune response. Another way to enhance immunogenicity is by using long peptides. The long peptide sequences (30 mers) are believed to be more efficient in generating effector T cells due to their extra length, thereby requiring only professional APCs that can process and present MCH molecules [112]. In addition, extra length may provide a place for tertiary structure, thus making the peptide not easily degradable by serum peptidases and tissue [113]. These make peptides suitable to be used in tandem with bioengineering applications and vaccine design, which significantly improves their effectiveness without jeopardizing the safety or quality of the peptide [114–116].

Due to this reason, the peptide-based vaccine is now gaining a great deal of scientific attention and has stepped significantly forward in the field of cancer immunotherapy. Despite using protein, peptide within the protein is more feasible with a distinctive individual epitope that can induce protective immunity. Multiple immunogenic epitopes, for example, containing T and B cell epitopes, can be linked covalently to form stable and strong linear complexes of peptide sequences, thus providing the platform necessary for immune cell recognition. The vaccines destroying cancer cells depend on unique peptide antigens obtained from the TAAs and TSAs and enable the immune system, especially cytotoxic T lymphocytes (CTLs), of the host to recognize them. The overexpression of TAAs and TSAs on the tumor surface increases the exposure of tumor cells to targeted therapeutics, and interestingly, the receptor-mediated tumor-targeting ligands are a type of protein. These ligands allow targeted delivery of peptides of interest to the tumor site either by direct coupling or through a carrier delivery system such as liposomes, micelles, or nanoparticles, thus triggering the specific tumor immune response.

5.2. Immunostimulatory Adjuvants

5.2.1. Toll-Like Receptors (TLRs) Ligands Based

TLRs belong to a class of pattern recognition receptors (PRRs) that are located at the surface or intracellular compartments of endosomal and cytoplasmic membranes, and they bind with the antigens before further creating intracellular signaling pathways that evoke immune system responses. Generally, TLRs resemble similar structural features of pathogen-associated molecular patterns (PAMPs) and have been easily recognized by PRRs. The ligand binding between TLR and antigen would activate the transcription factors nuclear factor kappa B (NF-B), interferon, and the release of cytokines such as IL6, IL1, IL8, IL12, TNF, and other molecules such as CD40, CD80, and CD86, which would lead to the recruitment of immune cells and, subsequently, induce killing mechanisms in cancer cells [117]. Examples of types of TLRs used as adjuvants in breast cancer vaccines are TLR9 ligand (CpG-oligonucleotides), TLR5 ligand (flagellin), TLR4 ligand (bacterial lipopolysaccharide (LPS)), and TLR3 ligand (Poly-ICLC).

TLRs have been found to be convincing immunostimulatory and are often co-delivered within the vaccine. TLR agonists were designed together within the peptide to overcome the poor immunogenicity of the peptide-based vaccine. To date, the incorporation of TLR agonists into a peptide has provided promising approaches for the development of an efficacious vaccine by targeting APC uptake and PRRs, thus allowing vaccines to achieve potent cell-mediated and humoral immunities [118]. As shown in many studies, the use of TLR-adjuvants helps in a way to increase the immunogenicity of a peptide [118–120]. In an in vitro study combining CpG oligonucleotides with cage protein, EP2 was able to increase MHC 1 display and induce CD8 T cell activation at a dose lower than required for DC maturation. The combination also increased MHC I display and CD8 T cell activation relative to unbound forms of the individual components.

Adjuvants need to follow pre-selected requirements, including being non-toxic, capable of protecting peptides from rapid degradation, stimulating a good humoral/T-cell response, and not causing autoimmune or allergic reactions [120–122]. Examples of widely used adjuvants include Montanide ISA-51 (IFA), poly I: C, GpG ODN, TLR-dependent, AS15, Freund's complete adjuvant (FCA), and GM-CSF. A combination of adjuvants may eventually be essential as it can modulate a potent immune response and intensify T helper cells as if one adjuvant were low. CpG-ODN induces the proliferation of B cells, activation of macrophages, and thus stimulation of the immune system, as in tetanus toxoid. CpGODN co-administration with antigens creates potent immunogens in vivo and in vitro for novel vaccine delivery [123].

Recent clinical trials on fifty-one patients reported by Melssen and team show LPS and poly-ICLC are safe and effective vaccine adjuvants even when combined with incomplete Freund's adjuvant (IFA), thus conferring protection against melanoma [124]. In fact, the results demonstrated that a multi-peptide vaccine with TLR agonists enhanced T-cell responses against the pathogen, which is otherwise difficult to achieve, especially without an adjuvant. The multi-peptide-based vaccines can also plausibly be designed to include multiple epitopes from more than one antigen.

To date, four TLRs agonists have been approved by the FDA for cancer treatment which are monophosphoryl lipid A (MPLA), IMQ, and REQ [125]. Among them, Bacillus Calmette-Guerin (BCG) was the first FDA-licensed TLR-based adjuvant used in bladder cancer vaccines. BCG is a mixture of TLR2/TLR4 agonists derived from *Mycobacterium bovis* bacteria that was previously used against tuberculosis. The licensed adjuvants are considered few despite extensive research and technology, and surprisingly, there is no information about standard adjuvants used for peptide-based vaccines against breast cancer. Hence, there is a necessity to explore the standard combination of vaccine-adjuvants that is effective.

Conjugation of the peptide sequence with immunostimulatory molecules such as TLRs and encapsulation with cancer antigens would enhance killing mechanisms against the tumor and improve its immunological effectiveness by targeting APC-expressed PRRs and allowing vaccines to achieve effective and long-term immunity [126]. For instance, it also improves the stability and reproducibility of vaccines besides amplifying the onset of an immune response [127,128].

5.2.2. Granulocyte Macrophage-Colony Stimulating Factor (GM-CSF)

GM-CSF is a type of cytokine derived from the activation of several types of cells, including T cells, B cells, macrophages, fibroblasts, monocytes, mast cells, and endothelial cells [129]. GM-CSF improves the role of APCs by activating, maturing, and conscripting dendritic cells (DCs), macrophages, eosinophils, and monocytes [130]. GM-CSF is a potent chemotactic factor that increases the expression of pro-inflammatory mediators such as IL-1 and TNF-, which in turn up-regulates GM-CSF itself [131]. Owing to this reason, GM-CSF has the potential to be explored as an immunostimulatory substance in various conditions such as autoimmunity, inflammation, and cancer [132–135]. In preclinical studies, the use of GM-CSF as an immunostimulant agent has been shown to evoke strong cell-mediated immune responses, thus suppressing tumor growth [136,137].

5.2.3. Keyhole Limpet Hemocyanin (KLH)

Keyhole limpet hemocyanin (KLH) is a potent immunogenic protein carrier that is able to induce both T cell and B cell production in animals and humans [138,139]. It is derived from the hemolymph of the inedible sea mollusk, *Megathura crenulata*. KLH was first introduced to patients in 1967 to determine the immunologic responsiveness of individuals undergoing immunosuppressive therapy with amethopterin or azathioprine [140,141]. KLH has also been used clinically as a carrier and adjuvant for the vaccine. Riggs and colleagues have demonstrated that KLH itself is able to inhibit the growth of ZR75-1, MCF-7, and PANC-1 cancer cell lines in vitro by an average of more than 30%. KLH acts as an immune stimulant when used as a conjugate vaccine, especially for peptide- and carbohydrate-based vaccines [142]. A current study from Wimmers et al. [138] monitored B cell responses to KLH in Stage III melanoma patients. The study found that a massive >100-fold expansion of CD19+ B cells was observed in all patients analyzed. Based on many previous studies, it can be concluded that KLH is a promising immunostimulant agent.

6. Selection of Main TAA-Derived Peptide Antigens

Peptides applied in the breast cancer vaccines have described in the Table 1. The peptides include GP2 peptide, peptide I-6, P5 peptide, MUC1-specific peptide vaccine sequence APGSTAPPA and SAPDTRPAP, E75 peptide, p5 HER-2/neu derived peptide and long peptide (conjugating SU18 peptide with SU22 peptide using glycine linker). Table 2 summarized the selected breast cancer peptides in the clinical trials.

Table 1. Peptides applied in breast cancer vaccines.

Peptides	Mechanism of Action	Types of Study	Results	Ref.
GP2 Peptide	Stimulates helper T cells, cytotoxic T lymphocytes, and antibodies	In vivo study in xenograft mice using TUBO cells	GP2 peptide alone did not have a significant therapeutic and prophylactic effect in mice	[143]
Peptide I-6	Targets MAGE-1 on breast cancer, thus, inducing the antitumor effect from CTLs	In vitro study: MDA-MB-231 cells In vivo study: MCF-7 cells	I-6 induced cytotoxic activity against MDA-MB-231 cells by activating CD8þ T lymphocytes	[32]
P5 peptide (HER-2 derived peptide)	P5 peptide releases a high amount of IFN-γ and IL-10, therefore, inducing a potent CTL immune response	In vivo: induce TUBO cells in BALB/c mice	P5 peptide conjugated with maleimide-PEG2000-DSPE incorporation into liposomes stimulate immunogenicity and anti-tumour activities more potent than P5 peptide alone	[109]
MUC1-specific peptide vaccine sequence APGSTAPPA and SAPDTRPAP	The peptide induces IFN-γ-producing T cells	In vitro MTag cell lines In vivo: Mammary gland tumors from PyV MT mice PyV MT mice	Immunosuppression within the tumor microenvironment hinders the immune response to anti-cancer vaccines	[144]

Table 1. Cont.

Peptides	Mechanism of Action	Types of Study	Results	Ref.
E75 Peptide, also known as p369 peptide	Ability to bind specific CD8$^+$ TL clones that could lyse HER2-positive tumor cells	In vitro breast cancer cell lines; MCF-7, MDA-MB-231 In vivo mice model Clinical trial	Two Phase-II clinical trials on patients resulted in remission after breast cancer but were considered at high risk of recurrence.	[52,145–147]
p5 HER-2/neu derived peptide	Induce a high level of CD8+ CTL, which is capable of killing tumor cells via recognizing the TAAs epitopes presented on the surface of cancer cells in association with MHC I molecules.	In vitro: TUBO cell In vivo: Female BALB/c mice were subcutaneously administered at the right flank	Free p5 peptide showed weak antitumor and CTLs response activities compared to Liposome–DOPE–p5 + CpG-ODN formulation	[65]
Long peptide (conjugating SU18 peptide with SU22 peptide using glycine linker)	The long peptides (containing T helper and killer epitope) targeted the overexpression of Survivin antigens in breast cancer cells.	Clinical trial (Phase 1). The vaccine was given every two weeks for 4 times.	A customized peptide with multiple epitopes and containing a long sequence of amino acids provide superior and innovative cancer vaccine designs, which are capable of inducing both Th1 and Th2 immune responses in cancer patients.	[148]

Table 2. Summary selected breast cancer peptides in clinical trials.

Agent	Phase	Adjuvant	Enrolment	Regime of Treatment	Ref.
NeuVax™ (Nelipepimut-S or E75)	III	Leukine® [sargramostim, GM-CSF]	758 patients	Once a month, for six consecutive months, and then booster for every six months total of 36 months	[52]
HER-2/neu ECD & ICD Peptides	I	Granulocyte-macrophage colony-stimulating factor (GM-CSF)	8 patients	Once a month for 2–6 months, intradermally	[149]
Folate Receptor Alpha (FRα) peptide vaccine	II	GM-CSF	80 patients	Single ID administration—monthly vaccinations repeated six times, followed by boosters every six months until recurrence.	[150–152]
MUC-1 peptide vaccine	I	poly-ICLC	29 patients	Subcutaneous (SC) injection in weeks 0, 2, and 10.	[153]
AE37 Peptide Vaccine	II	GMCSF	600 patients	Intradermally (ID) injection every 3–4 weeks for a total of up to 6 inoculations followed up every 3 months for the first 2 years.	[154]
E39 and J65 peptide vaccine	I	GMCSF	39 patients	Receive six monthly injections of peptide + GM-CSF booster inoculation within 1–2 weeks of their 6-month period	[155]
hTERT/Survivin Multi-Peptide Vaccine	1	-	11 patients	Receive subcutaneous injection every two weeks four times, then monthly up to 28 vaccinations, then every six months	[156]
WT1 peptide-based	I	Montanide ISA51	2 patients	Receive WT1 peptide intradermally three times at 2-week intervals	[157,158]
Ii-Key hybrid HER-2/neu peptide (AE37) vaccine	I	GM-CSF	15 patients	Receive vaccine via intradermal injection for six months	[159]

7. Nanoparticles as Peptide Vaccine Delivery Platform

Conjugation at the peptide terminals or encapsulation with liposomes, nanoparticles, immune-stimulating complexes (ISCOMs), and hydrogel has been extensively explored as an ideal delivery platform that can protect the peptide from degradation without jeopardizing its efficacy and safety. This effort could be synchronized with adjuvants to develop successful vaccines. Moreover, these manipulations are capable of increasing the APCs uptake, resulting in improved binding to MHC or T cell receptor sensitivity (TCR) and improving the secretion of interleukin-2 (IL-2) molecules and other mediators for CD8+ and CD4+ activation. As a consequence, the use of conjugation and encapsulation of nanoparticles tends to enhance the biostability and delivery system at the target site and make the vaccine more potent.

Liposomes are phospholipids or lipid bilayers that form closed membrane vesicles. Many researchers have proposed the use of liposomes in peptide-based vaccine development due to their good encapsulation efficiency, biodegradability, biocompatibility, and non-toxicity. Liposomes can be integrated or intercalated with lipophilic and hydrophilic peptides [160]. Moreover, liposomes have a long circulation in the body, and this provides an advantage for cellular uptake with APCs.

Integration of peptides onto the surface of the liposomes has become a famous approach in vaccine delivery due to its effectiveness. With regard to breast cancer vaccines, Razazan and colleagues developed a vaccine from HER2/neu-derived peptides using liposomes as a carrier delivery [143]. This lipopeptide vaccine induces a high CTL immune response and prolonged survivability in the BALB/c mouse model.

In another study, the use of cationic liposomes anchored with proteins conjugated with CpGODN elicited CD4+ T-cell responses. The positive charge benevolence formation of the depot action at the injection site is followed by a sustained release to the draining lymph nodes [123,161].

Nanoparticles (NPs) serve as a delivery platform in many vaccine formulations. The particle size, surface charge, and antigen loading mode of NPs greatly influence the adjuvanticity and immunoreactivity effects of potent and smooth vaccine delivery systems. Examples of famous nanomaterials that have continuously gained attention in the preparation of peptide-based vaccines are PLGA and chitosan. The strategy of using PLGA or chitosan as a vaccine delivery system has gained popularity among researchers. Both are types of polymers that are well-known as being non-harmful, non-toxic, biodegradable, and biocompatible. These nano-polymers, when incorporated with a peptide vaccine, can protect the peptide from degradation by proteolytic enzymes, thus providing more effective uptake and delivery to the lymphatic system [162]. Moreover, it could enhance and regulate immune responses.

Encapsulation of peptide antigens in polymeric particulates provides greater access to barrier compartments such as endothelial cell junctions, including the blood–brain barrier, when designed in a nanosize range of 120 nm; thus, this can be exploited for vaccine formulations, while sizes outside of this range are more likely to be blocked and cleared from the circulation [163,164]. PLGA, composed of two monomers, polylactic acid (LA) and polyglycolic acid, are the most ardent and ideal transport options for biomolecules as they are moist, biodegradable, biocompatible polymers and sturdy materials for medicinal and cosmetic applications [165]. They are metabolized to H_2O and CO_2 as end products before they are eliminated from the body.

To date, it is the most favored polymer to be used for drug delivery and is capable of protecting peptides from enzymatic degradation due to its ability to avoid the endosomal pouch. Peptide encapsulation into PLGA nanoparticles would provide a solid shield and efficiently improve drug release. Kroll et al. [166] demonstrated the ability of membrane-coated nanoparticle cancer cells to cause multi-antigenic antitumor immunity and extend the lives of mice with melanoma.

Gu et al. [167] reported that peptides coated with PLGA could protect peptide antigens from degradation by proteases and lysosomes and co-deliver the peptide antigen uptake by the APCs. Peptide antigen encapsulated in PLGA displayed slow antigen release. This process elicits immunological responses due to adequate antigen uptake by APCs, which express the fragments on their surface and then cross-present them through the MHC-binding complex pathway. Binding with MHC class 1 mostly activates cell-mediated immunity, while antigen binding with MHC class II optimally triggers an antibody response [168]. PLGA has been chosen as a compound delivery vehicle due to its wide safety profile and has been licensed by the FDA for medical applications [169,170]. Co-delivery of vaccine formulations in PLGA has shown promising results. For instance, Ma and co-researchers have developed a DC-based vaccine loaded with PLGA-NPs encapsulating the mSTEAP peptide against adenocarcinoma in C57BL/6 mice. The result was that PLGA NPs mediated good platform delivery and elicited strong immune responses in vivo [171].

Chen and colleagues have developed a combination of PLGA-ICG-R837, which is photothermal therapy with checkpoint blockade, adjuvant, and nanoparticles, against 4T1 mouse mammary carcinoma in BALB/c mice [172]. Synergistic anticancer effects were triggered by the combination of PLGA-ICG-R837 and checkpoint blockage, which induce potent immune responses from CTLs, thus suppressing tumor growth. This result indicated the benefit of using that combination, which may not just suppress cancer growth but also confer lifelong protection to prevent tumor relapse. Interestingly, the NPs and adjuvants used are FDA-approved ones, which have a good prospect for clinical translation.

Chu et al. [173] reported the potential of using chitosan NPs as a vaccine carrier for peptide-based vaccines due to their adjuvant effects as well as their stability properties in an acidic environment [174,175], which are suitable in breast cancer environments. Jadidi-Niaragh demonstrated that a dendritic cell vaccine incorporated with chitosan-lactate nanoparticles (ChLa NPs) inhibits metastasis and suppresses tumor growth, thus improving mice's survival [176]. The incorporation of chitosan NPs into vaccine formulation was proven to promote DC activation and trigger robust cell-mediated immunity, which can seemingly be applied to cancer therapy [177]. With this evidence, such approaches can be employed and exploited for the purpose of fighting breast cancer.

As a consequence, the use of peptide nanoparticles camouflaged with cancer cell membranes has become a new and interesting strategy in vaccinology. Fang et al. [178] developed a cancer vaccine by utilizing biodegradable polymeric nanoparticles cloaked with the MDA-MB-435 cell membrane, and a monophosphoryl lipid A-TLR-4 activator was used as an adjuvant. The cancer cell membrane-cloaked NPs vaccine demonstrated high dendritic cell maturation, which resulted in higher secretion of IFN. It is believed that MDA-MB-435 overexpressed galectin-3 and CEA, thus triggering this vaccine's antitumor response via the homologous binding mechanism. The idea was further explored by Chen et al. [172], which encapsulated PLGA-NPs with MDA-MB-231 and Hela cell line fragments. Doxorubicin and PD-L1 siRNA checkpoint inhibitors were loaded. The anticancer drug is capable of detecting cancer with an accurate and sensitive CTL response. As is currently the case, the restrictions on the use of most anticancer drugs do not provide long-lasting protection and have limited circulatory duration. Instead of using anticancer drugs, perhaps the methods can be shifted to peptide-based materials, which also have anticancer activity.

8. Future Direction: Rational Vaccine Design

8.1. Biomimetic Nano-Peptide Vaccine

The comprehensive advancement of the nanoparticle delivery system has driven significant improvements in cancer therapy. Biomimetic nanotechnology has recently attracted attention with its notion of wrapping polymeric nanoparticles (NPs) with cell membranes; they were extracted from breast cancer cells. This has resulted in multi-immune responses to different tumor antigens attributable to receptor-ligand interactions in surface cells [179,180], thereby enhancing biological adhesion and immune clearance [181]. This follows the weak efficacy of a single antigenic determinant vaccine, which lacks immune defenses, particularly when confronted with mutated and heterogeneous cancer cells.

The cell membrane is coated to preserve the biological function of the vaccine in inducing a robust and precise immune response. 'Ghost cancer cell membrane' is a term that has been used to extract intracellular cell material and preserve the cell membrane. The contents can be critically extracted by incubation in a hypotonic solution accompanied by sonication and by co-extrusion of the cell membrane and nanoparticle with the extruder. The illustration is shown in Figure 3.

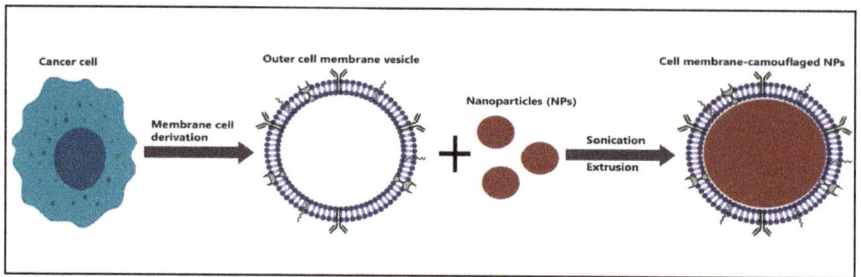

Figure 3. Schematic illustration of the preparation of cancer cell membrane-camouflaged nanoparticles. The intracellular contents of the cell can be extracted to form a 'ghost cancer cell' by applying hypotonic treatment. The NPs can be fused through sonication followed by extrusion.

This biomimetic concept can be adopted and further improved for peptide-based vaccine technology. Research by Jin et al. [182] showed the applicability of this biomimetic development to cancer immunotherapy-coated primary human glioblastoma cell membrane fractions (U87 and U87-CXCR4 cell lines) in PLGA NPs. The results showed that CD4+ and CD8+ T lymphocytes in the lymph nodes and spleens of the Balb/cc mouse model were activated, and the metastatic burden was significantly reduced. Future insights on the use of biomimetic technologies for breast cancer vaccinations are highly exciting and yet emerging.

8.2. Combining Immune Checkpoint Blockade Agents with Peptide-Based Vaccine

Similar to peptide-based cancer vaccines, to date, none of the currently available vaccinations can provide complete protection. With few side effects, it may be useful in cases of early cancer detection for preventing relapse or enhancing survival. Therefore, combinatorial treatments, on the other hand, may be able to successfully treat even advanced cancers while also overcoming immune escape mechanisms and tumor-mediated immunosuppression issues [183,184]. For example, by combining with immune checkpoint blockade therapies such as programmed cell death-1 (PD-1) and CTLA-4 targeted antibodies.

The vital importance of an efficient immune response for controlling cancerous cells was discussed when scientists found out that breast cancer is also a type of immunogenic disease [26,185]. Additionally, there is strong evidence indicating a link between a beneficial outcome in different malignancies and tumor-infiltrating lymphocytes (TILs) in tumor tissue [186,187]. The PD-1 receptor interacts with its ligand (either PD-L1 or PD-L2) on cancer cells, resulting in immune checkpoint pathway activation. When PD-1 is overexpressed on T cells, B cells, and NK cells, those immune cells are suppressed and deactivated, hijacked by tumors [188,189]. Anti-PD-1 agents have shown promising results in the metastatic environment, while combination strategies tend to enhance more responses [188]. In animal models, a combination of a peptide-based vaccine with CTLA-4, specific to T-cell inhibitory receptors, has shown remarkable activation of tumor-specific T cells against cancerous cells [190].

Several studies pertaining to combinational therapy with cancer vaccines and PD-1 were reported. The explanation for this is that some cancer patients do not respond to anti-PD-1 agents due to a lack of TILs and cancer vaccines that induce effector T-cell infiltration into the tumors. Thus, this provides a convincing concept that anti-PD-1/PD-L1 agents may act synergistically to induce a stronger immune response against the tumor [191].

9. Conclusions

Overall, the development of peptide-derived cancer antigen vaccines is encouraging because of their advantages in identifying target antigens and their ability to select appropriate adjuvants to combine to enhance immunogenicity. It opens a window for future breast cancer therapy. Further focus should be given to their experimental translation and their clinical implementation. The weak immunogenicity issue of the peptide can be improved with coadministration with an immunostimulatory agent such as KLH, GM-CSF, CpG-ODN, or poly:IC because the most important value of using peptides from an experimental point of view is their effectiveness with a wide safety margin.

Additionally, the peptide antigen itself can be chemically modified. For example, the lipopeptides derived from microbes can self-assemble into nanoparticles due to their amphiphilic properties. Changes in the microenvironment of breast cancer (poor tumor vasculature, hypoxia, acidic area) plus immunosuppression of the patient require improvement in peptide-based therapy that perhaps has antimicrobial properties, preventing secondary bacterial infections.

Intercalating the modification with multiple antigens and immunostimulatory substances would offer a potent immune response from the innate and adaptive immune systems. Moreover, multiple antigenic and chimeric peptide approaches offer the opportunity for antigenic peptides from various TAAs and CTL peptide epitopes to be covalently combined in a single system. In order to render their biological functions, the designated multiple peptides, antibodies, and nanoparticles can be further incorporated into the functionalized eukaryotic cell membrane. Owing to the cancer cell's antigenic diversity, membrane-coated nanoparticles are expected to be promising and can be used specifically to target breast cancer. Numerous cancer-associated moieties—proteins, ligands, and receptors—are found in the vaccine formulation from the use of peptides extracted from breast cancer. Among others, there must be a few or an entire group that is responsible for effective targeting, which can kill two birds with one stone. The concept of using a peptide-based vaccine for breast cancer therapy is fashionable and possible to accomplish. The more complex and comprehensive the system, the less chance there is for the cancer cells to escape.

Author Contributions: Conceptualization, M.L.N., A.K.A. and F.A.; validation, M.L.N. and F.A.; investigation, M.L.N., A.K.A. and A.H.N.; writing—original draft preparation, M.L.N., A.K.A., A.H.N., W.N., P.Y.N., Z.A.Z. and F.A.; writing—review and editing, M.L.N., A.K.A., A.H.N., W.N., P.Y.N., K.Y., N.A., Z.A.Z., N.I., F.A. and K.P.L.; funding acquisition, A.K.A. and N.I. All authors have read and agreed to the published version of the manuscript.

Funding: This research was financially supported by the Fundamental Research Grant Scheme (FRGS) funded by the Ministry of Higher Education, Malaysia, grant number FRGS/1/2019/SKK06/UKM/02/5.

Institutional Review Board Statement: Not applicable.

Informed Consent Statement: Not applicable.

Data Availability Statement: Data is contained within the article.

Acknowledgments: We would like to express our gratitude to the Faculty of Pharmacy, National University of Malaysia, Faculty of Veterinary Medicine, Universiti Malaysia Kelantan, and Ministry of Higher Education Malaysia, which provided financial support from the Fundamental Research Grant Scheme (FRGS) via grant number FRGS/1/2019/SKK06/UKM/02/5. We would like to thank the BioRender software (https://biorender.com, accessed on 10 March 2022) for the idea of the illustration.

Conflicts of Interest: The authors declare no conflict of interest.

References

1. World Health Organization. Breast Cancer: Prevention and Control. Available online: https://www.who.int/cancer/prevention/diagnosis-screening/breast-cancer/en/ (accessed on 10 December 2020).
2. Reddington, R.; Galer, M.; Hagedorn, A.; Liu, P.; Barrack, S.; Husain, E.; Sharma, R.; Speirs, V.; Masannat, Y. Incidence of Male Breast Cancer in Scotland over a Twenty-Five-Year Period (1992–2017). *Eur. J. Surg. Oncol.* **2020**, *46*, 1546–1550. [CrossRef] [PubMed]
3. Giordano, S.H. Breast Cancer in Men. *N. Engl. J. Med.* **2018**, *378*, 2311–2320. [CrossRef] [PubMed]
4. Gargiulo, P.; Pensabene, M.; Milano, M.; Arpino, G.; Giuliano, M.; Forestieri, V.; Condello, C.; Lauria, R.; De Placido, S. Long-Term Survival and BRCA Status in Male Breast Cancer: A Retrospective Single-Center Analysis. *BMC Cancer* **2016**, *16*, 375. [CrossRef] [PubMed]
5. Gaddam, S.; Heller, S.L.; Babb, J.S.; Gao, Y. Male Breast Cancer Risk Assessment and Screening Recommendations in High-Risk Men Who Undergo Genetic Counseling and Multigene Panel Testing. *Clin. Breast Cancer* **2021**, *21*, e74–e79. [CrossRef]
6. Siegel, R.L.; Miller, K.D.; Jemal, A. Cancer Statistics, 2015. *CA Cancer J. Clin.* **2015**, *65*, 5–29. [CrossRef]
7. Shah, R.; Rosso, K.; David Nathanson, S. Pathogenesis, Prevention, Diagnosis and Treatment of Breast Cancer. *World J. Clin. Oncol.* **2014**, *5*, 283–298. [CrossRef]
8. Ly, D.; Forman, D.; Ferlay, J.; Brinton, L.A.; Cook, M.B. An International Comparison of Male and Female Breast Cancer Incidence Rates. *Int. J. Cancer* **2013**, *132*, 1918–1926. [CrossRef]
9. Chakraborty, S.; Rahman, T. The Difficulties in Cancer Treatment. *Ecancermedicalscience* **2012**, *6*, ed16. [CrossRef]
10. Ekici, S.; Jawzal, H. Breast Cancer Diagnosis Using Thermography and Convolutional Neural Networks. *Med. Hypotheses* **2020**, *137*, 109542. [CrossRef]
11. Sánchez-Bermúdez, A.I.; Sarabia-Meseguer, M.D.; García-Aliaga, A.; Marín-Vera, M.; Macías-Cerrolaza, J.A.; Henaréjos, P.S.; Guardiola-Castillo, V.; la Peña, F.A.-D.; Alonso-Romero, J.L.; Noguera-Velasco, J.A.; et al. Mutational Analysis of RAD51C and RAD51D Genes in Hereditary Breast and Ovarian Cancer Families from Murcia (Southeastern Spain). *Eur. J. Med. Genet.* **2018**, *61*, 355–361. [CrossRef]
12. Karami, F.; Mehdipour, P. A Comprehensive Focus on Global Spectrum of BRCA1 and BRCA2 Mutations in Breast Cancer. *Biomed. Res. Int.* **2013**, *2013*, 928562. [CrossRef]
13. Ayala de la Peña, F.; Andrés, R.; Garcia-Sáenz, J.A.; Manso, L.; Margelí, M.; Dalmau, E.; Pernas, S.; Prat, A.; Servitja, S.; Ciruelos, E. SEOM Clinical Guidelines in Early Stage Breast Cancer (2018). *Clin. Transl. Oncol.* **2019**, *21*, 18–30. [CrossRef] [PubMed]
14. Suter, R.; Marcum, J.A. The Molecular Genetics of Breast Cancer and Targeted Therapy. *Biologics* **2007**, *1*, 241–258. [PubMed]
15. Townsend, D.M.; Tew, K.D. The Role of Glutathione-S-Transferase in Anti-Cancer Drug Resistance. *Oncogene* **2003**, *22*, 7369–7375. [CrossRef] [PubMed]
16. Aziz, M.H.; Ahmad, A. Epigenetic Basis of Cancer Drug Resistance. *Cancer Drug Resist.* **2020**, *3*, 113–116. [CrossRef] [PubMed]
17. Qin, S.; Xu, L.; Yi, M.; Yu, S.; Wu, K.; Luo, S. Novel Immune Checkpoint Targets: Moving beyond PD-1 and CTLA-4. *Mol. Cancer* **2019**, *18*, 155. [CrossRef]
18. Ji, X.; Lu, Y.; Tian, H.; Meng, X.; Wei, M.; Cho, W.C. Chemoresistance Mechanisms of Breast Cancer and Their Countermeasures. *Biomed. Pharmacother.* **2019**, *114*, 108800. [CrossRef]
19. Crespo, I.; Götz, L.; Liechti, R.; Coukos, G.; Doucey, M.A.; Xenarios, I. Identifying Biological Mechanisms for Favorable Cancer Prognosis Using Non-Hypothesis-Driven Iterative Survival Analysis. *NPJ Syst. Biol. Appl.* **2016**, *2*, 16037. [CrossRef]
20. Anfray, C.; Ummarino, A.; Andón, F.T.; Allavena, P. Current Strategies to Target Tumor-Associated-Macrophages to Improve Anti-Tumor Immune Responses. *Cells* **2020**, *9*, 46. [CrossRef]
21. Davis, R.J.; Van Waes, C.; Allen, C.T. Overcoming Barriers to Effective Immunotherapy: MDSCs, TAMs, and Tregs as Mediators of the Immunosuppressive Microenvironment in Head and Neck Cancer. *Oral Oncol.* **2016**, *58*, 59–70. [CrossRef]
22. Lin, Y.; Xu, J.; Lan, H. Tumor-Associated Macrophages in Tumor Metastasis: Biological Roles and Clinical Therapeutic Applications. *J. Hematol. Oncol.* **2019**, *12*, 76. [CrossRef]
23. Schmitt, N.; Bustamante, J.; Bourdery, L.; Bentebibel, S.E.; Boisson-Dupuis, S.; Hamlin, F.; Tran, M.V.; Blankenship, D.; Pascual, V.; Savino, D.A.; et al. IL-12 Receptor B1 Deficiency Alters in Vivo T Follicular Helper Cell Response in Humans. *Blood* **2013**, *121*, 3375–3385. [CrossRef]
24. Lindau, D.; Gielen, P.; Kroesen, M.; Wesseling, P.; Adema, G.J. The Immunosuppressive Tumour Network: Myeloid-Derived Suppressor Cells, Regulatory T Cells and Natural Killer T Cells. *Immunology* **2013**, *138*, 105–115. [CrossRef] [PubMed]
25. Li, S.; Yao, M.; Niu, C.; Liu, D.; Tang, Z.; Gu, C.; Zhao, H.; Ke, J.; Wu, S.; Wang, X.; et al. Inhibition of MCF-7 Breast Cancer Cell Proliferation by a Synthetic Peptide Derived from the C-Terminal Sequence of Orai Channel. *Biochem. Biophys. Res. Commun.* **2019**, *516*, 1066–1072. [CrossRef]
26. Gatti-Mays, M.E.; Balko, J.M.; Gameiro, S.R.; Bear, H.D.; Prabhakaran, S.; Fukui, J.; Disis, M.L.; Nanda, R.; Gulley, J.L.; Kalinsky, K.; et al. If We Build It They Will Come: Targeting the Immune Response to Breast Cancer. *NPJ Breast Cancer* **2019**, *5*, 37. [CrossRef]
27. Gun, S.Y.; Lee, S.W.L.; Sieow, J.L.; Wong, S.C. Targeting Immune Cells for Cancer Therapy. *Redox Biol.* **2019**, *25*, 101174. [CrossRef] [PubMed]
28. Zhang, Y.; Zhang, Z. The History and Advances in Cancer Immunotherapy: Understanding the Characteristics of Tumor-Infiltrating Immune Cells and Their Therapeutic Implications. *Cell Mol. Immunol.* **2020**, *17*, 807–821. [CrossRef] [PubMed]

29. Criscitiello, C.; Viale, G.; Curigliano, G. Peptide Vaccines in Early Breast Cancer. *Breast* **2019**, *44*, 128–134. [CrossRef]
30. Yamaguchi, Y.; Yamaue, H.; Okusaka, T.; Okuno, K.; Suzuki, H.; Fujioka, T.; Otsu, A.; Ohashi, Y.; Shimazawa, R.; Nishio, K.; et al. Guidance for Peptide Vaccines for the Treatment of Cancer. *Cancer Sci.* **2014**, *105*, 924–931. [CrossRef]
31. Waks, A.G.; Winer, E.P. Breast Cancer Treatment: A Review. *JAMA J. Am. Med. Assoc.* **2019**, *321*, 288–300. [CrossRef]
32. Shi, W.; Qiu, Q.; Tong, Z.; Guo, W.; Zou, F.; Feng, Z.; Wang, Y.; Huang, W.; Qian, H. Synthetic Tumor-Specific Antigenic Peptides with a Strong Affinity to HLA-A2 Elicit Anti-Breast Cancer Immune Response through Activating CD8+ T Cells. *Eur. J. Med. Chem.* **2020**, *189*, 112051. [CrossRef]
33. Nguyen, T.L.; Choi, Y.; Kim, J. Mesoporous Silica as a Versatile Platform for Cancer Immunotherapy. *Adv. Mater.* **2019**, *31*, e1803953. [CrossRef]
34. De Temmerman, M.L.; Rejman, J.; Demeester, J.; Irvine, D.J.; Gander, B.; De Smedt, S.C. Particulate Vaccines: On the Quest for Optimal Delivery and Immune Response. *Drug Discov. Today* **2011**, *16*, 569–582. [CrossRef]
35. Nejad, A.E.; Najafgholian, S.; Rostami, A.; Sistani, A.; Shojaeifar, S.; Esparvarinha, M.; Nedaeinia, R.; Javanmard, S.H.; Taherian, M.; Ahmadlou, M.; et al. The role of hypoxia in the tumor microenvironment and development of cancer stem cell: A novel approach to developing treatment. *Cancer Cell Intl.* **2021**, *21*, 62. [CrossRef]
36. Solinas, C.; Aiello, M.; Migliori, E.; Willard-Gallo, K.; Emens, L.A. Breast cancer vaccines: Heeding the lessons of the past to guide a path forward. *Cancer Treat. Rev.* **2020**, *84*, 101947. [CrossRef] [PubMed]
37. de Paula Peres, L.; da Luz, F.A.C.; dos Anjos Pultz, B.; Brígido, P.C.; de Araújo, R.A.; Goulart, L.R.; Silva, M.J.B. Peptide Vaccines in Breast Cancer: The Immunological Basis for Clinical Response. *Biotechnol. Adv.* **2015**, *33*, 1868–1877. [CrossRef]
38. Criscitiello, C. Tumor-Associated Antigens in Breast Cancer. *Breast Care* **2012**, *7*, 262–266. [CrossRef] [PubMed]
39. Thundimadathil, J. Cancer Treatment Using Peptides: Current Therapies and Future Prospects. *J. Amino Acids* **2012**, *2012*, 1–13. [CrossRef] [PubMed]
40. Blum, J.S.; Wearsch, P.A.; Cresswell, P. Pathways of Antigen Processing. *Annu. Rev. Immunol.* **2013**, *31*, 443–473. [CrossRef]
41. Zhang, L.; Huang, Y.; Lindstrom, A.R.; Lin, T.Y.; Lam, K.S.; Li, Y. Peptide-Based Materials for Cancer Immunotherapy. *Theranostics* **2019**, *9*, 7807–7825. [CrossRef]
42. Santoni, D. Viral Peptides-MHC Interaction: Binding Probability and Distance from Human Peptides. *J. Immunol. Methods* **2018**, *459*, 35–43. [CrossRef]
43. Jones, N.D.; Van Maurik, A.; Hara, M.; Spriewald, B.M.; Witzke, O.; Morris, P.J.; Wood, K.J. CD40-CD40 Ligand-Independent Activation of CD8+ T Cells Can Trigger Allograft Rejection. *J. Immunol.* **2000**, *165*, 1111–1118. [CrossRef] [PubMed]
44. Tay, N.Q.; Lee, D.C.P; Chua, Y.L.; Prabhu, N.; Gascoigne, N.R.J.; Kemeny, D.M. CD40L Expression Allows CD8+ T Cells to Promote Their Own Expansion and Differentiation through Dendritic Cells. *Front. Immunol.* **2017**, *8*, 1484. [CrossRef] [PubMed]
45. Wong, K.L.; Tang, L.F.M.; Lew, F.C.; Wong, H.S.K.; Chua, Y.L.; MacAry, P.A.; Kemeny, D.M. CD44high Memory CD8 T Cells Synergize with CpG DNA to Activate Dendritic Cell IL-12p70 Production. *J. Immunol.* **2009**, *183*, 41–50. [CrossRef] [PubMed]
46. Ikeda, H.; Shiku, H. Immunotherapy of Solid Tumor: Perspectives on Vaccine and Cell Therapy. *Nihon Rinsho* **2012**, *70*, 2043–2050.
47. Costa, R.L.B.; Czerniecki, B.J. Clinical Development of Immunotherapies for HER2+ Breast Cancer: A Review of HER2-Directed Monoclonal Antibodies and Beyond. *NPJ Breast Cancer* **2020**, *6*, 10. [CrossRef]
48. Lyu, H.; Huang, J.; He, Z.; Liu, B. Epigenetic Mechanism of Survivin Dysregulation in Human Cancer. *Sci. China Life Sci.* **2018**, *61*, 808–814. [CrossRef]
49. Genitsch, V.; Zlobec, I.; Thalmann, G.N.; Fleischmann, A. MUC1 Is Upregulated in Advanced Prostate Cancer and Is an Independent Prognostic Factor. *Prostate Cancer Prostatic. Dis.* **2016**, *19*, 242–247. [CrossRef]
50. Malonis, R.J.; Lai, J.R.; Vergnolle, O. Peptide-Based Vaccines: Current Progress and Future Challenges. *Chem. Rev.* **2020**, *120*, 3210–3229. [CrossRef]
51. Kim, I.; Sanchez, K.; McArthur, H.L.; Page, D. Immunotherapy in Triple-Negative Breast Cancer: Present and Future. *Curr. Breast Cancer Rep.* **2019**, *11*, 259–271. [CrossRef]
52. Mittendorf, E.A.; Lu, B.; Melisko, M.; Hiller, J.P.; Bondarenko, I.; Brunt, A.M.; Sergii, G.; Petrakova, K.; Peoples, G.E. Efficacy and Safety Analysis of Nelipepimut-S Vaccine to Prevent Breast Cancer Recurrence: A Randomized, Multicenter, Phase III Clinical Trial. *Clin. Cancer Res.* **2019**, *25*, 4248–4254. [CrossRef] [PubMed]
53. Jackson Doreen, O.; Francois, T.A.; Travis, C.G.; Vreeland Timothy, J.; Peace Kaitlin, M.; Hale Diane, F.; Litton Jennifer, K.; Murray James, L.; Perez Sonia, A.; Michael, P.; et al. Effects of HLA Status and HER2 Status on Outcomes in Breast Cancer Patients at Risk for Recurrence—Implications for Vaccine Trial Design. *Clin. Immunol.* **2018**, *195*, 28–35. [CrossRef]
54. Patil, R.; Clifton, G.T.; Holmes, J.P.; Amin, A.; Carmichael, M.G.; Gates, J.D.; Benavides, L.H.; Hueman, M.T.; Ponniah, S.; Peoples, G.E. Clinical and Immunologic Responses of HLA-A3+ Breast Cancer Patients Vaccinated with the HER2/Neu-Derived Peptide Vaccine, E75, in a Phase I/II Clinical Trial. *J. Am. Coll. Surg.* **2010**, *210*, 140–147. [CrossRef] [PubMed]
55. Gourley, C.; Thornton, C.; Massie, C.; Prescott, R.J.; Turner, M.; Leonard, R.C.F.; Kilpatrick, D.C. Is There a Relationship between HLA Type and Prognostic Factors in Breast Cancer? *Anticancer Res.* **2003**, *23*, 633–638.
56. Mahjoubin-Tehran, M.; Rezaei, S.; Jalili, A.; Aghaee-Bakhtiari, S.H.; Orafai, H.M.; Jamialahmadi, T.; Sahebkar, A. Peptide Decoys: A New Technology Offering Therapeutic Opportunities for Breast Cancer. *Drug Discov. Today* **2020**, *25*, 593–598. [CrossRef]
57. Farzad, N.; Barati, N.; Momtazi-Borojeni, A.A.; Yazdani, M.; Arab, A.; Razazan, A.; Shariat, S.; Mansourian, M.; Abbasi, A.; Saberi, Z.; et al. P435 HER2/Neu-Derived Peptide Conjugated to Liposomes Containing DOPE as an Effective Prophylactic Vaccine Formulation for Breast Cancer. *Artif. Cells Nanomed. Biotechnol.* **2019**, *47*, 664–672. [CrossRef]

58. Furrer, D.; Sanschagrin, F.; Jacob, S.; Diorio, C. Advantages and Disadvantages of Technologies for HER2 Testing in Breast Cancer Specimens: Table 1. *Am. J. Clin. Pathol.* **2015**, *144*, 686–703. [CrossRef]
59. Slamon, D.J.; Clark, G.M.; Wong, S.G.; Levin, W.J.; Ullrich, A.; McGuire, W.L. Human Breast Cancer: Correlation of Relapse and Survival with Amplification of the HER-2/Neu Oncogene. *Science* **1987**, *235*, 177–182. [CrossRef]
60. Ayoub, N.M.; Al-Shami, K.M.; Yaghan, R.J. Immunotherapy for HER2-Positive Breast Cancer: Recent Advances and Combination Therapeutic Approaches. *Breast Cancer Targets Ther.* **2019**, *11*, 53–69. [CrossRef]
61. Krasniqi, E.; Barchiesi, G.; Pizzuti, L.; Mazzotta, M.; Venuti, A.; Maugeri-Saccà, M.; Sanguineti, G.; Massimiani, G.; Sergi, D.; Carpano, S.; et al. Immunotherapy in HER2-Positive Breast Cancer: State of the Art and Future Perspectives. *J. Hematol. Oncol.* **2019**, *12*, 1–26. [CrossRef]
62. Katzorke, N.; Rack, B.K.; Haeberle, L.; Neugebauer, J.K.; Melcher, C.A.; Hagenbeck, C.; Forstbauer, H.; Ulmer, H.U.; Soeling, U.; Kreienberg, R.; et al. Prognostic Value of HER2 on Breast Cancer Survival. *J. Clin. Oncol.* **2013**, *31*, 640. [CrossRef]
63. Clynes, R.A.; Towers, T.L.; Presta, L.G.; Ravetch, J.V. Inhibitory Fc Receptors Modulate in Vivo Cytoxicity against Tumor Targets. *Nat. Med.* **2000**, *6*, 443–446. [CrossRef]
64. Wang, J.; Xu, B. Targeted Therapeutic Options and Future Perspectives for Her2-Positive Breast Cancer. *Signal Transduct. Target. Ther.* **2019**, *4*, 3305. [CrossRef]
65. Mansourian, M.; Badiee, A.; Jalali, S.A.; Shariat, S.; Yazdani, M.; Amin, M.; Jaafari, M.R. Effective Induction of Anti-Tumor Immunity Using P5 HER-2/Neu Derived Peptide Encapsulated in Fusogenic DOTAP Cationic Liposomes Co-Administrated with CpG-ODN. *Immunol. Lett.* **2014**, *162*, 87–93. [CrossRef] [PubMed]
66. Costa, R.L.B.; Soliman, H.; Czerniecki, B.J. The Clinical Development of Vaccines for HER2+ Breast Cancer: Current Landscape and Future Perspectives. *Cancer Treat. Rev.* **2017**, *61*, 107–115. [CrossRef]
67. Richman, C.M.; DeNardo, S.J. Systemic Radiotherapy in Metastatic Breast Cancer Using 90Y-Linked Monoclonal MUC-1 Antibodies. *Crit. Rev. Oncol. Hematol.* **2001**, *38*, 25–35. [CrossRef]
68. Nath, S.; Mukherjee, P. MUC1: A Multifaceted Oncoprotein with a Key Role in Cancer Progression. *Trends Mol. Med.* **2014**, *20*, 332–342. [CrossRef]
69. Kovjazin, R.; Horn, G.; Smorodinsky, N.I.; Shapira, M.Y.; Carmon, L. Cell Surface-Associated Anti-MUC1-Derived Signal Peptide Antibodies: Implications for Cancer Diagnostics and Therapy. *PLoS ONE* **2014**, *9*, e85400. [CrossRef] [PubMed]
70. Lakshminarayanan, V.; Thompson, P.; Wolfert, M.A.; Buskas, T.; Bradley, J.M.; Pathangey, L.B.; Madsen, C.S.; Cohen, P.A.; Gendler, S.J.; Boons, G.J. Immune Recognition of Tumor-Associated Mucin MUC1 Is Achieved by a Fully Synthetic Aberrantly Glycosylated MUC1 Tripartite Vaccine. *Proc. Natl. Acad. Sci. USA* **2012**, *109*, 261–266. [CrossRef]
71. Xing, P.X.; Michael, M.; Apostolopoulos, V.; Prenzoska, J.; Marshall, C.; Bishop, J.; McKenzie, I.F.C. Phase I Study of Synthetic MUC1 Peptides in Breast Cancer. *Int. J. Oncol.* **1995**, *6*, 1283–1289. [CrossRef]
72. Apostolopoulos, V.; Pietersz, G.A.; Tsibanis, A.; Tsikkinis, A.; Drakaki, H.; Loveland, B.E.; Piddlesden, S.J.; Plebanski, M.; Pouniotis, D.S.; Alexis, M.N.; et al. Pilot Phase III Immunotherapy Study in Early-Stage Breast Cancer Patients Using Oxidized Mannan-MUC1 [ISRCTN71711835]. *Breast Cancer Res.* **2006**, *8*, R27. [CrossRef]
73. Jeong, S.; Park, M.J.; Song, W.; Kim, H.S. Current Immunoassay Methods and Their Applications to Clinically Used Biomarkers of Breast Cancer. *Clin. Biochem.* **2020**, *78*, 43–57. [CrossRef]
74. Brantley-Sieders, D.M.; Zhuang, G.; Hicks, D.; Wei, B.F.; Hwang, Y.; Cates, J.M.M.; Coffman, K.; Jackson, D.; Bruckheimer, E.; Muraoka-Cook, R.S.; et al. The Receptor Tyrosine Kinase EphA2 Promotes Mammary Adenocarcinoma Tumorigenesis and Metastatic Progression in Mice by Amplifying ErbB2 Signaling. *J. Clin. Investig.* **2008**, *118*, 64–78. [CrossRef]
75. Gökmen-Polar, Y.; Toroni, R.A.; Hocevar, B.A.; Badve, S.; Zhao, Q.; Shen, C.; Bruckheimer, E.; Kinch, M.S.; Miller, K.D. Dual Targeting of EphA2 and ER Restores Tamoxifen Sensitivity in ER/EphA2-Positive Breast Cancer. *Breast Cancer Res. Treat.* **2011**, *127*, 375–384. [CrossRef] [PubMed]
76. Scarberry, K.E.; Dickerson, E.B.; McDonald, J.F.; Zhang, Z.J. Magnetic Nanoparticle-Peptide Conjugates for in Vitro and in Vivo Targeting and Extraction of Cancer Cells. *J. Am. Chem. Soc.* **2008**, *130*, 10258–10262. [CrossRef] [PubMed]
77. Salem, A.F.; Wang, S.; Billet, S.; Chen, J.F.; Udompholkul, P.; Gambini, L.; Baggio, C.; Tseng, H.R.; Posadas, E.M.; Bhowmick, N.A.; et al. Reduction of Circulating Cancer Cells and Metastases in Breast-Cancer Models by a Potent EphA2-Agonistic Peptide-Drug Conjugate. *J. Med. Chem.* **2018**, *61*, 2052–2061. [CrossRef]
78. Guo, Z.; He, B.; Yuan, L.; Dai, W.; Zhang, H.; Wang, X.; Wang, J.; Zhang, X.; Zhang, Q. Dual Targeting for Metastatic Breast Cancer and Tumor Neovasculature by EphA2-Mediated Nanocarriers. *Int. J. Pharm.* **2015**, *493*, 380–389. [CrossRef] [PubMed]
79. Jha, K.; Shukla, M.; Pandey, M. Survivin Expression and Targeting in Breast Cancer. *Surg. Oncol.* **2012**, *21*, 125–131. [CrossRef] [PubMed]
80. Altieri, D.C. Survivin, Cancer Networks and Pathway-Directed Drug Discovery. *Nat. Rev. Cancer.* **2008**, *8*, 61–70. [CrossRef]
81. Ryan, B.M.; Konecny, G.E.; Kahlert, S.; Wang, H.J.; Untch, M.; Meng, G.; Pegram, M.D.; Podratz, K.C.; Crown, J.; Slamon, D.J.; et al. Survivin Expression in Breast Cancer Predicts Clinical Outcome and Is Associated with HER2, VEGF, Urokinase Plasminogen Activator and PAI-1. *Ann. Oncol.* **2006**, *17*, 597–604. [CrossRef]
82. Tanaka, K.; Kanazawa, T.; Horiuchi, S.; Ando, T.; Sugawara, K.; Takashima, Y.; Seta, Y.; Okada, H. Cytoplasm-responsive nanocarriers conjugated with a functional cell-penetrating peptide for systemic siRNA delivery. *Intl. J. Pharm.* **2013**, *455*, 40–47. [CrossRef] [PubMed]

83. Rodel, F.; Sprenger, T.; Kaina, B.; Liersch, T.; Rodel, C.; Fulda, S.; Hehlgans, S. Survivin as a Prognostic/Predictive Marker and Molecular Target in Cancer Therapy. *Curr. Med. Chem.* **2012**, *19*, 3679–3688. [CrossRef]
84. Garg, H.; Suri, P.; Gupta, J.C.; Talwar, G.P.; Dubey, S. Survivin: A Unique Target for Tumor Therapy. *Cancer Cell Int.* **2016**, *16*, 49. [CrossRef] [PubMed]
85. Tsuda, N.; Murayama, K.; Ishida, H.; Matsunaga, K.; Komiya, S.; Itoh, K.; Yamada, A. Expression of a Newly Defined Tumor-Rejection Antigen SART3 in Musculoskeletal Tumors and Induction of HLA Class I-Restricted Cytotoxic T Lymphocytes by SART3-Derived Peptides. *J. Orthop. Res.* **2001**, *19*, 346–351. [CrossRef] [PubMed]
86. Miyagi, Y.; Sasatomi, T.; Mine, T.; Isomoto, H.; Shirouzu, K.; Yamana, H.; Imai, N.; Yamada, A.; Katagiri, K.; Muto, A.; et al. Induction of Cellular Immune Responses to Tumor Cells and Peptides in Colorectal Cancer Patients by Vaccination with SART3 Peptides. *Clin. Cancer Res.* **2001**, *7*, 3950–3962. [PubMed]
87. Sherman, E.J.; Mitchell, D.C.; Garner, A.L. The RNA-Binding Protein SART3 Promotes MiR-34a Biogenesis and G1 Cell Cycle Arrest in Lung Cancer Cells. *J. Biol. Chem.* **2019**, *294*, 17188–17196. [CrossRef]
88. Timani, K.A.; Gyorffy, B.; Liu, Y.; Mohammad, K.S.; He, J.J. Tip110/SART3 Regulates IL-8 Expression and Predicts the Clinical Outcomes in Melanoma. *Mol. Cancer* **2018**, *17*, 124. [CrossRef]
89. Lee, J.H.; Lee, S.W. The Roles of Carcinoembryonic Antigen in Liver Metastasis and Therapeutic Approaches. *Gastroenterol. Res. Pract.* **2017**, *2017*, 7521987. [CrossRef]
90. Turriziani, M.; Fantini, M.; Benvenuto, M.; Izzi, V.; Masuelli, L.; Sacchetti, P.; Modesti, A.; Bei, R. Carcinoembryonic Antigen (CEA)-Based Cancer Vaccines: Recent Patents and Antitumor Effects from Experimental Models to Clinical Trials. *Recent. Pat. Anticancer Drug Discov.* **2012**, *7*, 265–296. [CrossRef]
91. Ojima, T.; Iwahashi, M.; Nakamura, M.; Matsuda, K.; Nakamori, M.; Ueda, K.; Naka, T.; Ishida, K.; James Primus, F.; Yamaue, H. Successful Cancer Vaccine Therapy for Carcinoembryonic Antigen (CEA)-Expressing Colon Cancer Using Genetically Modified Dendritic Cells That Express CEA and T Helper-Type 1 Cytokines in CEA Transgenic Mice. *Int. J. Cancer* **2007**, *120*, 585–593. [CrossRef]
92. Gulley, J.L.; Arlen, P.M.; Tsang, K.-Y.; Yokokawa, J.; Palena, C.; Poole, D.J.; Remondo, C.; Cereda, V.; Jones, J.L.; Pazdur, M.P.; et al. Pilot study of vaccination with recombinant CEA-MUC-1-TRICOM poxviral-based vaccines in patients with metastatic carcinoma. *Clin. Cancer Res.* **2008**, *14*, 3060–3069. [CrossRef] [PubMed]
93. Liu, D.; Guo, P.; McCarthy, C.; Wang, B.; Tao, Y.; Auguste, D. Peptide Density Targets and Impedes Triple Negative Breast Cancer Metastasis. *Nat. Commun.* **2018**, *9*, 2612. [CrossRef] [PubMed]
94. Zhang, Q.; Bergman, J.; Wiman, K.G.; Bykov, V.J.N. Role of Thiol Reactivity for Targeting Mutant P53. *Cell Chem. Biol.* **2018**, *25*, 1219–1230. [CrossRef] [PubMed]
95. Bauer, M.R.; Joerger, A.C.; Fersht, A.R. 2-Sulfonylpyrimidines: Mild Alkylating Agents with Anticancer Activity toward P53-Compromised Cells. *Proc. Natl. Acad. Sci. USA* **2016**, *113*, E5271–E5280. [CrossRef] [PubMed]
96. Synnott, N.C.; Bauer, M.R.; Madden, S.; Murray, A.; Klinger, R.; O'Donovan, N.; O'Connor, D.; Gallagher, W.M.; Crown, J.; Fersht, A.R.; et al. Mutant P53 as a Therapeutic Target for the Treatment of Triple-Negative Breast Cancer: Preclinical Investigation with the Anti-P53 Drug, PK11007. *Cancer Lett.* **2018**, *414*, 99–106. [CrossRef]
97. Nijman, H.W.; Vermeij, R.; Leffers, N.; Van Der Burg, S.H.; Melief, C.J.; Daemen, T. Immunological and Clinical Effects of Vaccines Targeting P53-Overexpressing Malignancies. *J. Biomed. Biotechnol.* **2011**, *2011*, 702146. [CrossRef]
98. Vijayan, V.; Mohapatra, A.; Uthaman, S.; Park, I.K. Recent Advances in Nanovaccines Using Biomimetic Immunomodulatory Materials. *Pharmaceutics* **2019**, *11*, 534. [CrossRef]
99. Chianese-Bullock, K.A.; Lewis, S.T.; Sherman, N.E.; Shannon, J.D.; Slingluff, C.L. Multi-Peptide Vaccines Vialed as Peptide Mixtures Can Be Stable Reagents for Use in Peptide-Based Immune Therapies. *Vaccine* **2009**, *27*, 1764–1770. [CrossRef]
100. Oka, Y.; Tsuboi, A.; Nakata, J.; Nishida, S.; Hosen, N.; Kumanogoh, A.; Oji, Y.; Sugiyama, H. Wilms' Tumor Gene 1 (WT1) Peptide Vaccine Therapy for Hematological Malignancies: From CTL Epitope Identification to Recent Progress in Clinical Studies Including a Cure-Oriented Strategy. *Oncol. Res. Treat.* **2017**, *40*, 682–690. [CrossRef]
101. Qi, X.W.; Zhang, F.; Wu, H.; Liu, J.L.; Zong, B.G.; Xu, C.; Jiang, J. Wilms' Tumor 1 (WT1) Expression and Prognosis in Solid Cancer Patients: A Systematic Review and Meta-Analysis. *Sci. Rep.* **2015**, *5*, srep08924. [CrossRef]
102. Coosemans, A.; Moerman, P.; Verbist, G.; Maes, W.; Neven, P.; Vergote, I.; Van Gool, S.W.; Amant, F. Wilms' Tumor Gene 1 (WT1) in Endometrial Carcinoma. *Gynecol. Oncol.* **2008**, *111*, 502–508. [CrossRef] [PubMed]
103. Zhang, J.; Guo, F.; Wang, L.; Zhao, W.; Zhang, D.; Yang, H.; Yu, J.; Niu, L.; Yang, F.; Zheng, S.; et al. Screening and Identification of Non-Inflammatory Specific Protein Markers in Wilms' Tumor Tissues. *Arch. Biochem. Biophys.* **2019**, *676*, 108112. [CrossRef] [PubMed]
104. Goldstein, N.S.; Uzieblo, A. WT1 Immunoreactivity in Uterine Papillary Serous Carcinomas Is Different from Ovarian Serous Carcinomas. *Am. J. Clin. Pathol.* **2002**, *117*, 541–545. [CrossRef] [PubMed]
105. Iiyama, T.; Udaka, K.; Takeda, S.; Takeuchi, T.; Adachi, Y.C.; Ohtsuki, Y.; Tsuboi, A.; Nakatsuka, S.I.; Elisseeva, O.A.; Oji, Y.; et al. WT1 (Wilms' Tumor 1) Peptide Immunotherapy for Renal Cell Carcinoma. *Microbiol. Immunol.* **2007**, *51*, 519–530. [CrossRef]
106. O'Hagan, D.T.; Valiante, N.M. Recent Advances in the Discovery and Delivery of Vaccine Adjuvants. *Nat. Rev. Drug Discov.* **2003**, *2*, 727–735. [CrossRef]

107. Ghaffari-Nazari, H.; Tavakkol-Afshari, J.; Jaafari, M.R.; Tahaghoghi-Hajghorbani, S.; Masoumi, E.; Jalali, S.A. Improving Multi-Epitope Long Peptide Vaccine Potency by Using a Strategy That Enhances CD4+ T Help in BALB/c Mice. *PLoS ONE* **2015**, *10*, e0142563. [CrossRef]
108. Wu, C.Y.; Monie, A.; Pang, X.; Hung, C.F.; Wu, T.C. Improving Therapeutic HPV Peptide-Based Vaccine Potency by Enhancing CD4+ T Help and Dendritic Cell Activation. *J. Biomed. Sci.* **2010**, *17*, 88. [CrossRef]
109. Zamani, P.; Teymouri, M.; Nikpoor, A.R.; Navashenaq, J.G.; Gholizadeh, Z.; Darban, S.A.; Jaafari, M.R. Nanoliposomal Vaccine Containing Long Multi-Epitope Peptide E75-AE36 Pulsed PADRE-Induced Effective Immune Response in Mice TUBO Model of Breast Cancer. *Eur. J. Cancer* **2020**, *129*, 80–96. [CrossRef]
110. Yazdani, Z.; Rafiei, A.; Yazdani, M.; Valadan, R. Design an Efficient Multi-Epitope Peptide Vaccine Candidate against SARS-CoV-2: An in Silico Analysis. *Infect. Drug Resist.* **2020**, *13*, 3007–3022. [CrossRef]
111. Nezafat, N.; Ghasemi, Y.; Javadi, G.; Khoshnoud, M.J.; Omidinia, E. A Novel Multi-Epitope Peptide Vaccine against Cancer: An in Silico Approach. *J. Theor. Biol.* **2014**, *349*, 121–134. [CrossRef]
112. Bijker, M.S.; van den Eeden, S.J.F.; Franken, K.L.; Melief, C.J.M.; van der Burg, S.H.; Offringa, R. Superior Induction of Anti-Tumor CTL Immunity by Extended Peptide Vaccines Involves Prolonged, DC-Focused Antigen Presentation. *Eur. J. Immunol.* **2008**, *38*, 1033–1042. [CrossRef]
113. Slingluff, C.L. The Present and Future of Peptide Vaccines for Cancer: Single or Multiple, Long or Short, Alone or in Combination? *Cancer J.* **2011**, *17*, 343–350. [CrossRef] [PubMed]
114. Eskandari, S.; Guerin, T.; Toth, I.; Stephenson, R.J. Recent Advances in Self-Assembled Peptides: Implications for Targeted Drug Delivery and Vaccine Engineering. *Adv. Drug Deliv. Rev.* **2017**, *110–111*, 169–187. [CrossRef] [PubMed]
115. Mohit, E.; Hashemi, A.; Allahyari, M. Breast Cancer Immunotherapy: Monoclonal Antibodies and Peptide-Based Vaccines. *Expert Rev. Clin. Immunol.* **2014**, *10*, 927–961. [CrossRef] [PubMed]
116. Mufson, R.A. Tumor Antigen Targets and Tumor Immunotherapy. *Front. Biosci.* **2006**, *11*, 337–343. [CrossRef]
117. Kawasaki, T.; Kawai, T. Toll-like Receptor Signaling Pathways. *Front. Immunol.* **2014**, *5*, 461. [CrossRef]
118. Hos, B.J.; Tondini, E.; van Kasteren, S.I.; Ossendorp, F. Approaches to Improve Chemically Defined Synthetic Peptide Vaccines. *Front. Immunol.* **2018**, *9*, 884. [CrossRef]
119. Bartnik, A.; Nirmal, A.J.; Yang, S.-Y. Peptide Vaccine Therapy in Colorectal Cancer. *Vaccines* **2012**, *1*, 1–16. [CrossRef]
120. Tsuruma, T.; Hata, F.; Furuhata, T.; Ohmura, T.; Katsuramaki, T.; Yamaguchi, K.; Kimura, Y.; Torigoe, T.; Sato, N.; Hirata, K. Peptide-Based Vaccination for Colorectal Cancer. *Expert Opin. Biol. Ther.* **2005**, *5*, 799–807. [CrossRef]
121. Calvo Tardón, M.; Allard, M.; Dutoit, V.; Dietrich, P.-Y.; Walker, P.R. Peptides as Cancer Vaccines. *Curr. Opin. Pharmacol.* **2019**, *47*, 20–26. [CrossRef]
122. Azmi, F.; Fuaad, A.A.H.A.; Skwarczynski, M.; Toth, I. Recent Progress in Adjuvant Discovery for Peptide-Based Subunit Vaccines. *Hum. Vaccin. Immunother.* **2014**, *10*, 778–796. [CrossRef] [PubMed]
123. Chatzikleanthous, D.; Schmidt, S.T.; Buffi, G.; Paciello, I.; Cunliffe, R.; Carboni, F.; Romano, M.R.; O'Hagan, D.T.; D'Oro, U.; Woods, S.; et al. Design of a Novel Vaccine Nanotechnology-Based Delivery System Comprising CpGODN-Protein Conjugate Anchored to Liposomes. *J. Control. Release* **2020**, *323*, 125–137. [CrossRef]
124. Melssen, M.M.; Petroni, G.R.; Chianese-Bullock, K.A.; Wages, N.A.; Grosh, W.W.; Varhegyi, N.; Smolkin, M.E.; Smith, K.T.; Galeassi, N.V.; Deacon, D.H.; et al. A Multipeptide Vaccine plus Toll-like Receptor Agonists LPS or PolyICLC in Combination with Incomplete Freund's Adjuvant in Melanoma Patients. *J. Immunother. Cancer* **2019**, *7*, 163. [CrossRef] [PubMed]
125. Yang, Y.; Feng, R.; Wang, Y.Z.; Sun, H.W.; Zou, Q.M.; Li, H.B. Toll-like Receptors: Triggers of Regulated Cell Death and Promising Targets for Cancer Therapy. *Immunol. Lett.* **2020**, *223*, 1–9. [CrossRef] [PubMed]
126. Temizoz, B.; Kuroda, E.; Ishii, K.J. Combination and Inducible Adjuvants Targeting Nucleic Acid Sensors. *Curr. Opin. Pharmacol.* **2018**, *41*, 104–113. [CrossRef] [PubMed]
127. Van Doorn, E.; Liu, H.; Huckriede, A.; Hak, E. Safety and Tolerability Evaluation of the Use of Montanide ISATM51 as Vaccine Adjuvant: A Systematic Review. *Hum. Vaccin. Immunother.* **2016**, *12*, 159–169. [CrossRef]
128. Belnoue, E.; Di Berardino-Besson, W.; Gaertner, H.; Carboni, S.; Dunand-Sauthier, I.; Cerini, F.; Suso-Inderberg, E.M.; Wälchli, S.; König, S.; Salazar, A.M.; et al. Enhancing Antitumor Immune Responses by Optimized Combinations of Cell-Penetrating Peptide-Based Vaccines and Adjuvants. *Mol. Ther.* **2016**, *24*, 1675–1685. [CrossRef]
129. Ponomarev, E.D.; Shriver, L.P.; Maresz, K.; Pedras-Vasconcelos, J.; Verthelyi, D.; Dittel, B.N. GM-CSF Production by Autoreactive T Cells Is Required for the Activation of Microglial Cells and the Onset of Experimental Autoimmune Encephalomyelitis. *J. Immunol.* **2007**, *178*, 39–48. [CrossRef]
130. Yu, T.W.; Chueh, H.Y.; Tsai, C.C.; Lin, C.T.; Qiu, J.T. Novel GM-CSF-Based Vaccines: One Small Step in GM-CSF Gene Optimization, One Giant Leap for Human Vaccines. *Hum. Vaccin. Immunother.* **2016**, *12*, 3020–3028. [CrossRef]
131. Bhattacharya, P.; Budnick, I.; Singh, M.; Thiruppathi, M.; Alharshawi, K.; Elshabrawy, H.; Holterman, M.J.; Prabhakar, B.S. Dual Role of GM-CSF as a Pro-Inflammatory and a Regulatory Cytokine: Implications for Immune Therapy. *J. Interferon Cytokine Res.* **2015**, *35*, 585–599. [CrossRef]
132. Borriello, F.; Galdiero, M.R.; Varricchi, G.; Loffredo, S.; Spadaro, G.; Marone, G. Innate Immune Modulation by GM-CSF and IL-3 in Health and Disease. *Int. J. Mol. Sci.* **2019**, *20*, 834. [CrossRef]
133. Zhao, W.; Zhao, G.; Wang, B. Revisiting GM-CSF as an Adjuvant for Therapeutic Vaccines. *Cell. Mol. Immunol.* **2018**, *15*, 187–189. [CrossRef]

134. Shiomi, A.; Usui, T. Pivotal Roles of GM-CSF in Autoimmunity and Inflammation. *Mediat. Inflamm.* **2015**, *2015*, 1–13. [CrossRef]
135. Di Gregoli, K.; Johnson, J.L. Role of Colony-Stimulating Factors in Atherosclerosis. *Curr. Opin. Lipidol.* **2012**, *23*, 412–421. [CrossRef]
136. Kim, I.K.; Koh, C.H.; Jeon, I.; Shin, K.S.; Kang, T.S.; Bae, E.A.; Seo, H.; Ko, H.J.; Kim, B.S.; Chung, Y.; et al. GM-CSF Promotes Antitumor Immunity by Inducing Th9 Cell Responses. *Cancer Immunol. Res.* **2019**, *7*, 498–509. [CrossRef]
137. Decker, W.K.; Safdar, A. Cytokine Adjuvants for Vaccine Therapy of Neoplastic and Infectious Disease. *Cytokine Growth Factor Rev.* **2011**, *22*, 177–187. [CrossRef] [PubMed]
138. Wimmers, F.; De Haas, N.; Scholzen, A.; Schreibelt, G.; Simonetti, E.; Eleveld, M.J.; Brouwers, H.M.L.M.; Beldhuis-Valkis, M.; Joosten, I.; De Jonge, M.I.; et al. Monitoring of Dynamic Changes in Keyhole Limpet Hemocyanin (KLH)-Specific B Cells in KLHvaccinated Cancer Patients. *Sci. Rep.* **2017**, *7*, srep43486. [CrossRef]
139. Bi, S.; Bailey, W.; Brisson, C. Performance of Keyhole Limpet Hemocyanin (KLH) as an Antigen Carrier for Protein Antigens Depends on KLH Property and Conjugation Route. *J. Immunol.* **2016**, *196*, 76.16. [CrossRef]
140. Aarntzen, E.H.J.G.; De Vries, I.J.M.; GöErtz, J.H.; Beldhuis-Valkis, M.; Brouwers, H.M.L.M.; Van De Rakt, M.W.M.M.; Van Der Molen, R.G.; Punt, C.J.A.; Adema, G.J.; Tacken, P.J.; et al. Humoral Anti-KLH Responses in Cancer Patients Treated with Dendritic Cell-Based Immunotherapy Are Dictated by Different Vaccination Parameters. *Cancer Immunol. Immunother.* **2012**, *61*, 2003–2011. [CrossRef] [PubMed]
141. Swanson, M.A.; Schwartz, R.S. Immunosuppressive Therapy. *N. Engl. J. Med.* **1967**, *277*, 163–170. [CrossRef] [PubMed]
142. Slovin, S.F.; Ragupathi, G.; Musselli, C.; Fernandez, C.; Diani, M.; Verbel, D.; Danishefsky, S.; Livingston, P.; Scher, H.I. Thomsen-Friedenreich (TF) Antigen as a Target for Prostate Cancer Vaccine: Clinical Trial Results with TF Cluster (c)-KLH plus QS21 Conjugate Vaccine in Patients with Biochemically Relapsed Prostate Cancer. *Cancer Immunol. Immunother.* **2005**, *54*, 694–702. [CrossRef] [PubMed]
143. Razazan, A.; Behravan, J.; Arab, A.; Barati, N.; Arabi, L.; Gholizadeh, Z.; Jaafari, M.R. Conjugated nanoliposome with the HER2/neu-derived peptide GP2 as an effective vaccine against breast cancer in mice xenograft model. *PLoS ONE* **2017**, *12*, e0185099. [CrossRef] [PubMed]
144. Curry, J.M.; Besmer, D.M.; Erick, T.K.; Steuerwald, N.; Das Roy, L.; Grover, P.; Rao, S.; Nath, S.; Ferrier, J.W.; Reid, R.W.; et al. Indomethacin Enhances Anti-Tumor Efficacy of a MUC1 Peptide Vaccine against Breast Cancer in MUC1 Transgenic Mice. *PLoS ONE* **2019**, *14*, e0224309. [CrossRef] [PubMed]
145. Ladjemi, M.Z.; Jacot, W.; Chardès, T.; Pèlegrin, A.; Navarro-Teulon, I. Anti-HER2 Vaccines: New Prospects for Breast Cancer Therapy. *Cancer Immunol. Immunother.* **2010**, *59*, 1295–1312. [CrossRef]
146. Knutson, K.L.; Schiffman, K.; Cheever, M.A.; Disis, M.L. Immunization of Cancer Patients with a HER-2/Neu, HLA-A2 Peptide, P369-377, Results in Short-Lived Peptide-Specific Immunity. *Clin. Cancer Res.* **2002**, *8*, 1014–1018.
147. Anderson, B.W.; Peoples, G.E.; Murray, J.L.; Gillogly, M.A.; Gershenson, D.M.; Ioannides, C.G. Peptide Priming of Cytolytic Activity to HER-2 Epitope 369-377 in Healthy Individuals. *Clin. Cancer Res.* **2000**, *6*, 4192–4200.
148. Ohtake, J.; Ohkuri, T.; Togashi, Y.; Kitamura, H.; Okuno, K.; Nishimura, T. Identification of Novel Helper Epitope Peptides of Survivin Cancer-Associated Antigen Applicable to Developing Helper/Killer-Hybrid Epitope Long Peptide Cancer Vaccine. *Immunol. Lett.* **2014**, *161*, 20–30. [CrossRef]
149. Disis, M.L.; Gooley, T.A.; Rinn, K.; Davis, D.; Piepkorn, M.; Cheever, M.A.; Knutson, K.L.; Schiffman, K. Generation of T-Cell Immunity to the HER-2/Neu Protein after Active Immunization with HER-2/Neu Peptide-Based Vaccines. *J. Clin. Oncol.* **2002**, *20*, 2624–2632. [CrossRef]
150. Kalli, K.R.; Block, M.S.; Kasi, P.M.; Erskine, C.L.; Hobday, T.J.; Dietz, A.; Padley, D.; Gustafson, M.P.; Shreeder, B.; Puglisi-Knutson, D.; et al. Folate Receptor Alpha Peptide Vaccine Generates Immunity in Breast and Ovarian Cancer Patients. *Clin. Cancer Res.* **2018**, *24*, 3014–3025. [CrossRef]
151. Farran, B.; Pavitra, E.; Kasa, P.; Peela, S.; Rama Raju, G.S.; Nagaraju, G.P. Folate-Targeted Immunotherapies: Passive and Active Strategies for Cancer. *Cytokine Growth Factor Rev.* **2019**, *45*, 45–52. [CrossRef]
152. Folate Receptor Alpha Peptide Vaccine with GM-CSF in Patients with Triple Negative Breast Cancer. Available online: https://clinicaltrials.gov/ct2/show/NCT02593227 (accessed on 3 December 2020).
153. MUC1 Vaccine for Triple-Negative Breast Cancer. Available online: https://www.clinicaltrials.gov/ct2/show/NCT00986609 (accessed on 3 December 2020).
154. Vaccine Therapy in Treating Patients with Breast Cancer. Available online: https://clinicaltrials.gov/ct2/show/NCT00524277 (accessed on 3 December 2020).
155. Phase Ib Trial of Two Folate Binding Protein Peptide Vaccines (E39 and J65) in Breast and Ovarian Cancer Patients (J65). Available online: https://clinicaltrials.gov/ct2/show/NCT02019524 (accessed on 3 December 2020).
156. Multipeptide Vaccine for Advanced Breast Cancer. Available online: https://clinicaltrials.gov/ct2/show/NCT00573495 (accessed on 3 December 2020).
157. Oka, Y.; Tsuboi, A.; Taguchi, T.; Osaki, T.; Kyo, T.; Nakajima, H.; Elisseeva, O.A.; Oji, Y.; Kawakami, M.; Ikegame, K.; et al. Induction of WT1 (Wilms' Tumor Gene)-Specific Cytotoxic T Lymphocytes by WT1 Peptide Vaccine and the Resultant Cancer Regression. *Proc. Natl. Acad. Sci. USA* **2004**, *101*, 13885–13890. [CrossRef] [PubMed]
158. Oka, Y.; Tsuboi, A.; Oji, Y.; Kawase, I.; Sugiyama, H. WT1 Peptide Vaccine for the Treatment of Cancer. *Curr. Opin. Immunol.* **2008**, *20*, 211–220. [CrossRef] [PubMed]

159. Holmes, J.P.; Benavides, L.C.; Gates, J.D.; Carmichael, M.G.; Hueman, M.T.; Mittendorf, E.A.; Murray, J.L.; Amin, A.; Craig, D.; Von Hofe, E.; et al. Results of the First Phase I Clinical Trial of the Novel Ii-Key Hybrid Preventive HER-2/Neu Peptide (AE37) Vaccine. *J. Clin. Oncol.* **2008**, *26*, 3426–3433. [CrossRef]
160. Nevagi, R.J.; Toth, I.; Skwarczynski, M. Peptide-Based Vaccines. In *Peptide Applications in Biomedicine, Biotechnology and Bioengineering*; Elsevier: Amsterdam, The Netherlands, 2018; pp. 327–358.
161. Roces, C.B.; Khadke, S.; Christensen, D.; Perrie, Y. Scale-Independent Microfluidic Production of Cationic Liposomal Adjuvants and Development of Enhanced Lymphatic Targeting Strategies. *Mol. Pharm.* **2019**, *16*, 4372–4386. [CrossRef] [PubMed]
162. Salvador, A.; Igartua, M.; Hernández, R.M.; Pedraz, J.L. An Overview on the Field of Micro- and Nanotechnologies for Synthetic Peptide-Based Vaccines. *J. Drug Deliv.* **2011**, *2011*, 181646. [CrossRef]
163. Ohta, S.; Kikuchi, E.; Ishijima, A.; Azuma, T.; Sakuma, I.; Ito, T. Investigating the Optimum Size of Nanoparticles for Their Delivery into the Brain Assisted by Focused Ultrasound-Induced Blood–Brain Barrier Opening. *Sci. Rep.* **2020**, *10*, 18220. [CrossRef]
164. Koerner, J.; Horvath, D.; Groettrup, M. Harnessing Dendritic Cells for Poly (D,L-Lactide-Co-Glycolide) Microspheres (PLGA MS)-Mediated Anti-Tumor Therapy. *Front. Immunol.* **2019**, *10*, 707. [CrossRef]
165. Durán, V.; Yasar, H.; Becker, J.; Thiyagarajan, D.; Loretz, B.; Kalinke, U.; Lehr, C.M. Preferential Uptake of Chitosan-Coated PLGA Nanoparticles by Primary Human Antigen Presenting Cells. *Nanomedicine* **2019**, *21*, 102073. [CrossRef]
166. Kroll, A.V.; Fang, R.H.; Jiang, Y.; Zhou, J.; Wei, X.; Yu, C.L.; Gao, J.; Luk, B.T.; Dehaini, D.; Gao, W.; et al. Nanoparticulate Delivery of Cancer Cell Membrane Elicits Multiantigenic Antitumor Immunity. *Adv. Mater.* **2017**, *29*, 1–9. [CrossRef]
167. Gu, P.; Liu, Z.; Sun, Y.; Ou, N.; Hu, Y.; Liu, J.; Wu, Y.; Wang, D. Angelica Sinensis Polysaccharide Encapsulated into PLGA Nanoparticles as a Vaccine Delivery and Adjuvant System for Ovalbumin to Promote Immune Responses. *Int. J. Pharm.* **2019**, *554*, 72–80. [CrossRef]
168. Boraschi, D.; Italiani, P. From Antigen Delivery System to Adjuvanticy: The Board Application of Nanoparticles in Vaccinology. *Vaccines* **2015**, *3*, 930–939. [CrossRef] [PubMed]
169. Rezvantalab, S.; Drude, N.I.; Moraveji, M.K.; Güvener, N.; Koons, E.K.; Shi, Y.; Lammers, T.; Kiessling, F. PLGA-Based Nanoparticles in Cancer Treatment. *Front. Pharmacol.* **2018**, *9*, 1260. [CrossRef]
170. Pandey, A.; Jain, D.S. Poly Lactic-Co-Glycolic Acid (PLGA) Copolymer and Its Pharmaceutical Application. In *Handbook of Polymers for Pharmaceutical Technologies*; Scrivener Publishing LLC.: Beverly, MA, USA, 2015; Volume 2, pp. 151–172.
171. Ma, W.; Chen, M.; Kaushal, S.; McElroy, M.; Zhang, Y.; Ozkan, C.; Bouvet, M.; Kruse, C.; Grotjahn, D.; Ichim, T.; et al. PLGA Nanoparticle-Mediated Delivery of Tumor Antigenic Peptides Elicits Effective Immune Responses. *Int. J. Nanomed.* **2012**, *7*, 1475–1487. [CrossRef] [PubMed]
172. Chen, Q.; Xu, L.; Liang, C.; Wang, C.; Peng, R.; Liu, Z. Photothermal Therapy with Immune-Adjuvant Nanoparticles Together with Checkpoint Blockade for Effective Cancer Immunotherapy. *Nat. Commun.* **2016**, *7*, 13193. [CrossRef]
173. Chu, B.Y.; Al Kobiasi, M.; Zeng, W.; Mainwaring, D.; Jackson, D.C. Chitosan-Based Particles as Biocompatible Delivery Vehicles for Peptide and Protein-Based Vaccines. *Procedia Vaccinol.* **2012**, *6*, 74–79. [CrossRef]
174. Singh, B.; Maharjan, S.; Sindurakar, P.; Cho, K.H.; Choi, Y.J.; Cho, C.S. Needle-Free Immunization with Chitosan-Based Systems. *Int. J. Mol. Sci.* **2018**, *19*, 3639. [CrossRef] [PubMed]
175. Singla, A.K.; Chawla, M. Chitosan: Some Pharmaceutical and Biological Aspects—An Update. *J. Pharm. Pharmacol.* **2010**, *53*, 1047–1067. [CrossRef]
176. Jadidi-Niaragh, F.; Atyabi, F.; Rastegari, A.; Kheshtchin, N.; Arab, S.; Hassannia, H.; Ajami, M.; Mirsanei, Z.; Habibi, S.; Masoumi, F.; et al. CD73 Specific SiRNA Loaded Chitosan Lactate Nanoparticles Potentiate the Antitumor Effect of a Dendritic Cell Vaccine in 4T1 Breast Cancer Bearing Mice. *J. Control. Release* **2017**, *246*, 46–59. [CrossRef]
177. Pei, M.; Liang, J.; Zhang, C.; Wang, X.; Zhang, C.; Ma, G.; Sun, H. Chitosan/Calcium Phosphates Nanosheet as a Vaccine Carrier for Effective Cross-Presentation of Exogenous Antigens. *Carbohydr. Polym.* **2019**, *224*, 115172. [CrossRef]
178. Fang, R.H.; Hu, C.M.J.; Luk, B.T.; Gao, W.; Copp, J.A.; Tai, Y.; O'Connor, D.E.; Zhang, L. Cancer Cell Membrane-Coated Nanoparticles for Anticancer Vaccination and Drug Delivery. *Nano Lett.* **2014**, *14*, 2181–2188. [CrossRef]
179. Harris, J.C.; Scully, M.A.; Day, E.S. Cancer Cell Membrane-Coated Nanoparticles for Cancer Management. *Cancers* **2019**, *11*, 1836. [CrossRef] [PubMed]
180. Ruoslahti, E.; Bhatia, S.N.; Sailor, M.J. Targeting of Drugs and Nanoparticles to Tumors. *J. Cell Biol.* **2010**, *188*, 759–768. [CrossRef]
181. Xuan, M.; Shao, J.; Li, J. Cell Membrane-Covered Nanoparticles as Biomaterials. *Natl. Sci. Rev.* **2019**, *6*, 551–561. [CrossRef] [PubMed]
182. Jin, J.; Krishnamachary, B.; Barnett, J.D.; Chatterjee, S.; Chang, D.; Mironchik, Y.; Wildes, F.; Jaffee, E.M.; Nimmagadda, S.; Bhujwalla, Z.M. Human Cancer Cell Membrane-Coated Biomimetic Nanoparticles Reduce Fibroblast-Mediated Invasion and Metastasis and Induce T-Cells. *ACS Appl. Mater. Interfaces* **2019**, *11*, 7850–7861. [CrossRef] [PubMed]
183. Duan, Q.; Zhang, H.; Zheng, J.; Zhang, L. Turning Cold into Hot: Firing up the Tumor Microenvironment. *Trends Cancer* **2020**, *6*, 605–618. [CrossRef]
184. Nelde, A.; Rammensee, H.G.; Walz, J.S. The Peptide Vaccine of the Future. *Mol. Cell. Proteom.* **2021**, *20*, 100022. [CrossRef]
185. Vinay, D.S.; Ryan, E.P.; Pawelec, G.; Talib, W.H.; Stagg, J.; Elkord, E.; Lichtor, T.; Decker, W.K.; Whelan, R.L.; Kumara, H.M.C.S.; et al. Immune Evasion in Cancer: Mechanistic Basis and Therapeutic Strategies. *Semin. Cancer Biol.* **2015**, *35*, S185–S198. [CrossRef]
186. Ahmadzadeh, M. OR.93. Tumor Antigen-Specific CD8 T Cells Infiltrating the Tumor Express High Levels of PD-1 and Are Functionally Impaired. *Clin. Immunol.* **2009**, *131*, 1537–1544. [CrossRef]

187. Colozza, M.; de Azambuja, E.; Personeni, N.; Lebrun, F.; Piccart, M.J.; Cardoso, F. Achievements in Systemic Therapies in the Pregenomic Era in Metastatic Breast Cancer. *Oncologist* **2007**, *12*, 253–270. [CrossRef]
188. Planes-Laine, G.; Rochigneux, P.; Bertucci, F.; Chrétien, A.S.; Viens, P.; Sabatier, R.; Gonçalves, A. PD-1/PD-L1 Targeting in Breast Cancer: The First Clinical Evidences Are Emerging. a Literature Review. *Cancers* **2019**, *11*, 1033. [CrossRef]
189. Kamphorst, A.O.; Pillai, R.N.; Yang, S.; Nasti, T.H.; Akondy, R.S.; Wieland, A.; Sica, G.L.; Yu, K.; Koenig, L.; Patel, N.T.; et al. Proliferation of PD-1+ CD8 T Cells in Peripheral Blood after PD-1-Targeted Therapy in Lung Cancer Patients. *Proc. Natl. Acad. Sci. USA* **2017**, *114*, 4993–4998. [CrossRef] [PubMed]
190. Hirayama, M.; Nishimura, Y. The Present Status and Future Prospects of Peptide-Based Cancer Vaccines. *Int. Immunol.* **2016**, *28*, 319–328. [CrossRef] [PubMed]
191. Kleponis, J.; Skelton, R.; Zheng, L. Fueling the Engine and Releasing the Break: Combinational Therapy of Cancer Vaccines and Immune Checkpoint Inhibitors. *Cancer Biol. Med.* **2015**, *12*, 201–208. [PubMed]

Disclaimer/Publisher's Note: The statements, opinions and data contained in all publications are solely those of the individual author(s) and contributor(s) and not of MDPI and/or the editor(s). MDPI and/or the editor(s) disclaim responsibility for any injury to people or property resulting from any ideas, methods, instructions or products referred to in the content.

Article

A Peptide Vaccine Design Targeting KIT Mutations in Acute Myeloid Leukemia

Minji Kim [1], Kush Savsani [2] and Sivanesan Dakshanamurthy [3],*

1. College of Human Ecology, Cornell University, Ithaca, NY 14850, USA
2. College of Humanities and Sciences, Virginia Commonwealth University, Richmond, VA 23284, USA
3. Department of Oncology, Lombardi Comprehensive Cancer Center, Georgetown University Medical Center, Washington, DC 20057, USA
* Correspondence: sd233@georgetown.edu

Abstract: Acute myeloid leukemia (AML) is a leading blood cancer subtype that can be caused by 27 gene mutations. Previous studies have explored potential vaccine and drug treatments against AML, but many were proven immunologically insignificant. Here, we targeted this issue and applied various clinical filters to improve immune response. KIT is an oncogenic gene that can cause AML when mutated and is predicted to be a promising vaccine target because of its immunogenic responses when activated. We designed a multi-epitope vaccine targeting mutations in the KIT oncogene using CD8+ and CD4+ epitopes. We selected the most viable vaccine epitopes based on thresholds for percentile rank, immunogenicity, antigenicity, half-life, toxicity, IFNγ release, allergenicity, and stability. The efficacy of data was observed through world and regional population coverage of our vaccine design. Then, we obtained epitopes for optimized population coverage from PCOptim-CD, a modified version of our original Java-based program code PCOptim. Using 24 mutations on the KIT gene, 12 CD8+ epitopes and 21 CD4+ epitopes were obtained. The CD8+ dataset had a 98.55% world population coverage, while the CD4+ dataset had a 65.14% world population coverage. There were five CD4+ epitopes that overlapped with the top CD8+ epitopes. Strong binding to murine MHC molecules was found in four CD8+ and six CD4+ epitopes, demonstrating the feasibility of our results in preclinical murine vaccine trials. We then created three-dimensional (3D) models to visualize epitope–MHC complexes and TCR interactions. The final candidate is a non-toxic and non-allergenic multi-epitope vaccine against KIT mutations that cause AML. Further research would involve murine trials of the vaccine candidates on tumor cells causing AML.

Keywords: vaccine design; acute myeloid leukemia (AML); KIT oncogene; artificial neural networks; immunoinformatics; epitopes; MHC I and MHC II molecules; epitope–MHC complexes; TCR binding; murine MHC molecules

Citation: Kim, M.; Savsani, K.; Dakshanamurthy, S. A Peptide Vaccine Design Targeting KIT Mutations in Acute Myeloid Leukemia. *Pharmaceuticals* 2023, 16, 932. https://doi.org/10.3390/ph16070932

Academic Editor: Dhimant Desai

Received: 13 April 2023
Revised: 6 June 2023
Accepted: 21 June 2023
Published: 27 June 2023

Copyright: © 2023 by the authors. Licensee MDPI, Basel, Switzerland. This article is an open access article distributed under the terms and conditions of the Creative Commons Attribution (CC BY) license (https://creativecommons.org/licenses/by/4.0/).

1. Introduction

Acute myeloid leukemia (AML) is a blood cancer subtype where an overproduction of abnormal myeloid cells causes improper development of platelets, red blood cells, white blood cells, and bone marrow failure [1]. AML is the leading acute leukemia subtype (80%) and is common among older individuals. Genetic mutations (point mutations and chromosomal translocations) are the root cause of AML. Other conditions, including myelodysplastic syndrome, aplastic anemia, myelofibrosis, Down syndrome, blood syndrome, and environmental exposures such as chemotherapy, benzene, tobacco, and radiation, have been proven to increase the risks of AML [2,3].

Common mutations that cause AML are in the genes Nucleophosmin 1 (*NPM1*), FMS-like tyrosine kinase 3 (*FLT3*), Runt-related transcription factor (*RUNX1*), and KIT, all of which are critical for hematopoiesis. The KIT gene encodes for a type III receptor tyrosine kinase that is critical for pathways involved in cell proliferation, survival, and differentiation of hematopoietic progenitor cells [4]. Synonyms for the KIT gene include c-kit, CD117, and mast/stem cell growth factor receptor (SCFR).

KIT is a proto-oncogene that can cause AML, most often core-binding factor acute myeloid leukemia (CBF-AML), and gastrointestinal stromal tumors [5,6]. CBF-AML is characterized by the chromosomal alterations t(8;21) and inv(16) [7], and 15–45% of patients suffer from mutations in the KIT gene [8]. We did not focus on CBF-AML because most of the clinically studied KIT mutations referenced in our data were not specific to CBF-AML. Still, most of the KIT mutations we analyze may be relevant to CBF-AML.

The KIT gene comprises six domains: Ig-like C2-type 1, Ig-like C2-type 2, Ig-like C2-type 3, Ig-like C2-type 4, Ig-like C2-type 5, and protein kinase. The protein kinase domain is intracellular, while the five other domains work on extracellular regions [9]. KIT mutations on the cell surface exhibit various ligand-induced activities, including ubiquitination, cell transformation, and greater sensitivity for basal tyrosine phosphorylation. Intracellular KIT mutations exhibit ligand-independent activities, including KIT activation, ubiquitination, and cell transformation [10].

AML is currently targeted with chemotherapy, drugs, and stem cell transplants [11]. Less aggressive treatments are needed because AML is an acute disease with a quick and poor prognosis, especially among the elderly. Research on CBF-AML treatments includes drug therapies such as cytarabine [12]. A successful preclinical murine trial was also conducted for a solid vaccine treatment made of polyethylene glycol and alginate [13]. Other peptide vaccine studies have targeted major mutated genes involved with AML that induce T-cell responses, including Wilms' tumor 1 (*WT1*), proteinase 3 (*PR3*), hyaluronic acid-mediated motility receptor (*RHAMM*), and mucinone1 protein (*MUC1*). However, these vaccines have had limited success in phase II clinical trials.

The KIT gene has not been targeted with vaccines to treat AML, but trials on the gene's effectiveness in drugs and vaccines for other conditions exist. KIT gene mutations are common targets for the treatment of various cancers because of the gene's role in cellular functions such as hematopoiesis, carcinogenesis, and melanogenesis [14,15]. Furthermore, KIT is predicted to be a strong target for drugs and vaccines owing to its immunogenicity. C-kit ligation is associated with the release of cytokines and other pro-inflammatory mediators, and c-kit signaling can impact adaptive immunity [15]. Completed studies on treatments targeting the KIT gene include anti-drug conjugates against small cell lung cancer and a DNA vaccine targeting ligand attachment to fight tumor growth [14,16].

This study relied on inducing an immunogenic response to point mutations on the KIT oncogene that have been found in AML patients. Single amino acid changes in genes can change the chemical properties of peptide sequences and have been associated with several cancers [17]. The KIT gene was our chosen target because of its significance in cancers such as AML and its critical functions in hematopoiesis. Point mutations on KIT can lead to the development of ubiquitinated tumor-specific antigens (TSAs), which are cleaved into epitopes in proteasomes. Transporter-associated antigen processing (TAP) protein complexes direct the epitopes to bind with MHC molecules at the endoplasmic reticulum. After epitope–MHC complexes are transported to the surface of tumor cells, antigen-presenting cells help induce CD8+ and CD4+ T cells. Through their T-cell receptors (TCRs), an immune response is initiated to respond to and attack the antigen [18]. Combining CD8+ and CD4+ immune pathways may create a potent vaccine because CD4+ T cells improve the immune response of cytotoxic T cells.

Our design was a multi-epitope AML vaccine predicted to induce effective CD8+ and CD4+ T-cell responses by targeting the intracellular and extracellular domains of the KIT oncogene. We used epitopes derived from common AML-inducing point mutations in the KIT gene and overlapping CD8+ and CD4+ epitopes. Our epitope vaccine design elicits T-cell immune responses by releasing epitopes cleared through the following clinically relevant variables: HLA binding affinity, immunogenicity, antigenicity, half-life, instability, toxicity, IFNγ release, allergenicity, population coverage, and murine MHC binding affinity. CD8+ T cells respond to endogenous antigens and participate in a cytotoxic activity, while CD4+ T cells respond to exogenous antigens and are helper T cells that induce a more refined immune response. Implementing CD8+ and CD4+ epitopes into cancer vaccines is predicted to safely induce an immune response that removes tumor cells expressing the same epitopes. This study is the first to target AML with a vaccine against KIT gene mutations.

2. Results

We targeted AML-inducing KIT mutations because of the gene's critical roles in hematopoietic cell survival, proliferation, and differentiation (Figure 1) [19–22]. When bound to a stem cell factor, KIT facilitates multiple intracellular signaling pathways, which helps maintain normal hematopoietic cell activity. KIT mutations are also involved in immunogenic responses such as cytokine release and adaptive immunity [15]. Mutations in the KIT gene can cause improper differentiation and growth in hematopoietic cells, causing harmful conditions such as AML. The KIT gene plays critical roles in tumor cell activity, but AML vaccines targeting KIT have yet to be explored. We designed a multi-epitope vaccine combining CD8+ and CD4+ epitopes that is predicted to induce a safe immune response against mutations in the KIT gene.

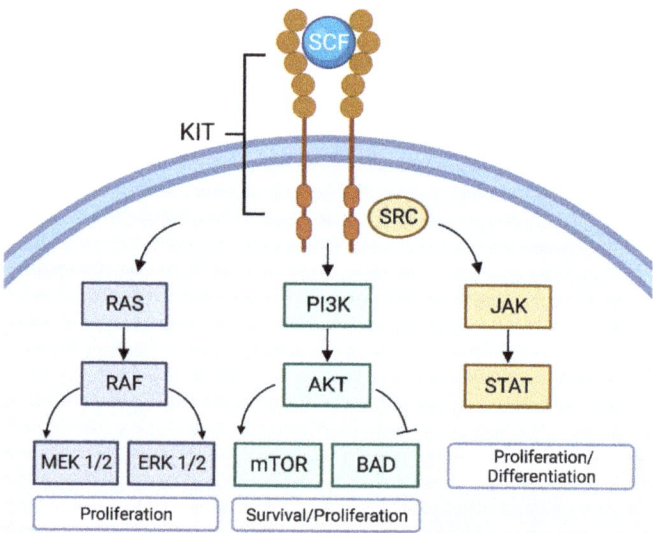

Figure 1. Intracellular signaling pathways that KIT is involved in. The KIT receptor tyrosine kinase can facilitate multiple signaling pathways when bound to a stem cell factor ligand. The RAS/RAF/MEK/ERK pathway guides cell proliferation. The PI3K-Akt pathway helps determine cell survival and proliferation. The JAK-STAT pathway plays a role in cell proliferation and differentiation/development (Created with BioRender.com accessed on 28 July 2022).

Figure 2 provides a workflow of the methodology we used to design the multi-epitope AML vaccine. We started by choosing a cancer subtype, researching common gene mutations, and obtaining the mutated peptide sequences. Then we computed percentile rank, binding affinity, immunogenicity, antigenicity, half-life, instability, toxicity, IFNγ, and allergenicity values of the epitopes and filtered through the data based on specific thresholds. Optimized epitopes were obtained through a modified version of PCOptim called PCOptim-CD, which finds epitopes for optimal population coverage for both CD8+ and CD4+ datasets. Steps five through eight were repeated for both the CD8+ and CD4+ data. Finally, we modeled the top epitope–MHC complexes and their binding with TCR complexes.

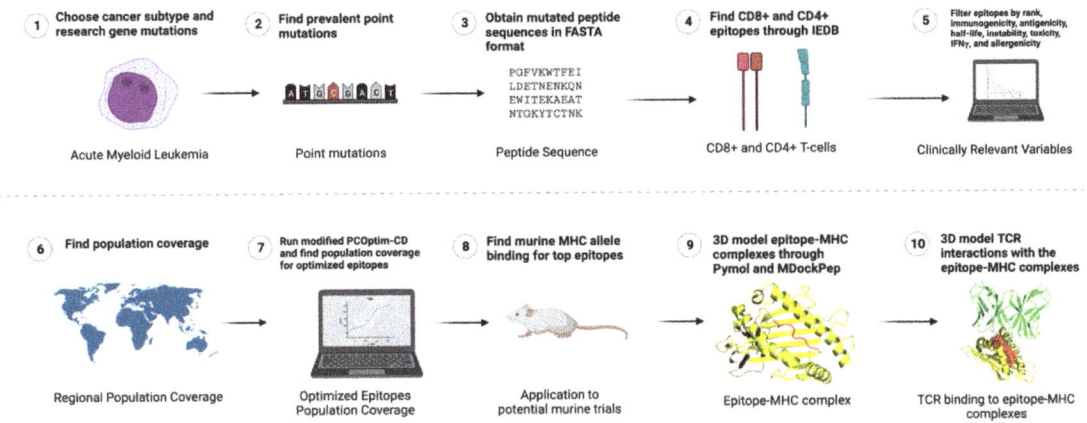

Figure 2. Workflow diagram of the study methodology. This process was used to develop the vaccine design; Created with BioRender.com (accessed on 18 February 2023).

2.1. Filtration of CD8+ Epitopes

The final CD8+ dataset included 12 ninemer epitopes filtered for rank (<10), immunogenicity (>0), antigenicity (>0.4), half-life (>1 h), toxicity (non-toxin), allergenicity (non-allergen), and instability (<40). CD8+ epitopes were not filtered for IFNγ release. IFNγ release was not prioritized for the CD8+ epitopes because CD8+ T cells are cytotoxic and less involved in releasing IFNγ than CD4+ helper T cells. Table 1 lists the top CD8+ epitopes, their respective mutations, and their binding HLA alleles. Specific values for clinically relevant variables (rank, immunogenicity, antigenicity, half-life, toxicity, allergenicity, instability, and IFNγ release) of the top CD8+ epitopes are in Supplementary Table S7.

2.2. Population Coverage for CD8+ Epitopes

Next, we determined that the world population coverage for MHC Class I binding CD8+ epitopes was 98.55% (Figure 3). Regions with high population coverage included East Asia (98.18%), Europe (99.68%), East Africa (98.18%), West Indies (98.98%), and North America (99.06%). However, Central America had an especially low population coverage of 7.76%. Population coverage for all regions is listed in Supplementary Table S1.

Table 1. Top CD8+ Epitopes and Murine Binding.

Mutation	Epitope	HLA Alleles	Strong H2 Allele Restriction	Weak H2 Allele Restriction
I571L	INGNNYVYL	HLA-A*24:02, HLA-B*08:01, HLA-A*23:01, HLA-A*68:02	H-2-Db, H-2-Dd, H-2-Kb	H-2-Ld
K550N	NPMYEVQWK	HLA-A*68:01, HLA-B*35:01, HLA-A*33:01, HLA-B*53:01, HLA-A*11:01, HLA-A*03:01, HLA-B*07:02	Not available	Not available

Table 1. Cont.

Mutation	Epitope	HLA Alleles	Strong H2 Allele Restriction	Weak H2 Allele Restriction
R49H	GKSDLIVHV	HLA-A*02:06, HLA-A*02:03, HLA-A*68:02, HLA-A*02:01, HLA-B*40:01 HLA-A*30:01, HLA-B*44:03, HLA-B*51:01, HLA-B*44:02, HLA-A*30:02, HLA-A*26:01, HLA-B*15:01	Not available	Not available
R49H	KSDLIVHVG	HLA-B*58:01, HLA-B*57:01, HLA-A*01:01	Not available	Not available
R49H	VHVGDEIRL	HLA-A*23:01, HLA-B*40:01, HLA-A*24:02, HLA-B*44:03, HLA-B*35:01, HLA-B*44:02, HLA-B*53:01	Not available	H-2-Kd
V399I	SDINAAIAF	HLA-B*44:03, HLA-B*44:02, HLA-B*40:01, HLA-B*15:01, HLA-B*35:01, HLA-A*26:01, HLA-A*30:02, HLA-B*53:01, HLA-A*01:01, HLA-B*07:02, HLA-A*32:01, HLA-A*23:01, HLA-A*24:02, HLA-B*58:01	H-2-Qa2	H-2-Kk, H-2-Ld
V399I	SNSDINAAI	HLA-A*68:02, HLA-B*51:01, HLA-A*02:06, HLA-B*40:01, HLA-A*30:02, HLA-A*02:03, HLA-A*26:01, HLA-B*07:02, HLA-B*58:01, HLA-A*32:01, HLA-B*44:02, HLA-B*44:03, HLA-A*01:01, HLA-B*53:01, HLA-B*35:01, HLA-A*23:01, HLA-A*24:02	Not available	H-2-Kk

Table 1. Cont.

Mutation	Epitope	HLA Alleles	Strong H2 Allele Restriction	Weak H2 Allele Restriction
V399I	NSDINAAIA	HLA-A*01:01, HLA-B*51:01, HLA-A*68:02, HLA-B*35:01	Not available	H-2-Db
D760V	AIMEDVELA	HLA-A*02:06, HLA-A*02:01, HLA-A*02:03, HLA-A*68:02, HLA-A*30:02, HLA-A*26:01, HLA-A*01:01, HLA-A*32:01, HLA-A*11:01	Not available	Not available
C809R	GRITKIRDF	HLA-B*08:01, HLA-A*30:02, HLA-B*15:01, HLA-A*23:01, HLA-A*26:01, HLA-B*44:03, HLA-A*32:01, HLA-A*24:02, HLA-B*44:02, HLA-B*40:01	Not available	Not available
C809R	ITKIRDFGL	HLA-B*08:01, HLA-B*57:01, HLA-A*30:01, HLA-B*58:01, HLA-A*68:02, HLA-A*32:01, HLA-A*02:06, HLA-B*07:02, HLA-A*30:02, HLA-B*51:01, HLA-A*02:03, HLA-A*31:01, HLA-B*15:01, HLA-A*33:01, HLA-A*24:02, HLA-A*23:01	Not available	Not available
C809R	THGRITKIR	HLA-A*33:01, HLA-A*31:01, HLA-A*68:01	Not available	Not available

2.3. Murine MHC Binding for CD8+ Epitopes

We used the default thresholds provided by NetMHCpan-4.0 to determine strong- and weak-binding epitopes to murine MHC molecules. Strong-binding epitopes had a threshold of 0.5% and weak-binding epitopes had a threshold of 2%. There were four strong binders and six weak binders. Top CD8+ epitopes had strong binding to the murine MHC alleles H-2-Db, H-2-Dd, H-2-Kb, and H-2-Qa2. Table 1 lists strong- and weak-binding murine MHC alleles for the top CD8+ epitopes.

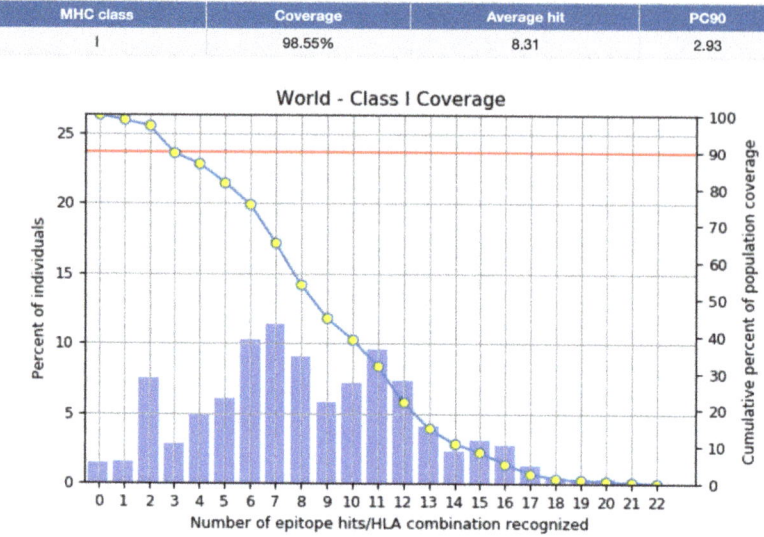

Figure 3. World Population Coverage for top CD8+ epitopes. World population coverage for the top CD8+ epitopes was 98.55%. Epitopes included in the calculation were filtered for percentile rank/binding affinity, immunogenicity, antigenicity, half-life, toxicity, allergenicity, and stability. Greater variety in HLA alleles resulted in higher population coverage.

2.4. Optimized Data for CD8+ Epitopes

PCOptim-CD was used on CD8+ epitopes filtered for rank, immunogenicity, antigenicity, half-life, toxicity, allergenicity, and stability. The resulting dataset with optimal population coverage included four CD8+ epitopes (Supplementary Table S2). One optimized epitope matched a top CD8+ epitope (SNSDINAAI) from Table 1. Population coverage rates of the final CD8+ epitopes and the optimized CD8+ epitopes were both 98.55% because PCOptim-CD was run on the same epitopes as the final filtered dataset.

2.5. Filtration of CD4+ Epitopes

The final CD4+ dataset included 21 epitopes filtered for rank (<10), immunogenicity (<50), antigenicity (>0.4), half-life (>1 h), toxicity (non-toxin), IFNγ (positive), allergenicity (non-allergen), and instability (<40). Two epitopes were 15-mers, two were 16-mers, five were 17-mers, and 12 were 18-mers. Thus, longer length epitopes had higher potency for MHC class II binding in our vaccine design. Table 2 lists the mutations, lengths, and binding HLA alleles of our top CD4+ epitopes. Specific values for clinically relevant variables (rank, immunogenicity, antigenicity, half-life, toxicity, allergenicity, instability, and IFNγ release) of the top CD8+ epitopes are in Supplementary Table S8. There were five CD4+ epitopes overlapping with top CD8+ epitopes. The C809R mutation resulted in four CD4+ epitopes (AARNILLTHGRITKI**R**DF, ARNILLTHGRITKI**R**DF, ARNILLTHGRITKI**R**DFG, ILLTHGRITKI**R**DFGLAR) overlapping with the CD8+ epitope G**R**ITKI**R**DF from the same mutation. The K550N mutation resulted in the CD4+ epitope TYKYLQ**N**PMYEVQWK overlapping with the CD8+ epitope **N**PMYEVQWK from the same mutation.

2.6. Population Coverage for CD4+ Epitopes

We determined that the world population coverage for MHC Class II-binding CD4+ epitopes was 65.14% (Figure 4). Regions with highest population coverage included South Asia (62.22%), Europe (71.47%), and North America (73.34%). Regions with the lowest population coverage were Southeast Asia (29.2%), Southwest Asia (33.7%), South

Africa (5.91%), and Oceania (37.6%). Population coverage for all regions is listed in Supplementary Table S3.

Table 2. Top CD4+ Epitopes.

Mutation	Length	Epitope	HLA Alleles	Strong H2 Allele Restriction	Weak H2 Allele Restriction
D816H	18	FGLARHIKNDSNYVVKGN	HLA-DRB1*13:02, HLA-DRB3*02:02, HLA-DRB3*01:01	Not available	Not available
D816H	18	GLARHIKNDSNYVVKGNA	HLA-DRB1*13:02, HLA-DRB3*02:02, HLA-DRB3*01:01	Not available	Not available
D816H	18	LARHIKNDSNYVVKGNAR	HLA-DRB1*13:02, HLA-DRB3*02:02, HLA-DRB3*01:01	Not available	Not available
D816V	18	VIKNDSNYVVKGNARLPV	HLA-DRB1*13:02, HLA-DRB3*02:02, HLA-DRB1*08:02, HLA-DRB1*15:01	Not available	H-2-IEd
D816Y	17	DFGLARYIKNDSNYVVK	HLA-DRB3*02:02, HLA-DRB1*13:02, HLA-DRB3*01:01, HLA-DRB1*08:02, HLA-DRB1*04:01, HLA-DRB1*15:01	Not available	H-2-IEd
D816Y	18	DFGLARYIKNDSNYVVKG	HLA-DRB3*02:02, HLA-DRB1*13:02, HLA-DRB3*01:01, HLA-DRB1*08:02, HLA-DRB1*04:01, HLA-DRB1*15:01	Not available	H-2-IEd
D816Y	16	FGLARYIKNDSNYVVK	HLA-DRB3*02:02, HLA-DRB1*13:02, HLA-DRB3*01:01, HLA-DRB1*15:01, HLA-DRB1*04:01, HLA-DRB1*08:02	H-2-IEd	Not available
D816Y	17	FGLARYIKNDSNYVVKG	HLA-DRB3*02:02, HLA-DRB1*13:02, HLA-DRB3*01:01, HLA-DRB1*08:02, HLA-DRB1*04:01, HLA-DRB1*15:01	H-2-IEd	Not available
D816Y	18	FGLARYIKNDSNYVVKGN	HLA-DRB3*02:02, HLA-DRB1*13:02, HLA-DRB3*01:01, HLA-DRB1*08:02, HLA-DRB1*04:01, HLA-DRB1*15:01	Not available	H-2-IEd
D816Y	17	GLARYIKNDSNYVVKGN	HLA-DRB3*02:02, HLA-DRB1*13:02, HLA-DRB3*01:01, HLA-DRB1*08:02, HLA-DRB1*04:01, HLA-DRB1*15:01	H-2-IEd	Not available

Table 2. *Cont.*

Mutation	Length	Epitope	HLA Alleles	Strong H2 Allele Restriction	Weak H2 Allele Restriction
D816Y	18	GLARYIKNDSNYVVKGNA	HLA-DRB3*02:02, HLA-DRB1*13:02, HLA-DRB3*01:01, HLA-DRB1*08:02, HLA-DRB1*04:01, HLA-DRB1*15:01	Not available	H-2-IEd
D816Y	17	LARYIKNDSNYVVKGNA	HLA-DRB3*02:02, HLA-DRB1*13:02, HLA-DRB3*01:01, HLA-DRB1*08:02, HLA-DRB1*04:01, HLA-DRB1*15:01	Not available	H-2-IEd
D816Y	18	LARYIKNDSNYVVKGNAR	HLA-DRB3*02:02, HLA-DRB1*13:02, HLA-DRB3*01:01, HLA-DRB1*08:02, HLA-DRB1*04:01, HLA-DRB1*15:01	Not available	H-2-IEd
N822K	15	DSKYVVKGNARLPVK	HLA-DRB3*02:02, HLA-DRB1*13:02, HLA-DRB5*01:01, HLA-DRB1*01:01, HLA-DRB1*08:02, HLA-DRB1*11:01, HLA-DRB1*15:01	H-2-IEd, H-2-IEk	Not available
N822K	16	NDSKYVVKGNARLPVK	HLA-DRB3*02:02, HLA-DRB1*13:02, HLA-DRB5*01:01	H-2-IEd	H-2-IEk
K550N	15	TYKYLQNPMYEVQWK	HLA-DRB3*02:02 HLA-DRB1*04:05, HLA-DRB1*04:01, HLA-DPA1*01:03/DPB1*04:01	Not available	Not available
C809R	18	AARNILLTHGRITKIRDF	HLA-DRB1*07:01	Not available	Not available
C809R	17	ARNILLTHGRITKIRDF	HLA-DRB1*07:01	Not available	Not available
C809R	18	ARNILLTHGRITKIRDFG	HLA-DRB1*07:01	Not available	Not available
C809R	18	ILLTHGRITKIRDFGLAR	HLA-DRB1*07:01	Not available	Not available
T417D & Y418F	18	AAIAFNVYVNTKPEILDF	HLA-DRB1*07:01, HLA-DRB3*02:02	Not available	Not available

2.7. Murine MHC Binding for CD4+ Epitopes

We used the default thresholds provided by NetMHCIIpan-4.0 to determine strong- and weak-binding epitopes to murine MHC molecules. Strong-binding epitopes had a threshold of 1% and weak-binding epitopes had a threshold of 5%. There were six strong-binding epitopes and eight weak-binding epitopes. The top CD4+ epitopes had strong binding to the murine MHC alleles H-2-IEd and H-2-IEk. Table 2 lists strong- and weak-binding murine MHC alleles for the top CD4+ epitopes.

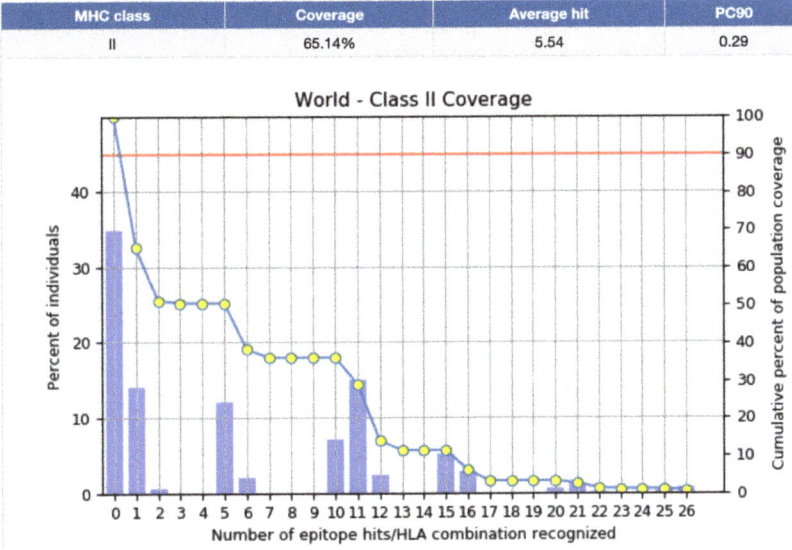

Figure 4. World Population Coverage for top CD4+ epitopes. World population coverage for the top CD8+ epitopes was 65.14%. Epitopes included in the calculation were filtered for percentile rank/binding affinity, immunogenicity, antigenicity, half-life, toxicity, IFNγ release, allergenicity, and stability. HLA-DRB3*02:02, HLA-DRB3*01:01, HLA-DRB1*04:05, HLA-DRB5*01:01, and HLA-DPA1*01:03/DPB1*04:01 were removed from the population coverage calculations because IEDB did not contain these alleles in their dataset.

2.8. Optimized Data for CD4+ Epitopes

PCOptim-CD was used on CD4+ epitopes filtered for rank, immunogenicity, and antigenicity. The resulting dataset included six CD4+ epitopes (Supplementary Table S4) with a world population coverage of 99.68%. There was no overlap between the optimized epitopes and the top CD4+ epitopes from Table 2, indicating weaker results for CD4+ data compared to CD8+ data. Regions with the highest population coverage for the optimized CD4+ dataset were Northeast Asia (99.39%), South Asia (99.74%), Europe (99.98%), East Africa (99.98%), West Africa (99.94%), Central Africa (99.88%), Central America (99.5%), South America (99.99%), and Oceania (99.54%). South Africa had the lowest regional population coverage (32.1%). HLA-DRB3*01:01, HLA-DRB3*02:02, and HLA-DRB5*01:01 were disregarded from the CD4+ optimized epitopes population coverage because the IEDB dataset did not include these alleles. Supplementary Table S5 provides the world and regional population coverage of the optimized CD4+ epitopes.

2.9. Population Coverage for Combined Class I and Class II Molecules

We combined the final filtered dataset for Class I and Class II MHC binding epitopes and used the IEDB population coverage tool to obtain 99.49% world population coverage. Population coverage rates for specific regions are listed in Supplementary Table S6. HLA-DRB3*01:01, HLA-DRB3*02:02, HLA-DRB1*04:05, HLA-DPA1*01:03/DPB1*04:01, HLA-DRRB5*01:01, HLA-B*40:01, and HLA-A*30:01 were excluded from the combined population coverage because the IEDB dataset did not contain data for those alleles.

2.10. 3D Modeling for Peptide–MHC Complexes and TCR Interactions

We modeled four top epitope–MHC complexes using MDockPep, CABS-dock, and PyMOL. We created 3D models for SDINAAIAF binding to HLA-A*01:01, GKSDLIVHV binding to HLA-A*02:06, GLARYIKNDSNYVVKGN binding to HLA-DRB1*04:01, and FGLARYIKNDSNYVVK binding to HLA-DRB3*01:01 (Figure 5). The 3D models for TCR interactions with peptide–MHC complexes were obtained using TCRModel (Figure 6). The A6 TCR is specific to HLA-A2 and was thus used to model an immune response to HLA-A*02:06 and the CD8+ epitope GKSDLIVHV [23]. The HA1.7 TCR is specific to HLA-DRB1*04:01 and was thus used to model an immune response to HLA-DRB1*04:01 and the CD4+ epitope GLARYIKNDSNYVVKGN [24]. Supplementary Figure S1 includes superimposed images of our epitope–MHC complexes with sample peptides from the RCSB Protein Data Bank [25] to validate the binding affinity of our epitopes to select MHC molecules.

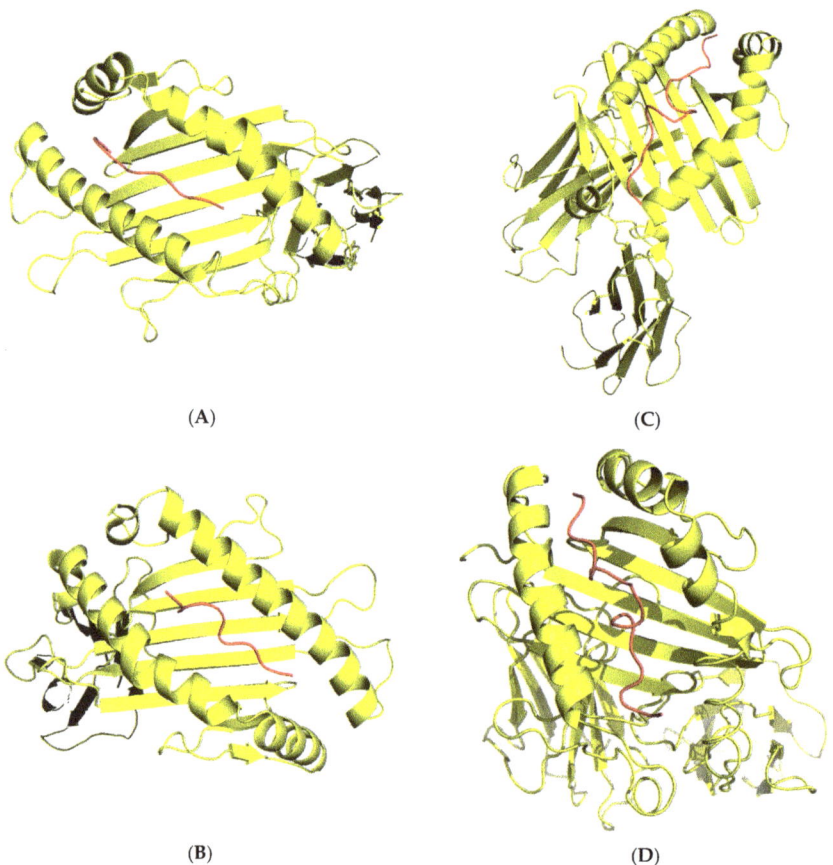

Figure 5. 3D models for peptide–MHC complexes. SDINAAIAF binding to MHC Class I molecule HLA-A*01:01 (RCSB PDB: 6MPP) (**A**). GKSDLIVHV binding to MHC Class I molecule HLA-A*02:06 (RCSB PDB: 3OXR) (**B**). GLARYIKNDSNYVVKGN binding to MHC Class II molecule HLA-DRB1*04:01 (RCSB PDB: 5JLZ) (**C**). FGLARYIKNDSNYVVK binding to MHC Class II molecule HLA-DRB3*01:01 (RCSB PDB: 2Q6W) (**D**). Yellow represents HLA alleles, and red represents epitopes.

2.11. 3D Modeling of Epitopes on KIT Gene

We selected a total of 33 CD8+ and CD4+ epitopes based on the filters: binding affinity/percentile rank, immunogenicity, antigenicity, half-life, toxicity, IFNγ release, allergenicity, and stability (12 CD8+ epitopes and 21 CD4+ epitopes). The KIT gene's protein kinase domain, which affects intracellular signaling pathways, holds 26 of our top epitopes. Figure 7 locates our top epitopes in a 3D model of the KIT gene.

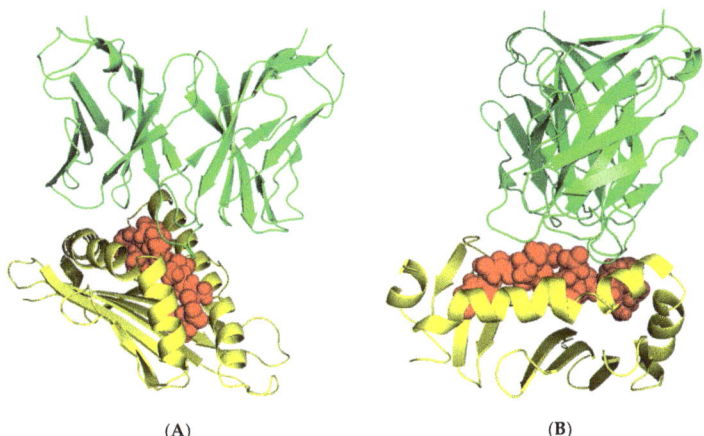

Figure 6. 3D models for TCR binding to top epitope–MHC complexes. GKSDLIVHV epitope and MHC Class I HLA-A*02:06 binding with the A6 TCR complex (RCSB PBD: 3QH3) (**A**). GLARYIKND-SNYVVKGN epitope and MHC Class II HLA-DRB1*04:01 binding with the HA1.7 TCR complex (RCSB PDB: 4GKZ) (**B**). Green represents the TCR complex specific to the HLA allele, yellow represents the HLA allele, and red represents the epitope.

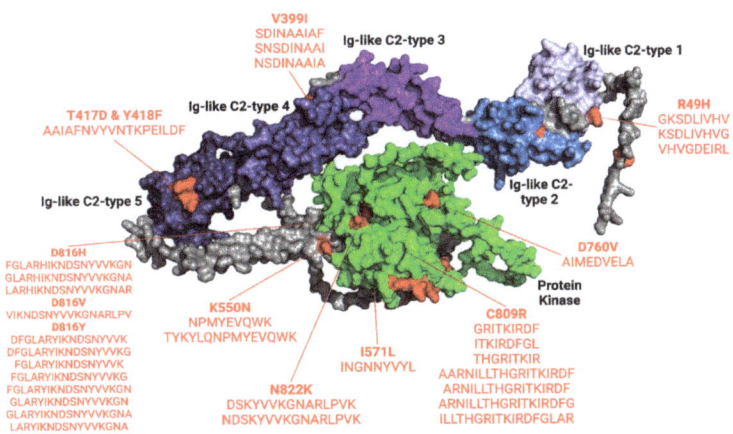

Figure 7. 3D structure of KIT marked with the locations of our top filtered epitopes. Three of our top CD8+ epitopes (GKSDLIVHV, KSDLIVHVG, VHVGDEIRL) are on the Ig-like C2-type 1 domain, and three (SDINAAIAF, SNSDINAAI, NSDINAAIA) are on the Ig-like C2-type 4 domain. Six CD8+ epitopes are on the protein kinase domain: NPMYEVQWK, INGNNYYVL, GRITKIRDF, ITKIRDFGL, THGRITKIR, AIMEDVELA. One CD4+ epitope (AAIAFNVYVNTKPEILDF) is located on the Ig-like C2-type five domain. The protein kinase holds 20 of our CD4+ epitopes: FGLARHIKNDSNYVVKGN, GLARHIKNDSNYVVKGNA, LARHIKNDSNYVVKGNAR, VIKNDSNYVVKGNARLPV, DFGLARYIKNDSNYVVK, DFGLARYIKNDSNYVVKG, FGLARYIKNDSNYVVK, FGLARYIKNDSNYVVKG, FGLARYIKNDSNYVVKGN, GLARYIKNDSNYVVKGN, GLARYIKNDSNYVVKGNA, LARYIKNDSNYVVKGNA, LARYIKNDSNYVVKGNAR, TYKYLQNPMYEVQWK, DSKYVVKGNARLPVK, NDSKYVVKGNARLPVK, AARNILLTHGRITKIRDF, ARNILLTHGRITKIRDF, ARNILLTHGRITKIRDFG, ILLTHGRITKIRDFGLAR.

3. Discussion

There is limited research on treatments for AML that target the KIT gene. Instead, peptide vaccines and dendritic cell vaccines have targeted other tumor-associated antigens (TAAs), including WT1, PR3, RHAMM, and MUC1. Limited MHC allele interactions with epitopes have been tested in WT1 vaccines, indicating potentially low population coverage. Additionally, a WT1 vaccine restricted in HLA-A*02 had no immunological significance in its phase II clinical trial owing to minimal vaccine benefits and low sample size. A vaccine targeting OCV-501 with MHC class II molecules resulted in insignificant immunological improvements in its phase II clinical trial. TAAs are less effective than TSAs in eliciting safe immune responses to cancer cells. TSAs are only present in cancer cells and have a higher affinity to MHC molecules and TCRs, making them better candidates for anticancer vaccines. TAAs are more widely studied, but their potential toxicity and lack of specificity for tumors indicate that targeting TSAs may be an improved approach. Clinical trials with TSA-based anticancer vaccines have also been successful [18]. Further research is needed to treat AML patients with vaccines targeting TSAs, but existing trials have shown the potential use of peptide vaccines in treating AML [26].

Clinical trials for CD8+ and CD4+ epitope vaccines against AML exist, but with limited success. One such vaccine targeting the WT1 gene reached phase II of clinical trials but did not develop strong immunological memory. We addressed this issue in our vaccine design by only selecting epitopes with high antigenicity scores. However, another vaccine targeting mutated WT1 peptides resulted in improved survival. Future trials for AML vaccines must prioritize targeting TSAs instead of TAAs to ensure proper and safe immune responses [26]. In this study, we targeted the proto-oncogene KIT and identified top epitopes predicted to elicit safe immunogenicity by selecting those with high binding affinity, immunogenicity, antigenicity, half-life, toxicity, IFNγ release, allergenicity, and population coverage.

Current studies on treatments for AML that target the KIT gene emphasize drug therapy, such as combined treatment with nilotinib and chemotherapy [27] and midostaurin on patients with (8;21) translocation AML. Patients in these studies had mutations in the KIT or FLT3-ITD genes, and similar to our study, the effects of midostaurin are being observed on mut-KIT8 and mut-KIT17 [28].

Our vaccine design follows in silico methods predicted to safely induce CD8+ and CD4+ immunogenic responses. We demonstrated predicted vaccine efficacy by filtering epitopes through clinically relevant variables such as immunogenicity, antigenicity, toxicity, and allergenicity, to obtain top epitopes. Designing vaccines through bioinformatics offers a quick and cost-effective method of developing anti-cancer treatments before murine or pre-clinical trials. We identified four CD8+ and six CD4+ epitopes that were strong binders to murine MHC molecules, demonstrating potential use of our vaccine design in further research including murine trials.

We filtered out many potential CD8+ and CD4+ epitopes because of low immunogenicity scores. In the CD8+ dataset, 50% of epitopes that passed the percentile rank filter also passed the immunogenicity filter. All top CD8+ epitopes failed to pass the IFNγ filter because IFNepitope was only developed for CD4+ epitopes. Thus, the IFNγ filter was disregarded for MHC I binding molecules. For the CD4+ dataset, IFNγ and allergenicity filtered out most of the epitopes in addition to immunogenicity.

Our vaccine design was strengthened by the five CD4+ epitopes overlapping with the top CD8+ epitopes. Overlapping epitopes emphasizes their strength and our vaccine's potential to elicit high immunogenic responses involving both cytotoxic and helper T cells. Population coverage for overlapping epitopes alone remains low, but the two potential immunogenic pathways that may be induced by the overlapping CD8+ and CD4+ epitopes indicate high potency for attacking cancerous cells. Further research on increasing population coverage of overlapping epitopes can help improve the vaccine's effectiveness. Additionally, combined usage of CD8+ and CD4+ epitopes increase the likelihood of stability despite the short peptide lengths of CD8+ epitopes. CD8+ epitopes were limited to

9-mers, but CD4+ data included epitopes of up to 18-mers. Still, previous studies indicated that CD8+ and CD4+ immunogenic responses are inducible with vaccines using 9- or 10-mer peptides in patients with solid tumors [29].

The protein kinase domain of the KIT gene held 26 out of 33 top CD8+ and CD4+ epitopes. Each epitope was critical for our vaccine design. Still, the intracellular signaling pathways that the KIT gene is involved in, such as those outlined in Figure 2, are mainly instigated in the protein kinase domain. The protein kinase's critical role in hematopoietic cell growth, proliferation and development makes the domain an important location for our top epitopes. Cancerous activities caused by mutations in the protein kinase domain can be primarily targeted by having most of our target mutations in this domain. Our study was unique in targeting a proto-oncogene for which not many have studied AML vaccine therapies. The 33 combined CD8+ and CD4+ epitopes induced a population coverage of 99.49%, ensuring that our vaccine may effectively improve AML prognosis for a large population. For both CD8+ and CD4+ epitopes, we determined population coverage based on HLA alleles that the peptides could bind to and the frequency of those alleles among various regions worldwide. High population coverage was optimal because more patients could effectively be treated with the vaccine. However, regions including Central America had lower population coverage for CD8+ and CD4+ epitopes. Large differences in population coverage such as between Central America and Europe were due to varying frequencies of HLA alleles in different populations. Each population has a unique frequency of HLA alleles, so the potency of our epitope design varies by region. In Central America, frequent HLA alleles include A*02:06:01, A*02, DQA1*05:01, and A*02:02 [30]. However, frequent alleles in Central America, such as DQA1*05:01, were still included in our top epitopes. This indicated that other discrepancies in the region's genetic makeup may have caused lower population coverage in this region. Vaccine design methods can be improved by filtering for top epitopes that specifically bind to alleles prevalent in regions with low population coverage found in our data to maximize efficacy. Limitations in IEDB's allele dataset also resulted in lower population coverage for certain regions, primarily with the CD4+ dataset.

Our vaccine design would be the most effective on AML patients within Asia, Europe, and North America, which included regions with the highest population coverage. HLA alleles that our top epitopes bind to were more prevalent in these regions. AML is most reported in North America, Western Europe, and South Asia, which validates our vaccine design, as our targeted population would be the most reactive to our vaccine [31].

Our data were weakly validated for population coverage owing to the minimal overlap between final epitope datasets and optimized epitopes from PCOptim-CD. However, PCOptim-CD was not as effective in our vaccine design as compared to other datasets—when PCOptim-CD was used on epitope data for a vaccine design targeting the HRAS gene for squamous cell carcinoma, the optimized dataset contained six epitopes [32]. PCOptim-CD analysis on CD8+ epitopes filtered for rank and immunogenicity only resulted in one optimal epitope. This demonstrated the high quality of our epitopes in the inputted dataset because it showed that maximum population coverage could be obtained with one epitope. However, to find more optimized epitopes, every filter had to be applied to the inputted data, making the optimized CD8+ epitope population coverage identical to that of the top CD8+ epitopes. Only one epitope from the optimized dataset matched one of our top CD8+ epitopes. Additionally, none of the epitopes in the CD4+ optimized dataset matched the top CD4+ epitopes. Therefore, population coverage of the CD4+ epitope dataset was weaker than that of CD8+. With CD4+ epitopes having a lower population coverage and less validity from PCOptim-CD, CD4+ T-cell response to our vaccine design was weak.

Peptide vaccine designs are a cost-effective method of developing treatments to target tumors and/or viruses. Computational methods also allow for large protein datasets to be quickly tested for vaccine efficacy. When compared to in vitro and in vivo studies, in silico methods are unable to reflect direct testing with living cells. To address these challenges, tools for in silico vaccine studies are constantly being developed to form optimal vaccine designs. IntegralVac is an example of this, where MHCSeqNet, DeepVacPred, and hemolytic/non-hemolytic peptide predictors were combined to improve vaccine design accuracy and safety [33]. Our data can also be used for future research to develop immunoinformatic methods to strengthen our epitope design.

Limitations of the Study

Compared to the population coverage of the CD8+ epitopes, the CD4+ epitopes had low coverage. Additional mutations were filtered through to find more epitopes, including combination mutations with double missense, but population coverage remained low. One potential reason was that IEDB had limitations in their HLA allele dataset—a few alleles in the final CD4+ dataset were excluded from the population coverage calculation. For example, HLA-DRB3*02:02 was not included in the CD4+ population coverage, but the allele could bind to 16 of the final CD4+ epitopes. HLA-DRB3*01:01 was also excluded from the population coverage but could bind to 12 final CD4+ epitopes. Thus, the accuracy of CD4+ population coverage was limited owing to the IEDB database.

CD4+ epitope population coverage may have also been low because studies show that the KIT gene does not often interact with CD4+ T cells [34]. The KIT gene is involved in CD8+ T-cell immunodominance, but the gene was not expressed in the presence of CD4+ T cells. KIT genes can induce CD4+ T-cell immune responses, but KIT gene expression is less involved with CD4+ T cells than CD8+ T cells. Past experiments that found minimal interaction with the KIT gene and CD4+ T cells indicated why population coverage may have been low. Furthermore, because most of our mutations were on the intracellular protein kinase domain of the KIT gene, CD8+ T cells are more susceptible to being instigated, as CD8+ T cells respond to endogenous antigens, while CD4+ T cells mainly respond to exogenous antigens.

4. Materials and Methods

4.1. Finding Prevalent Point Mutations on the KIT Gene

Common mutations of the KIT gene that cause AML are located on exons 17 and 8, and D816V is the most prevalent [1,35]. Mutations were chosen based on prevalence—CoDing Sequence (CDS) mutations of alanine to threonine were present in 48.20% of samples compiled in the COSMIC database. We used Y418F and D816V in this study. The CDS mutation of glycine to threonine was present in 15.51% of samples, including the point mutations W8C and D816Y used in our study. The CDS mutation of glycine to cysteine was present in 11.91% of samples, which included the point mutation D816H used in our study. Lastly, the CDS mutation of threonine to glycine was present in 9.14% of samples, including the point mutation N822K used in our study [36]. The most common mutations were located at point 816 on aspartic acid [36]. Additional point mutations found in past clinical trials [1,4,36], as well as those found in the COSMIC database (https://cancer.sanger.ac.uk/cosmic, (accessed on 2 June 2022)) [37] were used to obtain mutated KIT gene sequences. Figure 8 shows where the 24 mutations we observed are located on the KIT gene.

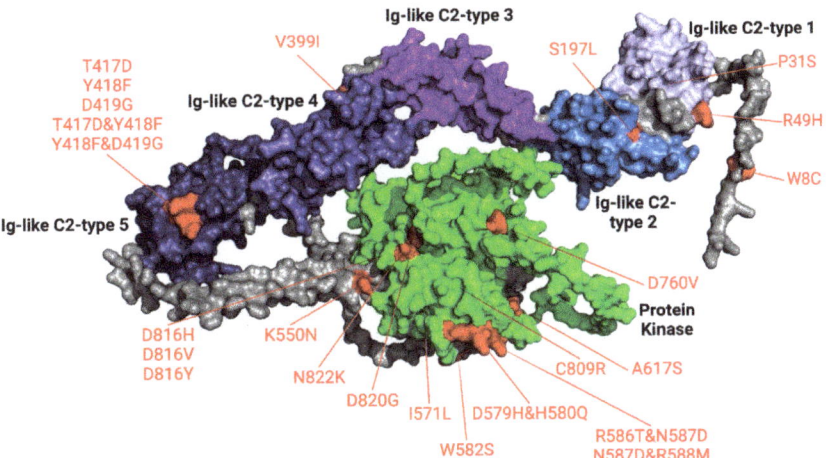

Figure 8. 3D structure of KIT with the locations of point mutations used in our study. Two mutations are on the Ig-like C2-type 1 domain (P31S, R49H), one mutation is on the Ig-like C2-type 2 domain (S197L), and one mutation is on the Ig-like C2-type 4 domain (V399I), and five mutations are on the Ig-like C2-type 5 domain (T417D, Y418F, D419G, T417D and Y418F, Y418F and D419G). The protein kinase domain holds 14 of the mutations we used in our study (K550N, D816H/V/Y, D820G, I571L, N822K, D579H and H580Q, R586T and N587D, N587D and R588M, C809R, A617S, D760V). The AlphaFold Protein Structure Database (https://alphafold.ebi.ac.uk/entry/P10721, (accessed on 31 July 2022) [9] was used to obtain the whole KIT gene structure, and UniProt was used to identify the domains [38].

4.2. Identifying Mutated Sequences

The "mast/stem cell growth factor receptor Kit" peptide sequence was obtained in FASTA format using UniProt (https://www.uniprot.org/, (accessed on 31 July 2022) [39]. Mutated peptide sequences were determined based on the point mutations labeled in Figure 1.

4.3. MHC Class I Binding Epitope Prediction

9-mer CD8+ epitopes for each point mutation were obtained using the IEDB T Cell Epitope Prediction Tool with MHC I Binding (http://tools.iedb.org/mhci/, (accessed on 7 June 2022) [40]. The prediction tool was trained to predict binding affinity for top HLA alleles in humans using binding affinity and eluted ligand data. IEDB calculated a percentile rank for each epitope's binding affinity to 27 HLA alleles: HLA-A*01:01, HLA-A*02:01, HLA-A*02:03, HLA-A*02:06, HLA-A*03:01, HLA-A*11:01, HLA-A*23:01, HLA-A*23:01, HLA-A*24:02, HLA-A*26:01, HLA-A*30:01, HLA-A*30:02, HLA-A*31:01, HLA-A*32:01, HLA-A*33:01, HLA-A*68:01, HLA-A*68:02, HLA-B*07:02, HLA-B*08:01, HLA-B*15:01, HLA-B*35:01, HLA-B*40:01, HLA-B*44:02, HLA-B*44:03, HLA-B*51:01, HLA-B*53:01, HLA-B*57:01, HLA-B*58:01. IEDB derived the percentile rank by comparing IC_{50} values of each peptide in the protein sequence with the IC_{50} values of other peptides found in the SWISSPROT database [41]. Lower percentages (above 0%) indicated higher binding affinity, and a maximum threshold of 10% was used for this filter.

Strong and stable epitope candidates were determined based on a variety of clinically relevant variables, including percentile rank (binding affinity), immunogenicity, antigenicity, half-life, instability, isoelectric point, aliphatic index, GRAVY score, toxicity, IFNγ release, and allergenicity. Only epitopes that passed these filters (IFNγ was disregarded for CD8+) were presented as top epitopes for our vaccine design.

Each epitope that passed the percentile rank filter was tested for immunogenicity using the IEDB Class I Immunogenicity tool (http://tools.iedb.org/immunogenicity/, (accessed on 13 June 2022) [42]. IEDB trained the tool to identify immunogenicity through a study of 600 immunogenic and 181 non-immunogenic peptide–MHC complexes. Further training included analysis of non-anchor positions (positions 4–6) in determining positions of high interaction with T-cell receptors (TCRs). Higher scores indicated greater immunogenicity of the epitopes, and the minimum threshold was set to 0 [43].

Antigenicity for top immunogenic epitopes was determined by VaxiJen v2.0 (http://www.ddg-pharmfac.net/vaxijen/VaxiJen/VaxiJen.html, (accessed on 13 June 2022) [44]. VaxiJen is up to 89% accurate and was developed with auto-cross covariance (ACC), turning protein sequences into vectors representing principal amino acid properties. VaxiJen v2.0 provides datasets for five organisms: bacteria, viruses, tumors, parasites, and fungi—we used tumors for our dataset. The minimum threshold used for antigenicity was 0.4—when VaxiJen was developed and tested on viral antigens, a threshold of 0.4 had 70% accuracy for external validation [44,45]. VaxiJen has been tested on multiple in silico vaccine designs, one of which identified T-cell and B-cell epitopes targeting the SARS-COV2 S protein [46].

Half-life, instability, isoelectric point, aliphatic index, and GRAVY score were determined through ProtParam (https://web.expasy.org/protparam/, (accessed on 14 June 2022). To calculate half-life, ProtParam analyzed each epitope's N-terminal amino acid [47]. Amino acids in mammals have a minimum half-life of 0.8–1 h. Thus, we used one hour as the minimum threshold for half-life [48]. ProtParam calculates the instability index based on dipeptides. ProtParam trained the program to calculate instability using a study of 400 dipeptides in test tubes that were given weight values based on dipeptides of known stable and unstable proteins. A maximum threshold of 40 was used by ProtParam and our study to distinguish instability [49,50].

While the epitopes were not filtered for isoelectric point, aliphatic index, and GRAVY score, these values demonstrate the physicochemical properties of our top epitopes. Isoelectric point indicates the pH when a peptide reaches a neutral charge [51]. The aliphatic index was calculated by ProtParam based on the volume of aliphatic side chains (alanine, valine, leucine, and isoleucine) in the epitopes. A higher aliphatic index indicates higher thermostability. The GRAVY (grand average of hydropathy) score reveals the hydropathy of peptides, with higher scores indicating higher hydrophobicity [49].

Toxicity was obtained with ToxinPred (https://webs.iiitd.edu.in/raghava/toxinpred/, (accessed on 14 June 2022) [52]. Toxicity is determined based on SVM scores, which ToxinPred calculates based on the amino acid and dipeptide composition, binary profile pattern, and motif-based profile. The main training dataset used to develop ToxinPred included 1805 toxic peptides and 3593 non-toxic peptides. Performance on the main training dataset resulted in 93.92% maximum accuracy from the amino acid-based SVM model, 94.50% accuracy from the dipeptide-based SVM model, and 91.63% accuracy from the binary profile-based SVM model [53].

IFNγ release was tested using IFNepitope (http://crdd.osdd.net/raghava/ifnepitope/, (accessed on 14 June 2022) [54]. IFNepitope determines IFNγ release with an accuracy of 82.10% based on motifs likely to release IFNγ. IFNepitope obtained 10,433 CD4+ epitopes from IEDB to develop the dataset—3705 resulted in positive IFNγ release, and 6728 resulted in negative IFNγ release [55]. The IFNγ filter was disregarded for MHC class I molecules because IFNepitope was only developed using MHC class II molecules. However, results were still obtained for MHC class I molecules.

Allergenicity was determined using AllerTOP v2.0 (https://www.ddg-pharmfac.net/AllerTOP/, (accessed on 14 June 2022) [56]. AllerTOP v2.0 also uses ACC to develop uniform vectors from proteins. Datasets in AllerTOP v2.0 were tested against known allergenic and non-allergenic peptides. Filtering out allergenic epitopes helps design a safe vaccine because certain proteins can induce abnormal immune responses, such as rashes, sneezing, and mucous membrane swelling [57].

Population coverage was calculated using the IEDB epitope analysis tool "Population Coverage" (http://tools.iedb.org/population/, (accessed on 21 June 2022)). We used population coverage to determine our vaccine's effectiveness on the world population and on the populations of 16 regions: East Asia, Northeast Asia, South Asia, Southeast Asia, Southwest Asia, Europe, East Africa, West Africa, Central Africa, North Africa, South Africa, West Indies, North America, Central America, South America, and Oceania. We observed population because HLA type representation varies by population and ethnicity, and maximum coverage is ideal for a vaccine design [58].

4.4. MHC Class II Binding Epitope Prediction

We used the same method for filtering through the CD4+ epitope dataset as we did for the CD8+ epitopes. However, percentile rank/binding affinity and immunogenicity were calculated with different tools. Percentile rank/binding affinity was obtained using the MHC II Binding Prediction tool on IEDB (http://tools.iedb.org/mhcii/, (accessed on 4 July 2022)) [59]. IEDB included 27 HLA alleles, and epitopes of length 12–18 mers were obtained. The HLA alleles studied for MHC class II molecules were HLA-DRB1*-1:-1, HLA-DRB1*03:01, HLA-DRB1*04:01, HLA-DRB1*04:05, HLA-DRB1*07:01, HLA-DRB1*08:02, HLA-DRB1*09:01, HLA-DRB1*11:01, HLA-DRB1*12:01, HLA-DRB1*13:02, HLA-DRB1*15:01, HLA-DRB3*01:01, HLA-DRB3*02:02, HLA-DRB4*01:01, HLA-DRB5*01:01, HLA-DQA1*05:01/DQB1*02:01, HLA-DQA1*05:01/DQB1*03:01, HLA-DQA1*03:01/DQB1*03:02, HLA-DQA1*04:01/DQB1*04:02, HLA-DQA1*01:01/DQB1*05:01, HLA-DQA1*01:02/DQB1*06:02, HLA-DPA1*02:01/DPB1*01:01, HLA-DPA1*01:03/DPB1*02:01, HLA-DPA1*01:02/DPB1*04:01, HLA-DPA1*03:01/DPB1*04:02, HLA-DPA1*02:01/DPB1*05:01, HLA-DPA1*02:01/DPB1*14:01. A percentile rank threshold of 10% was kept.

Immunogenicity was determined using the IEDB CD4+ T cell immunogenicity prediction tool (http://tools.iedb.org/CD4episcore/, (accessed on 4 July 2022)) [60]. This tool calculated an IEDB-recommended combined score, which is the combination of each epitope's immunogenicity and their HLA binding prediction scores. Combined scores had a maximum area under the ROC curve (AUC) score of 0.71 with a training dataset of 530 immunogenic peptides and 1758 non-immunogenic peptides [61]. We calculated the percent of MHC class I binding epitopes that passed the immunogenicity threshold of 0 to determine a threshold for CD4+ epitope immunogenicity. Half of the CD8+ epitopes passed the immunogenicity filter, so a maximum combined score of 50 (out of 100) was used for the CD4+ epitope immunogenicity threshold. A lower combined score indicated a better T-cell response.

4.5. Obtaining Optimized Population Coverage with PCOptim-CD

The final epitopes had several overlapping amino acid sequences. We developed PCOptim-CD to find an optimized epitope dataset with maximal population coverage to reduce redundancy in epitope selection. The original program, PCOptim, was only designed for CD8+ datasets. The modified version, PCOptim-CD, was programmed to obtain the optimized epitopes for the CD4+ dataset as well. PCOptim-CD (Supplementary Figure S1) was based on the console version, called PopCoverageOptimization. Therefore, it is text-based rather than GUI-based, and instructions for using the program can be found in the comments of the Java code.

Epitopes and their MHC-restricted alleles for optimal population coverage were obtained using PCOptim-CD [32]. We used CD8+ epitopes filtered by rank, immunogenicity, antigenicity, half-life, instability, toxicity, and allergenicity to find multiple optimal CD8+ epitopes. We used CD4+ epitopes filtered by rank, immunogenicity, and antigenicity to obtain the optimized CD4+ dataset. PCOptim-CD allowed us to identify epitopes from our full dataset that were likely to have optimal population coverage.

4.6. Murine MHC Binding

Strong- and weak-binding CD8+ epitopes to murine MHC molecules were identified using NetMHCpan-4.0 for peptide-MHC class I binding (https://services.healthtech.dtu.dk/service.php?NetMHCpan-4.0, (accessed on 28 June 2022) [62]. NetMHCpan-4.0 used artificial neural networks (ANNs) to give results for the following murine MHC alleles:

H-2-Db, H-2-Dd, H-2-Kb, H-2-Kd, H-2-Kk, H-2-Ld, H-2-Qa1, and H-2-Qa2. For CD4+ epitopes, NetMHCIIpan-4.0 for peptide–MHC class II binding was used (https://services.healthtech.dtu.dk/service.php?NetMHCIIpan-4.0, (accessed on 19 July 2022) [63], which gave results for the following murine MHC alleles: H-2-IAu, H-2-Ied, and H-2-IEk.

4.7. Three-Dimensional (3D) Modeling of Peptide–MHC Complex and TCR Interactions

We found PDB files for four HLA alleles that were restricted by several of our top epitopes on the RCSB Protein Data Bank (https://www.rcsb.org/, (accessed on 21 July 2022) [25]. HLA alleles were chosen based on the MHC restrictions presented for each CD8+ and CD4+ epitope listed by IEDB. Using MDockPeP (https://zougrouptoolkit.missouri.edu/mdockpep/, (accessed on 21 July 2022) [64–66] and CABS-dock [67], select epitopes from our final dataset were attached to binding grooves of HLA alleles to create four 3D models of peptide–MHC complexes. Both MDockPeP and CABS-dock generate top-scoring docking models with minimal binding energy. TCRModel (https://tcrmodel.ibbr.umd.edu/rtcrex/TCRSDM6_180718_160348, (accessed on 31 July 2022) [68] was used to create 3D models of TCR complex interactions with our peptide–MHC complexes. All 3D models were edited with PyMOL.

5. Conclusions

Several studies have investigated how to treat AML, including drug therapies, combination therapy (drugs and chemotherapy), stem cell transplants, and vaccines. However, many treatments, including AML vaccines that target the KIT gene, remain unexplored. The purpose of this study was to develop a vaccine design for AML using in silico methods that target missense mutations on the KIT oncogene. We applied several clinically relevant variables to our vaccine epitopes, including percentile rank, immunogenicity, antigenicity, half-life, toxicity, IFNγ release, allergenicity, and stability, to ensure the vaccine's safety and effectiveness. Then, population coverage demonstrated the broadness of our vaccine design's potential. Using this method, we found 12 CD8+ and 21 CD4+ epitopes from mutated KIT peptide sequences that can be implemented in a vaccine and potentially used in murine trials. The 12 CD8+ epitopes were immunogenic, antigenic, non-toxic, non-allergenic, and had long half-lives. In comparison, the 21 CD4+ epitopes were immunogenic, antigenic, non-toxic, non-allergenic, have long half-lives, and release IFNγ. The CD8+ epitopes had a high population coverage of 98.55%, while the CD4+ epitopes had a lower population coverage of 65.14% owing to limitations in our tools' datasets and minimal interactions between the KIT gene and CD4+ T-cells. PCOptim was modified into PCOptim-CD to analyze both CD8+ and CD4+ datasets for optimized population coverage. There was minimal overlap between the final filtered epitopes and the optimized epitopes from PCOptim-CD, proving that further research is needed to develop a stronger dataset with greater validity. The four CD8+ and six CD4+ epitopes that were strong binders to murine MHC alleles indicated that our results can lead to preclinical studies with vaccine trials on murine models. We designed a vaccine predicated to be safe and effective through in silico methods to help improve treatments for AML and develop cost-effective methods for vaccine designs before pre-clinical trials. Our data may be used to facilitate future studies in investigating the use of our vaccine design in murine and clinical trials and improving immunoinformatic tools. Murine trials with the peptide vaccine design would be the next step for advancing research on this treatment for AML. Using the top epitopes with strong binding to murine MHC molecules, hematopoietic and stem and progenitor cells from mice would be modified with genome editing in vitro. The treatment group would receive these cells intravenously (IV) in addition to radiation treatment [69] and IV-administered peptide vaccine, and the control group would receive normal saline administration. The study would include dosage testing to measure the appropriate dosage needed for the peptide vaccine. qPCR analysis may be conducted to measure the presence of the KIT gene as well as mutant KIT genes. RNA-sequence analysis would be used to measure prevalence of the single amino acid mutations found in our peptide vaccine. MHC-epitope binding complexes would be isolated with immunoprecipitation assays to confirm the success of epitope binding to target MHC allele. SCF binds to the KIT gene to induce various cellular pathways, and SCF-ELISA assay may be used to analyze antibody binding levels on KIT to assess KIT function. The results of these experiments with murine trials would determine whether the peptide vaccine can be tested further clinically.

Supplementary Materials: The following supporting information can be downloaded at: https://www.mdpi.com/article/10.3390/ph16070932/s1, Figure S1: Superimposed models of epitope–MHC complexes with sample peptides. SDINAAIAF binding to MHC Class I molecule HLA-A*01:01 superimposed with PDB ID: 6MPP (**A**). GKSDLIVHV binding to MHC Class I molecule HLA-A*02:06 superimposed with PDB ID: 3OXR (**B**). GLARYIKNDSNYVVKGN binding to MHC Class II molecule HLA-DRB1*04:01 superimposed with PDB ID: 5JLZ (**C**). FGLARYIKNDSNYVVK binding to MHC Class II molecule HLA-DRB3*01:01 superimposed with PDB ID: 2Q6W (**D**); Table S1: Population Coverage for CD8 Epitopes; Table S2: Optimized CD8 Epitopes; Table S3: Population Coverage for CD4 Epitopes; Table S4: Optimized CD4 Epitopes; Table S5: Population Coverage for Optimized CD4 Epitopes; Table S6: Combined Class I and Class II Population Coverage; Table S7: Top CD8+ Epitopes Clinically Relevant Variables; Table S8: Tope CD4+ Epitopes Clinically Relevant Variables.

Author Contributions: Conceptualization, S.D.; methodology, K.S., S.D. and M.K.; software, K.S. and S.D.; validation, M.K. and K.S.; formal analysis, M.K. and K.S.; investigation, K.S., S.D. and M.K.; resources, S.D.; data curation, M.K. and K.S.; writing—original draft preparation, M.K., K.S. and S.D.; writing—review and editing, M.K., K.S. and S.D.; visualization, M.K.; supervision, S.D.; project administration, S.D.; funding acquisition, S.D. All authors have read and agreed to the published version of the manuscript.

Funding: This research received no external funding.

Institutional Review Board Statement: Not applicable.

Informed Consent Statement: Not applicable.

Data Availability Statement: Data is contained within the article and Supplementary Material.

Acknowledgments: This work was supported in part by funding from Georgetown Lombardi's Comprehensive Cancer (LCCC) Research Training and Education Coordination (CRTEC), and the author M.K. and K.S. were part of the GLCCC Undergraduate Summer Research Program. The author D.S. acknowledges the support of the LCCC METRO PILOT Award.

Conflicts of Interest: The authors declare no conflict of interest.

References

1. DiNardo, C.D.; Cortes, J.E. Mutations in AML: Prognostic and therapeutic implications. *Hematol. Am. Soc. Hematol. Educ. Program* **2016**, *1*, 348–355. [CrossRef] [PubMed]
2. Vakiti, A.; Mewawalla, P. Acute Myeloid Leukemia. In *StatPearls [Internet]*; [Updated 17 August 2021]; StatPearls Publishing: Treasure Island, FL, USA, 2022.
3. Mayo Foundation for Medical Education and Research. Acute Myelogenous Leukemia. Mayo Clinic. 10 February 2021. Available online: https://www.mayoclinic.org/diseases-conditions/acute-myelogenous-leukemia/symptoms-causes/syc-20369109 (accessed on 21 July 2022).
4. Hussain, S.R.; Raza, S.T.; Babu, S.G.; Singh, P.; Naqvi, H.; Mahdi, F. Screening of C-kit gene Mutation in Acute Myeloid Leukaemia in Northern India. *Iran. J. Cancer Prev.* **2012**, *5*, 27–32.
5. Poklepovic, A.; Bose, P. Molecularly Targeted Therapy: Imatinib and Beyond Gastrointestinal stromal tumor. In *Gastrointestinal Stromal Tumor*; Lunevicius, R., Ed.; InTech: London, UK, 2012; Volume 49.
6. Tabone-Eglinger, S.; Subra, F.; El Sayadi, H.; Alberti, L.; Tabone, E.; Michot, J.-P.; Théou-Anton, N.; Lemoine, A.; Blay, J.-Y.; Emile, J.-F. KIT Mutations Induce Intracellular Retention and Activation of an Immature Form of the KIT Protein in Gastrointestinal Stromal Tumors. *Clin. Cancer Res.* **2008**, *14*, 2285–2294. [CrossRef] [PubMed]
7. Sangle, N.A.; Perkins, S.L. Core-Binding Factor Acute Myeloid Leukemia. *Arch. Pathol. Lab. Med.* **2011**, *135*, 1504–1509. [CrossRef]
8. Badr, P.; Elsayed, G.M.; Eldin, D.N.; Riad, B.Y.; Hamdy, N. Detection of KIT mutations in core binding factor acute myeloid leukemia. *Leuk. Res. Rep.* **2018**, *10*, 20–25. [CrossRef]
9. Mast/Stem Cell Growth Factor Receptor Kit: AlphaFold Structure Prediction. AlphaFold Protein Structure Database. Updated 1 June 2022. Available online: https://alphafold.ebi.ac.uk/entry/P10721 (accessed on 31 July 2022).

10. Shi, X.; Sousa, L.P.; Mandel-Bausch, E.M.; Tome, F.; Reshetnyak, A.V.; Hadari, Y.; Schlessinger, J.; Lax, I. Distinct cellular properties of oncogenic kit receptor tyrosine kinase mutants enable alternative courses of cancer cell inhibition. *Proc. Natl. Acad. Sci. USA* **2016**, *113*, E4784–E4793. [CrossRef]
11. Treating Acute Myeloid Leukemia (AML). American Cancer Society. Available online: https://www.cancer.org/cancer/acute-myeloid-leukemia/treating.html (accessed on 21 July 2022).
12. Paschka, P.; Konstanze, D. Core-binding factor acute myeloid leukemia: Can we improve on HiDAC consolidation? *Hematol. Am. Soc. Hematol. Educ. Program* **2013**, *1*, 209–219. [CrossRef]
13. Brownell, L. Solid Vaccine Eliminates Acute Myeloid Leukemia in Mice. Harvard Gazette. 14 January 2020. Available online: https://news.harvard.edu/gazette/story/2020/01/solid-vaccine-eliminates-acute-myeloid-leukemia-in-mice/ (accessed on 21 July 2022).
14. Kim, K.H.; Kim, J.O.; Park, J.Y.; Seo, M.D.; Park, S.G. Antibody-drug conjugate targeting c-KIT for the treatment of small cell lung cancer. *Int. J. Mol. Sci.* **2022**, *23*, 2264. [CrossRef] [PubMed]
15. Ray, P.; Krishnamoorthy, N.; Oriss, T.B.; Ray, A. Signaling of c-kit in dendritic cells influences adaptive immunity. *Ann. N. Y. Acad. Sci.* **2010**, *1183*, 104–122. [CrossRef]
16. Dentelli, P.; Cavallo, F.; Brizzi, M.F. Membrane-bound KIT ligand-targeting DNA vaccination inhibits mammary tumor growth. *Oncoimmunology* **2014**, *3*, e27259. [CrossRef]
17. Liu, J.J.; Yu, C.S.; Wu, H.W.; Chang, Y.J.; Lin, C.P.; Lu, C.H. The structure-based cancer-related single amino acid variation prediction. *Sci. Rep.* **2021**, *11*, 13599. [CrossRef] [PubMed]
18. Zhao, Y.; Baldin, A.V.; Isayev, O.; Werner, J.; Zamyatnin, A.A., Jr.; Bazhin, A.V. Cancer Vaccines: Antigen Selection Strategy. *Vaccines* **2021**, *9*, 85. [CrossRef]
19. Liang, J.; Wu, Y.L.; Chen, B.J.; Zhang, W.; Tanaka, Y.; Sugiyama, H. The c-kit receptor-mediated signal transduction and tumor-related diseases. *Int. J. Biol. Sci.* **2013**, *9*, 435–443. [CrossRef]
20. de Lartigue, J. The SCF/KIT Pathway's Roles: Interest in Therapeutic Targets is Growing. OncLive. Updated 1 September 2011. Available online: https://www.onclive.com/view/the-scfkit-pathways-roles-interest-in-therapeutic-targets-is-growing (accessed on 31 July 2022).
21. Feng, Z.C.; Riopel, M.; Popell, A.; Wang, R. A survival Kit for pancreatic beta cells: Stem cell factor and c-Kit receptor tyrosine kinase. *Diabetologia* **2015**, *58*, 654–665. [CrossRef]
22. Carlino, M.S.; Todd, J.R.; Rizos, H. Resistance to c-KIT inhibitors in melanoma: Insights for Future Therapies. *Oncoscience* **2014**, *1*, 423–426. [CrossRef] [PubMed]
23. Wang, Z.; Turner, R.; Baker, B.M.; Biddison, W.E. MHC Allele-Specific Molecular Features Determine Peptide/HLA-A2 Conformations That Are Recognized by HLA-A2-Restricted T Cell Receptors. *J. Immunol.* **2002**, *169*, 3146–3154. [CrossRef]
24. Ge, C.; Weisse, S.; Xu, B.; Dobritzsch, D.; Viljanen, J.; Kihlberg, J.; Do, N.-N.; Schneider, N.; Lanig, H.; Holmdahl, R.; et al. Key interactions in the trimolecular complex consisting of the rheumatoid arthritis-associated DRB1*04:01 molecule, the major glycosylated collagen II peptide and the T-cell receptor. *Ann. Rheum. Dis.* **2022**, *81*, 480–489. [CrossRef] [PubMed]
25. RCSB PDB. Available online: https://www.rcsb.org/ (accessed on 21 July 2022).
26. Barbullushi, K.; Rampi, N.; Serpenti, F.; Sciumè, M.; Fabris, S.; De Roberto, P.; Fracchiolla, N.S. Vaccination Therapy for Acute Myeloid Leukemia: Where Do We Stand? *Cancers* **2022**, *14*, 2994. [CrossRef]
27. Combination of Nilotinib (AMN107) and RAD001 in Patients with Acute Myeloid Leukemia. ClinicalTrials.gov. Updated 8 August 2012. Available online: https://clinicaltrials.gov/ct2/show/NCT00762632?term=KIT (accessed on 21 July 2022).
28. Trial to Assess the Efficacy of Midostaurin (PKC412) in Patients with c-KIT or FLT3-ITD Mutated t(8;21) AML (MIDOKIT). ClinicalTrials.gov. Updated 6 August 2020. Available online: https://clinicaltrials.gov/ct2/show/NCT01830361?term=KIT (accessed on 21 July 2022).
29. Gross, S.; Lennerz, V.; Gallerani, E.; Mach, N.N.; Böhm, S.; Hess, D.; von Boehmer, L.; Knuth, A.; Ochsenbein, A.; Gnad-Vogt, U.S.; et al. Short Peptide Vaccine Induces CD4+ T Helper Cells in Patients with Different Solid Cancers. *Cancer Immunol. Res.* **2016**, *4*, 18–25. [CrossRef]
30. Gonzalez-Galarza, F.F.; McCabe, A.; Melo dos Santos, E.J.; Jones, J.; Takeshita, L.; Ortega-Rivera, N.D.; Del Cid-Pavon, G.M.; Ramsbottom, K.; Ghattaoraya, G.; Alfirevic, A.; et al. Allele frequency net database (AFND) 2020 update: Gold-standard data classification, open access genotype data and new query tools. *Nucleic Acids Res.* **2020**, *48*, D783–D788. [CrossRef]
31. Yi, M.; Li, A.; Zhou, L.; Chu, Q.; Song, Y.; Wu, K. The global burden and attributable risk factor analysis of acute myeloid leukemia in 195 countries and territories from 1990 to 2017: Estimates based on the global burden of disease study 2017. *J. Hematol. Oncol.* **2020**, *13*, 72. [CrossRef]
32. Savsani, K.; Jabbour, G.; Dakshanamurthy, S. A New Epitope Selection Method: Application to Design a Multi-Valent Epitope Vaccine Targeting HRAS Oncogene in Squamous Cell Carcinoma. *Vaccines* **2022**, *10*, 63. [CrossRef] [PubMed]
33. Suri, S.; Dakshanamurthy, S. IntegralVac: A Machine Learning-Based Comprehensive Multivalent Epitope Vaccine Design Method. *Vaccines* **2022**, *10*, 1678. [CrossRef]
34. Frumento, G.; Zuo, J.; Verma, K.; Croft, W.; Ramagiri, P.; Chen, F.E.; Moss, P. CD117 (c-kit) is expressed during CD8+ T cell priming and stratifies sensitivity to apoptosis according to strength of TCR Engagement. *Front. Immunol.* **2019**, *10*, 468. [CrossRef] [PubMed]

35. Fuster, O.; Barragán, E.; Bolufer, P.; Cervera, J.; Larráyoz, M.J.; Jiménez-Velasco, A.; Martínez-López, J.; Valencia, A.; Moscardó, F.; Sanz, M. Rapid detection of KIT mutations in core-binding factor acute myeloid leukemia using high-resolution melting analysis. *J. Mol. Diagn.* **2009**, *11*, 458–463. [CrossRef] [PubMed]
36. KIT Gene–Somatic Mutations in Cancer. COSMIC. Available online: https://cancer.sanger.ac.uk/cosmic/gene/analysis?all_data=&coords=AA%3AAA&dr=&end=977&gd=&hn=haematopoietic_neoplasm&id=258193&ln=KIT&seqlen=977&sh=acute_myeloid_leukaemia&sn=haematopoietic_and_lymphoid_tissue&ss=NS&start=1#ts (accessed on 8 August 2022).
37. Catalogue of Somatic Mutations in Cancer. COSMIC. Updated 31 May 2022. Available online: https://cancer.sanger.ac.uk/cosmic (accessed on 24 July 2022).
38. P10721: KIT_HUMAN. UniProt. Available online: https://www.uniprot.org/uniprotkb/P10721/entry#family_and_domains (accessed on 31 July 2022).
39. Uniprot. UniProt. Available online: https://www.uniprot.org/ (accessed on 21 July 2022).
40. MHC-I Binding Predictions. IEDB Analysis Resource. Available online: http://tools.iedb.org/mhci/ (accessed on 7 June 2022).
41. MHC-I Binding Predictions–Tutorial. IEDB Analysis Resource. Available online: http://tools.iedb.org/mhci/help/ (accessed on 21 July 2022).
42. Class I Immunogenicity. IEDB Analysis Resource. Available online: http://tools.iedb.org/immunogenicity/ (accessed on 13 June 2022).
43. Calis, J.J.A.; Maybeno, M.; Greenbaum, J.A.; Weiskopf, D.; De Silva, A.D.; Sette, A.; Keşmir, C.; Peters, B. Properties of MHC Class I Presented Peptides That Enhance Immunogenicity. *PLoS Comput. Biol.* **2013**, *9*, e1003266. [CrossRef]
44. VaxiJen: Prediction of Protective Antigens and Subunit Vaccines. VaxiJen v2.0. Available online: http://www.ddg-pharmfac.net/vaxijen/VaxiJen/VaxiJen.html (accessed on 13 June 2022).
45. Doytchinova, I.A.; Flower, D.R. VaxiJen: A server for prediction of protective antigens, tumour antigens and subunit vaccines. *BMC Bioinform.* **2007**, *8*, 4. [CrossRef]
46. Chen, Z.; Ruan, P.; Wang, L.; Nie, X.; Ma, X.; Tan, Y. T and B Cell Epitope analysis of SARS-COV-2 S protein based on immunoinformatics and experimental research. *J. Cell. Mol. Med.* **2021**, *25*, 1274–1289. [CrossRef]
47. Varshavsky, A. The N-end rule pathway of protein degradation. *Genes Cells* **2003**, *2*, 13–28. [CrossRef]
48. Gonda, D.K.; Bachmair, A.; Wünning, I.; Tobias, J.W.; Lane, W.S.; Varshavsky, A. Universality and structure of the N-end rule. *J. Biol. Chem.* **1989**, *264*, 16700–16712. [CrossRef]
49. Wilkins, M.R.; Gasteiger, E.; Bairoch, A.; Sanchez, J.C.; Williams, K.L.; Appel, R.D.; Hochstrasser, D.F. Protein identification and analysis tools in the ExPASy server. *Methods Mol. Biol.* **1999**, *112*, 531–552. [CrossRef] [PubMed]
50. ProtParam Tool. Expasy. Available online: https://web.expasy.org/protparam/ (accessed on 14 June 2022).
51. Smoluch, M.; Mielczarek, P.; Drabik, A.; Silberring, J. 5–Online and Offline Sample Fractionation. In *Proteomic Profiling and Analytical Chemistry*, 2nd ed.; Ciborowski, P., Silberring, J., Eds.; Elsevier: Amsterdam, The Netherlands, 2016; pp. 63–99. [CrossRef]
52. Virtual Scanning of Toxic Peptides. ToxinPred: Designing and Prediction of Toxic Peptides. Available online: https://webs.iiitd.edu.in/raghava/toxinpred/ (accessed on 14 June 2022).
53. Gupta, S.; Kapoor, P.; Chaudhary, K.; Gautam, A.; Kumar, R.; Raghava, G.P.S. In Silico Approach for Predicting Toxicity of Peptides and Proteins. *PLoS ONE* **2013**, *8*, e73957. [CrossRef] [PubMed]
54. Epitope Prediction. IFNepitope: A Server for Predicting and Designing Interferon-Gamma Inducing Epitopes. Available online: http://crdd.osdd.net/raghava/ifnepitope/ (accessed on 14 June 2022).
55. Dhanda, S.K.; Vir, P.; Raghava, G.P. Designing of interferon-gamma inducing MHC class-II binders. *Biol. Direct.* **2013**, *8*, 30. [CrossRef] [PubMed]
56. Bioinformatics Tool for Allergenicity Prediction. AllerTop v. 2.0. Available online: https://www.ddg-pharmfac.net/AllerTOP/index.html (accessed on 14 June 2022).
57. Dmitrov, I.; Bangov, I.; Flower, D.R.; Doytchinova, I. AllerTOP v.2.0—A server for in silico prediction of allergens. *J. Mol. Model.* **2014**, *20*, 2278. [CrossRef]
58. Population Coverage. IEDB Analysis Resource. Available online: http://tools.iedb.org/population/ (accessed on 21 June 2022).
59. MHC-II Binding Predictions. IEDB Analysis Resource. Available online: http://tools.iedb.org/mhcii/ (accessed on 4 July 2022).
60. CD4 T Cell Immunogenicity Prediction. IEDB Analysis Resource–Labs. Available online: http://tools.iedb.org/CD4episcore/ (accessed on 4 July 2022).
61. Dhanda, S.K.; Karosiene, E.; Edwards, L.; Grifoni, A.; Paul, S.; Andreatta, M.; Weiskopf, D.; Sidney, J.; Nielsen, M.; Peters, B.; et al. Predicting HLA CD4 Immunogenicity in Human Populations. *Front. Immunol.* **2018**, *9*, 1369. [CrossRef]
62. NetMHCpan–4.0: Pan-Specific Binding of Peptides to MHC Class I Alleles of Known Sequence. DTU Health Tech. Available online: https://services.healthtech.dtu.dk/service.php?NetMHCpan-4.0 (accessed on 28 June 2022).
63. NetMHCpan–5.0: Pan-Specific Binding of Peptides to MHC Class II Alleles of Known Sequence. DTU Health Tech. Available online: https://services.healthtech.dtu.dk/service.php?NetMHCIIpan-4.0 (accessed on 19 July 2022).
64. MDockPeP Server. (n.d.). Updated 21 March 2019. Available online: https://zougrouptoolkit.missouri.edu/mdockpep/ (accessed on 21 July 2022).
65. Xu, X.; Yan, C.; Zou, X. MDockPeP: An ab-initio protein-peptide docking server. *J. Comput. Chem.* **2018**, *39*, 2409–2413. [CrossRef]

66. Yan, C.; Xu, X.; Zou, X. Fully Blind Docking at the Atomic Level for Protein-Peptide Complex Structure Prediction. *Structure* **2016**, *24*, 1842–1853. [CrossRef]
67. CABS-Dock: Server for Flexible Protein-Peptide Docking. Available online: http://biocomp.chem.uw.edu.pl/CABSdock (accessed on 8 August 2022).
68. TCR Model: Automated High Resolution Modeling of T Cell Receptors. Available online: https://tcrmodel.ibbr.umd.edu/rtcrex/TCRSDM6_180718_160348 (accessed on 31 July 2022).
69. Almosailleakh, M.; Schwaller, J. Murine Models of Acute Myeloid Leukaemia. *Int J Mol Sci.* **2019**, *20*, 453. [CrossRef]

Disclaimer/Publisher's Note: The statements, opinions and data contained in all publications are solely those of the individual author(s) and contributor(s) and not of MDPI and/or the editor(s). MDPI and/or the editor(s) disclaim responsibility for any injury to people or property resulting from any ideas, methods, instructions or products referred to in the content.

Article

Anti-Inflammatory Effects of Ang-(1-7) Bone-Targeting Conjugate in an Adjuvant-Induced Arthritis Rat Model

Sana Khajeh pour [1], Arina Ranjit [1], Emma L. Summerill [2] and Ali Aghazadeh-Habashi [1,*]

[1] College of Pharmacy, Idaho State University, Pocatello, ID 83209, USA
[2] College of Health, Idaho State University, Pocatello, ID 83209, USA
* Correspondence: habaali@isu.edu; Tel.: +1-(208)-282-1409

Abstract: Rheumatoid arthritis (RA) is a chronic inflammatory condition of synovial joints that causes disability and systemic complications. Ang-(1-7), one of the main peptides in the renin-angiotensin (Ang) system (RAS), imposes its protective effects through Mas receptor (MasR) signaling. It has a short half-life, limiting its feasibility as a therapeutic agent. In this study, we evaluated the anti-inflammatory effects of Ang-(1-7)'s novel and stable conjugate (Ang. Conj.) by utilizing its affinity for bone through bisphosphonate (BP) moiety in an adjuvant-induced arthritis (AIA) rat model. The rats received subcutaneous injections of vehicle, plain Ang-(1-7), or an equivalent dose of Ang. Conj. The rats' body weights, paws, and joints' diameters were measured thrice weekly. After 14 days, the rats were euthanized, and the blood and tissue samples were harvested for further analysis of nitric oxide (NO) and RAS components' gene and protein expression. The administration of Ang. Conj. reduced body weight loss, joint edema, and serum NO. Moreover, the Ang. Conj. treatment significantly reduced the classical arm components at peptide, enzyme, and receptor levels while augmenting them for the protective arm. The results of this study introduce a novel class of bone-targeting natural peptides for RA caused by an inflammation-induced imbalance in the activated RAS. Our results indicate that extending the half-life of Ang-(1-7) augments the RAS protective arm and exerts enhanced therapeutic effects in the AIA model in rats.

Keywords: rheumatoid arthritis; inflammation; renin-angiotensin system; Angiotensin-(1-7)

1. Introduction

Rheumatoid arthritis (RA) is a chronic and inflammatory autoimmune disorder affecting about 1% of the world population [1]. It is a chronically inflamed condition of synovial joints, eventually progressing to disability and systemic complexities leading to early fatality [2]. RA affects large and small joints through symmetrical polyarthritis. This disorder typically presents between the ages of 30 to 50 and is the most common inflammatory arthritis. The etiology of RA has not been fully defined, but genetic and environmental factors are considered the main initiators of the disease. Complications associated with RA include increased risk of developing cardiovascular disease, cerebrovascular and coronary artery atherosclerosis [3], and abnormal skeletal health [4]. Joint destruction and systemic complications can be mediated by complex interactions among multiple immune cell types, growth factors, and cytokines [5]. Multiple systems orchestrate the complex and dynamic situation of inflammation in the body, including the renin-angiotensin system (RAS) [6]. One of the factors responsible for RA's pathology and tissue damage is Ang II. Ang II can activate proinflammatory cytokines and initiate reactive oxygen species (ROS) production in vascular smooth muscle cells, neutrophils, and osteoclasts [7]. RA is associated with excess morbidity and mortality from cardiovascular complications [8,9]. Evidence suggests that common proinflammatory cytokines are involved in the development and progression of both atherosclerosis and cardiovascular complications [3,10]. Current therapeutic options, such as analgesics, disease-modifying antirheumatic drugs, non-steroidal anti-inflammatory drugs, and biologics, are associated with severe and broad-spectrum side effects. There is a dire need for safer and more effective therapeutic options for RA management.

The RAS consists of proinflammatory and anti-inflammatory axes balanced in the normal physiological condition (Figure 1). As one of the RAS main vasoactive peptides, Ang II is produced by Ang-Converting Enzyme (ACE) from Ang I peptide. It mediates proinflammatory actions by binding to the Ang II type 1 receptor (AT1R) [11]. Whereas the protective RAS axis component, Ang-(1-7), is produced by the action of ACE2. It binds to the G-protein coupled MasR and exerts anti-inflammatory and anti-pressor effects [12]. The ACE/Ang II/AT1R arm overactivation in some pathological conditions switches the balance off towards the inflammatory axis. It is justifiable to endogenously increase the Ang-(1-7) level and augment the protective arm to suppress the overactivation of the proinflammatory axis and restore the disturbed balance.

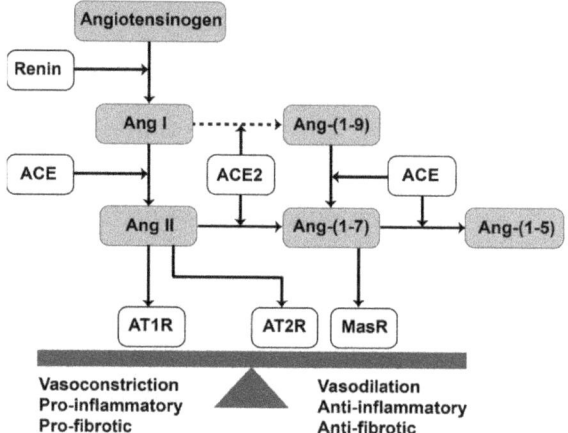

Figure 1. Schematic overview of the RAS. Ang—angiotensin, ACE—angiotensin-converting enzyme, ACE2—angiotensin-converting enzyme 2, MasR—Mas receptor, AT1R—angiotensin II type 1 receptor, AT2R—angiotensin II type 2 receptor.

Ang-(1-7) as a vasoactive peptide induces vasodilatory, anti-inflammatory, antifibrotic, antiangiogenic, and antihypertensive effects of MasR signaling. These beneficial actions of Ang-(1-7) make this peptide an attractive target for cardioprotective therapies. These protective effects have been demonstrated in numerous animal models of human diseases, including hypertension, diabetes mellitus, and atherosclerosis [13]. The administration of Ang-(1-7) is therapeutically effective in the experimental animal model of RA due to the activity mediated by the reduction of neutrophil accumulation, inhibition of cytokine release, and improvement of joint hyper nociception [12,14]. Additionally, Ang-(1-7) exhibits anti-inflammatory behavior by reducing cytokine release, tissue damage, leukocyte attraction, and fibrosis [7].

Activating the RAS induces inflammation by signaling AT1R by Ang II on leukocytes. After RA's initiation and development, inflammatory cells infiltrate the articular synovial tissues and consequently secrete inflammatory cytokines, such as TNF-α, IL-1, and IL-6 [15]. In rodent models of AIA through augmentation of the RAS protective arm by Ang-(1-7) or the Mas agonist AVE 0991, the AIA-induced neutrophil accumulation and production of proinflammatory mediators such as TNF-a, IL-1β, and CXCL1 in preartricular tissue were diminished, and paw histopathological markers were significantly reduced [12]. Ang-(1-7) also improved paw histological changes and normalized cytokine biomarkers in the Collagen-induced arthritis rat model [16]. Furthermore, increased levels of NO in plasma [17,18], articular [19] fluid, and a high expression of iNOS in hyperplastic synovium [17,20] and chondrocytes [21] have been reported in different models of AIA. This elevation of NO and ROS are related to the presence of macrophages in the inflamed tissues due to higher expression of iNOS induced by inflammatory cytokines. In a study using cultured human aortic smooth muscle cells (HASMC), the expression of inducible nitric oxide synthase (iNOS) and the release of nitric oxide (NO) were stimulated by both Ang II and IL-1β. Treatment with Ang-(1-7) inhibits the NO production, which Mas receptor antagonists block such effect [22]. These findings indicate that the Mas receptor activation by Ang-(1-7) prevents neutrophil influx, cytokine production, and No release and, as a result, significantly improves arthritis in rats and mice AIA models. However, Ang-(1-7) has an extremely short half-life (3–15 min) due to the rapid systemic clearance, which restricts its potential therapeutic benefits [23]. Therefore, introducing an appropriate drug delivery system that extends the systemic half-life of Ang-(1-7) could offer a most-needed safe [24] and effective therapeutic option for RA. As mentioned above, Ang-(1-7) presented protective effects by reducing proinflammatory cytokine levels in the plasma, HASMC, and joint tissues. In the current study, we further investigated the impact of inflammation on the RAS at the cellular level and focused on exploring the cardiovascular complication of RA. We additionally studied the protective effects of Ang. Conj. treatment, as a stable form of Ang-(1-7), on the RAS in the enzyme, peptide, and receptor levels of the heart, kidney, liver, and lung tissues.

We designed a bone-targeting peptide delivery system that actively targets the bone, makes a drug depot, slowly releases the active peptide, and prolongs its circulation half-life [25]. This approach seems promising in the delivery strategy for treating bone disorders like RA, osteoporosis, and cancers with bone complications [26]. In this targeted drug delivery, a drug is conjugated with a bone targeting moiety like BP with linkers like PEG. The BP has an intrinsic affinity to the hydroxyapatite of the bone, which serves as a drug reservoir and increases its metabolic stability [27]. Similar peptides such as parathyroid hormone [28], salmon calcitonin [29], and osteoprotegerin [30] showed higher therapeutic efficacy than their parent analog when delivered through a bone-targeting approach due to the half-life extension of conjugated peptide drugs. It is worth mentioning that contrary to the nitrogen-containing bisphosphonates, the BP moiety of the Ang. Conj. does not possess any significant pharmacological effect [28].

Using a gamma counter radioassay, we have shown that Ang. Conj. has a longer half-life [25]. The radioassay methods have some limitations in their application for the quantitative measurement of radiolabeled material. We are currently developing an LC-MS/MS method to quantify Ang. Conj. in plasma to confirm our previous PK results. Assuming a direct relationship between the plasma concentration of Ang-(1-7) and its anti-inflammatory effect, we investigated Ang. Conj. therapeutic effects using a rat model of RA compared to plain Ang-(1-7) and elucidated the impact of bone-targeting delivery and stability improvement strategy on enhancing their anti-inflammatory effects. To make the comparison more effective and the effect size more significant, we increased the dosing interval from the previously used daily regimen [16] to 3×/week.

In this study, we focused on investigating the cardiovascular complication of RA. We aimed to explore the impact of the inflammation on the RAS in the heart, kidney, liver, and lung tissues and investigate the protective effects of Ang. Conj. treatment. We did not include the paw tissue or inflammatory cytokines measurement, as it was previously shown by Liu et al. that Ang-(1-7) improves paw histological and normalized cytokine biomarkers changed in the Collagen-induced arthritis rat model [16]. Increased plasma and synovial tissue NO levels due to high expression of iNOS have been reported in patients with RA [31,32]. In the rat model of AIA, systemic administration of selective and nonselective iNOS halted the development of the disease [17,33]. Similarly, we evaluated Ang. Conj. systemic anti-inflammatory effects through assessment of serum concentration of NO.

2. Results

2.1. Ang. Conj. Treatment Improved Body Weight Gain and Reduced Paw and Joint Swelling

Adjuvant-induced Arthritis (AIA) emerged 8–10 days post-adjuvant injection. It manifested itself by redness of the paw and erythema of ankle joints, followed by the involvement of the metatarsal and interphalangeal joints. The symptoms spread progressively with time into other parts of the hind and forepaws. The weight, paw, and joint measurements were done thrice per week. The paw and joint diameters are reported as percentage change on day 24 compared to day 0.

The percentage change of animals' body weight gain was significantly reduced in AIA animals and treatment with Ang-(1-7) and Ang. Conj. restored the body weight gain over time (Figure 2A). The absolute mean ± SEM values of body weight (g) on day 24 were 409.6 ± 8.4 for control, 290.0 ± 2.8 for inflamed, 362.0 ± 19.4 for Ang-(1-7), and 373 ± 16.6 for Ang. Conj. groups. The weight gain percentage value in the inflamed (12.6 ± 5.5%) and Ang-(1-7)-treated group (31.6 ± 5.3%) was significantly lower than the control group (50.1 ± 2.5%). However, the value in the Ang. Conj-treated group (34.2 ± 4.5%) was comparable to that of the control group (Figure 2A). As a measure of signs and symptoms of arthritis, the arthritis index (AI) was significantly higher in the inflamed group indicating efficacious arthritis induction. The therapeutic efficacy of Ang-(1-7) and Ang. Conj. was noticeable after administration of three consecutive doses (~6 days), but it was statistically significant 10 days after treatment started (Figure 2A,B). The right and left joints' diameter percent changes at the end of the experiment compared to day 0 in the non-treated inflamed rats (29.0 ± 9.0% and 27.7 ± 7.4%) were significantly higher than in the control group (2.7 ± 1.2 and 3.8 ± 0.8%). The drug treatment impacted the right and left joints swelling as it was substantially lower in Ang. Conj. (3.6 ± 1.6% and 2.5 ± 1.4%) and Ang-(1-7) (9.6 ± 2.5% and 12.4 ± 7.4%) groups (Table 1 and Figure 2C,E). The absolute mean ± SEM values for the right joint diameter (mm) on day 24 were 6.7 ± 0.06, 8.6 ± 0.06, 7.2 ± 0.06, and 6.7 ± 0.08, and for the left joint diameter (mm) were 6.8 ± 0.15, 8.6 ± 0.03, 7.1 ± 0.08, and 6.8 ± 0.08 for the control, inflamed, Ang-(1-7)-treated, and Ang. Conj.-treated groups, respectively. The same trend was seen for the left and right paw diameter percent changes but did not reach a significant difference due to high variability (Table 1, Figure 2D,F). The absolute mean ± SEM values on day 24 for the right paw diameter (mm) were 4.2 ± 0.05, 5.3 ± 0.08, 4.6 ± 0.03, and 4.2 ± 0.06, and for the left paw diameter (mm) were 4.3 ± 0.04, 5.1 ± 0.04, 4.4 ± 0.02, and 4.3 ± 0.02 for the control, inflamed, Ang-(1-7)-treated, and Ang. Conj.-treated groups, respectively.

Table 1. The percentage change in paw and joint diameters in different treatment groups at the end of the experiment compared to day 0.

Animal Group	Paw Diameter Percentage Change Mean (SEM)		Joint Diameter Percentage Change Mean (SEM)	
	Left Hind	Right Hind	Left Hind	Right Hind
Control ($n = 6$)	2.7 (2.0) [a]	0.0 (1.3) [a]	2.7 (1.2) [a]	3.8 (0.8) [a]
Inflamed ($n = 6$)	22.7 (10.6) [a]	23.7 (10.5) [a]	29.0 (9.0) [b]	27.7 (7.4) [b]
Ang-(1-7) ($n = 5$)	4.8 (1.3) [a]	10.0 (10.7) [a]	9.6 (2.5) [ab]	12.4 (4.6) [a]
Ang. Conj. ($n = 5$)	3.1 (1.5) [a]	1.8 (0.6) [a]	3.6 (1.6) [a]	2.5 (1.4) [a]

Values are reported as mean ± SEM, and statistical analysis was done using one-way ANOVA with the Tukey multiple comparison post-hoc test. In each column, groups labeled with different letters (a or b) have significant differences between them, $p < 0.05$.

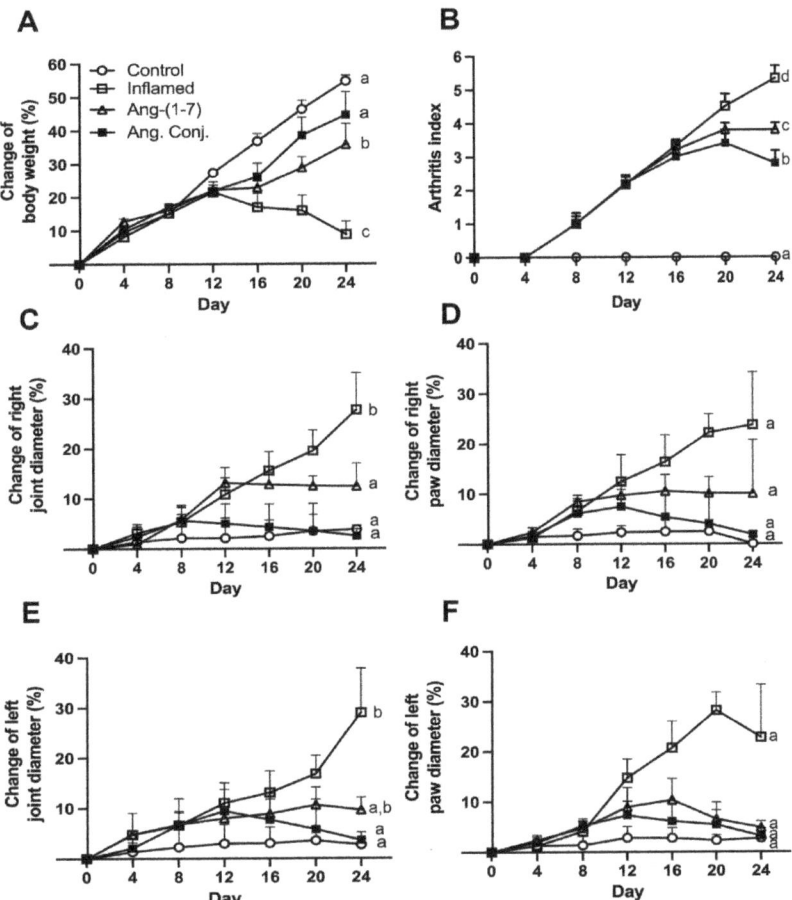

Figure 2. Effect of adjuvant-induced arthritis and treatment with Ang-(1-7) and Ang. Conj. on the percentage change of body weight (**A**) and the arthritis index values during the study period (**B**). The percentage change in the right joint (**C**), right paw (**D**), left joint (**E**), and left paw (**F**) diameters compared to day 0. The number of animals in control and inflamed groups were six and in Ang-(1-7) and Ang. Conj.-treated groups were five. The values are reported as mean ± SEM, and statistical analysis was done using one-way ANOVA with the Tukey multiple comparison post-hoc test. Groups labeled with different letters have significant differences between them ($p < 0.05$).

2.2. The Arthritis-Induced High Serum Nitrate and Nitrite Levels Were Reduced after Treatment with Ang-(1-7) and Ang. Conj.

The serum nitric oxide (NO) level was significantly elevated in inflamed animals (5.15 ± 1.02 ng/mL) compared to healthy control rats (0.22 ± 0.13 ng/mL). Ang. Conj. treatment significantly reduced the serum NO concentration to 0.95 ± 0.52 ng/mL, which was comparable to the control group. Although Ang-(1-7) treatment reduced the serum NO concentration to 2.35 ± 0.73 ng/mL, this reduction did not happen to the same extent as observed after treatment with Ang. Conj. The serum NO level in Ang-(1-7)-treated rats was not significantly different from the inflamed or control group (Figure 3).

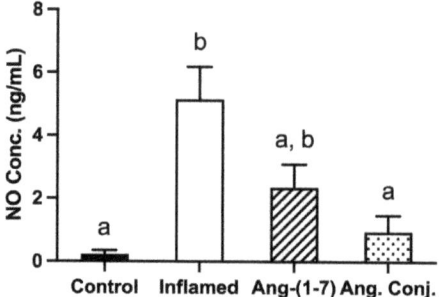

Figure 3. Effect of adjuvant-induced arthritis and Ang-(1-7) or Ang. Conj. treatment on serum NO concentrations at the end of the experiment. The number of animals in control and inflamed groups were six, and in Ang-(1-7) and Ang. Conj.-treated groups were five. The values are reported as mean ± SEM, and statistical analysis was done using one-way ANOVA with the Tukey multiple comparison post-hoc test. Groups labeled with different letters have significant differences between them ($p < 0.05$).

2.3. Treatment with Ang-(1-7) or Ang. Conj. Increased Ang-(1-7) and Reduced Ang II Peptides Levels in Plasma

The plasma level of Ang-(1-7) and Ang II peptides and their ratio are presented in Figure 4. These data confirmed the treatment of AIA rats with Ang. Conj. significantly increases the Ang-(1-7) plasma levels (1.45 ± 0.22 ng/mL) compared with the control (0.75 ± 0.08 ng/mL), inflamed (0.19 ± 0.05 ng/mL), and Ang-(1-7)-treatment (0.90 ± 0.17 ng/mL) groups (Figure 4A). The Ang II plasma concentrations, on the other hand, present a reverse trend. The AIA significantly elevates Ang II plasma levels in the inflamed group (1.70 ± 0.37 ng/mL), which reduced to a comparable plasma concentration of the control group (0.27 ± 0.02 ng/mL) after treatment with Ang-(1-7) (0.30 ± 0.036 ng/mL) or Ang. Conj. (0.21 ± 0.02 ng/mL) groups (Figure 4B).

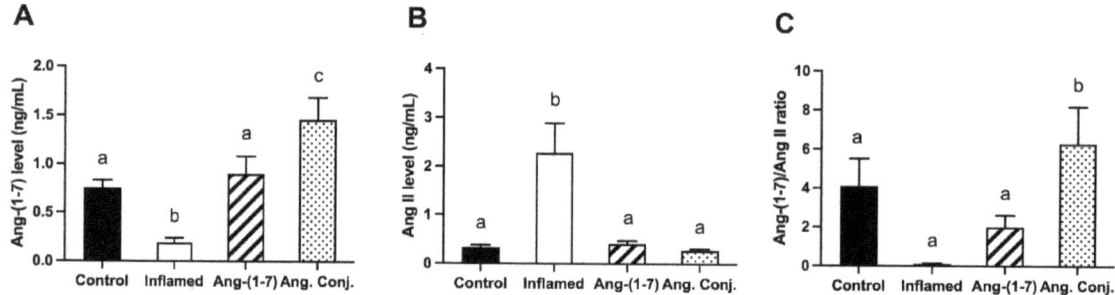

Figure 4. The plasma concentration of Ang-(1-7) (**A**), Ang II (**B**), and their ratio (**C**) in control ($n = 6$), inflamed ($n = 6$), Ang-(1-7)-treated ($n = 5$), and Ang. Conj.-treated ($n = 5$) groups. The values are reported as mean ± SEM, and statistical analysis was done using one-way ANOVA with the Tukey multiple comparison post-hoc test. Groups labeled with different letters have significant differences between them ($p < 0.05$).

The Ang-(1-7)/Ang II ratio was significantly higher in the Ang. Conj-treated group (6.30 ± 1.89) compared with the healthy-control (4.10 ± 1.42), inflamed (0.1 ± 0.03), and Ang-(1-7) (2.00 ± 0.60) groups (Figure 4C).

2.4. Treatment with Ang-(1-7) or Ang. Conj. Reversed the AIA-Induced Changes in ACE1, ACE2, MasR, and AT1R Gene Expression in Different Tissues

The mRNA expression of ACE1, ACE2, AT1R, and MasR in the heart, lung, liver, and kidney, are shown in Figure 5. ACE1 and AT1R gene expression levels significantly increased in all tested tissues of the inflamed rats, and ACE2 gene expression was reduced in the heart, lung, liver, and kidney. MasR gene expression increased in the heart and decreased in the lung and kidney, with no change observed in the liver. Treatment with Ang-(1-7) and Ang. Conj. (in a higher magnitude) significantly reversed all changes. However, in the heart tissue, the increased MasR expression due to AIA was further increased by Ang. Conj. ACE2/ACE1 and MasR/AT1R gene expression ratios in all tested tissues were reduced due to AIA, and treatment with Ang-(1-7) normalized them while Ang. Conj. increased them several folds.

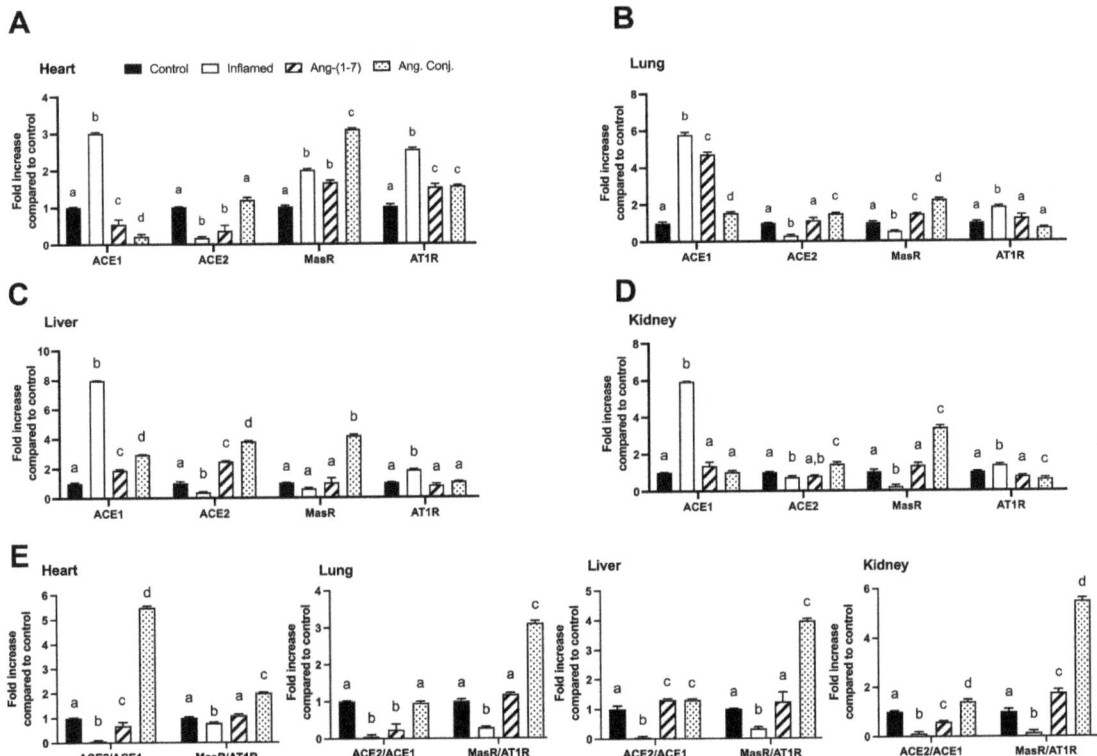

Figure 5. Gene expression of ACE1, ACE2, MasR, and AT1R levels in the heart (**A**), lung (**B**), liver (**C**), kidney (**D**) tissues, and the ratio of ACE2/ACE1 and MasR/AT1R in those tissues (**E**) in control ($n = 6$), inflamed ($n = 6$), Ang-(1-7)-treated ($n = 5$), and Ang. Conj.-treated ($n = 5$) groups. ACE1—angiotensin-converting enzyme 1, ACE2—angiotensin-converting enzyme 2, MasR—Mas receptor, AT1R—angiotensin II type 1 receptor. The values are reported as mean ± SEM, and statistical analysis was done using one-way ANOVA with the Tukey multiple comparison post-hoc test. Groups labeled with different letters have significant differences between them ($p < 0.05$).

2.5. Treatment with Ang-(1-7) or Ang. Conj. Reversed the AIA-Induced Changes in ACE1, ACE2, MasR, and AT1R Protein Expression in Different Tissues

Data shown in Figure 6 represent the significant changes in relative protein density of ACE1, ACE2, MasR, and AT1R in the heart, lung, liver, and kidney tissues due to AIA. These changes were normalized by Ang-(1-7) or Ang. Conj. treatment and the effects were more pronounced in the case of the latter. The individual proteins WB results indicate

(i) AIA caused a significant increase in the ACE1 protein expression in the heart, liver, and kidney tissues of the inflamed rats and Ang-(1-7) or Ang. Conj. treatment reversed it, which was more efficient in the latter case, and brought it back to the control group level. (ii) AIA significantly reduced the ACE2 expression in the heart and lung tissues, and treatment with Ang-(1-7) or Ang. Conj. normalized it. In the case of the liver and kidney, AIA resulted in a similar change trend but was not significant. (iii) MasR's expression was significantly reduced in inflamed animals' lung, liver, and kidney tissues, which were again normalized by Ang-(1-7) or Ang. Conj. treatments. (iv) The AT1R expression was significantly increased in all tissues other than the kidney in the inflamed group and Ang. Conj. treatment reversed the expression more efficiently than Ang-(1-7). (v) ACE2/ACE1 and MasR/AT1R protein expression ratios in all tested tissues were reduced by AIA, and treatment with Ang-(1-7) or Ang. Conj. normalized them.

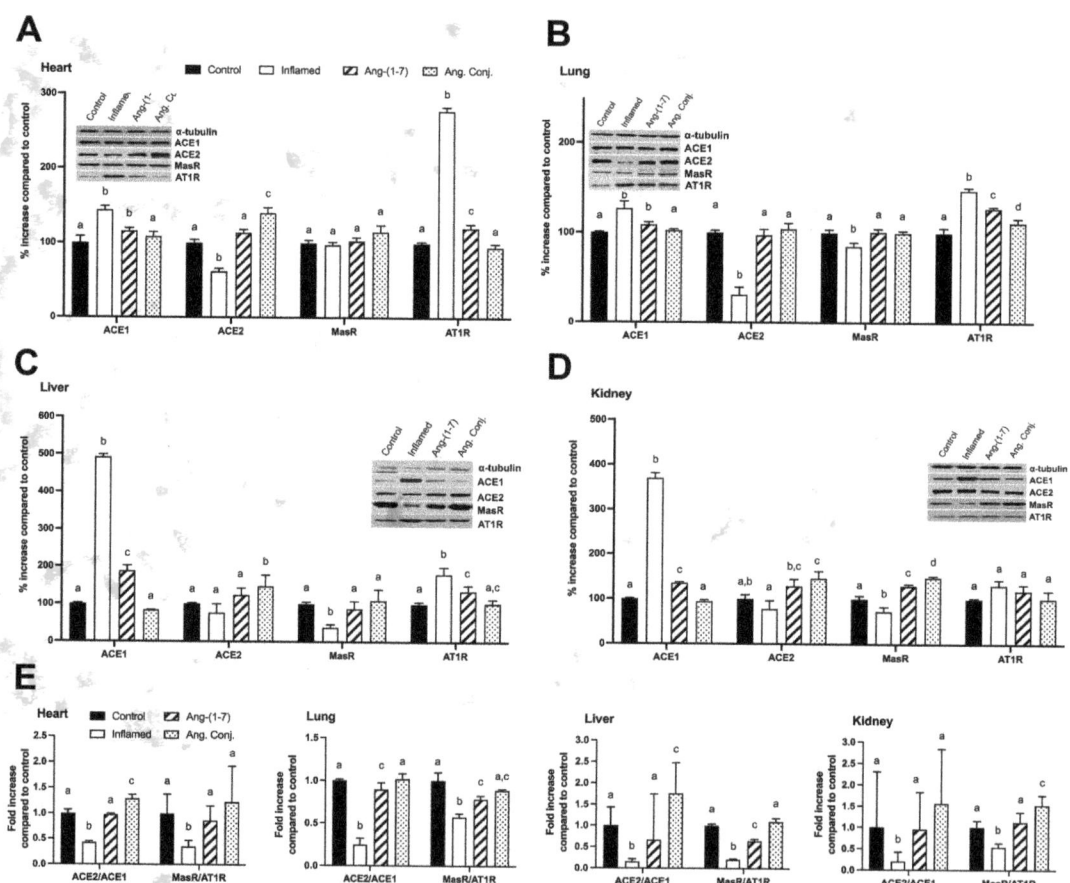

Figure 6. Protein expression of ACE1, ACE2, MasR, and AT1R, levels in the heart (**A**), lung (**B**), liver (**C**), kidney (**D**) tissues, and the ratio of ACE2/ACE1 and MasR/AT1R (**E**) in control ($n = 6$), inflamed ($n = 6$), Ang-(1-7)-treated ($n = 5$), and Ang. Conj.-treated ($n = 5$) groups. ACE1—angiotensin-converting enzyme 1, ACE2—angiotensin-converting enzyme 2, MasR—Mas receptor, AT1R—angiotensin II type 1 receptor. The values are reported as mean ± SEM, and statistical analysis was done using one-way ANOVA with the Tukey multiple comparison post-hoc test. Groups labeled with different letters have significant differences between them ($p < 0.05$).

3. Discussion

The administration of Ang-(1-7) or Ang. Conj. impacted the activated RAS, which was more pronounced in the Ang. Conj. case. The anti-inflammatory effects of Ang-(1-7) have been reported previously [12]. However, its short half-life hampers such an application's feasibility. The findings of this study, for the first time, indicate that it is feasible. We designed a bone-targeting drug conjugate of Ang-(1-7); Ang. Conj. and improved its pharmacokinetics, leading to an enhancement of pharmacodynamic properties [25].

The results of the present study demonstrate an augmentation of the RAS protective arm by exogenous administration of Ang-(1-7) as a stable conjugate, Ang. Conj., in an AIA model of RA. Such a boost exerts significant anti-inflammatory effects by restoring weight gain loss, reducing the paw and joint swelling, and diminishing the increased NO level and AI due to inflammation (Figures 2 and 3). The anti-inflammatory effect of Ang. Conj. reducing and normalizing the NO serum concentration followed the same trend of change in Ang-(1-7) and Ang II plasma levels. Although the NO serum levels in Ang-(1-7) or Ang Conj.-treated groups did not reduce to the exact same level as the control group; there was no significant difference between their NO values. Similarly, the therapeutic efficacy of Ang-(1-7) and Ang. Conj. on reducing AI was not significantly noticeable until ten days after treatment started (Figure 2A,B). This observation could be attributed to the time required for Ang-(1-7) plasma concentration to release from the conjugate and to reach an adequate steady-state concentration. The two weeks treatment period with a low dose of Ang-(1-7) or Ang. Conj. in this pilot study presents a promising positive outcome, which can be potentiated by increasing the dose and duration. These findings are mostly in line with reestablishing the disturbed balance between the classical and protective arms, which is presented as ratios of Ang-(1-7)/Ang II peptides (Figure 4), or ACE2/ACE1 and MasR/AT1R at the gene or protein levels (Figures 5 and 6).

In agreement with previous reports [34], the results of this study (Figures 4–6) indicate that induced inflammation in the AIA rats alters the balance of the RAS components in plasma peptides and enzyme and receptor levels in all tested tissues. The ACE and ACE2 enzyme expression alteration affect the plasma Ang-(1-7), Ang II, and their ratio (Figure 4). This observation indicates that the lower expression of ACE2 resulted in a lower plasma concentration of the vasodilator peptide, Ang-(1-7). This effect was reported in the heart and kidney tissues of the AIA rats and most likely applies to other tissues as well [34]. Considering the association of RAS with cardiovascular diseases, the observed RAS imbalance could be attributed to the well-known effect of inflammation in increasing cardiovascular risks [35]. It is notable that Ang II promotes proinflammatory outcomes and is elevated in many cardiovascular conditions, such as hypertension, atherosclerosis, and coronary heart disease, by stimulating the production of different inflammatory mediators and their migration into sites of tissue injury [36,37]. Consistent with the proinflammatory actions of Ang II, treatment with ACE inhibitors (ACEIs) and AT1R blockers (ARBs) diminishes the production and release of inflammatory mediators in models of inflammation [38–43]. ACE2, by degrading the Ang II to Ang-(1-7) and MasR signaling, counters the +Ang II proinflammatory effects [44,45]. There were no significant changes in the gene expression of MasR in any tested tissues due to AIA, but it was significantly increased in the case of AT1R in the heart and liver tissues (Figure 5). These gene expression changes variably translated to changes in protein expression as the lung, liver, and kidney tissues were presented with lower MasR receptor density and all other than the kidney tissue had higher AT1R expression.

Nevertheless, the MasR/AT1R protein ratio was lower in all tissues in non-treated AIA animals. This observation exerts a further change in balance in the activated RAS components toward the classical arm deleterious effects. The observed significant elevation of AT1R and reduced MasR/AT1R (Figure 6) support the notion that arthritis can change the vasodilation–vasoconstriction balance as there will be less target protein, MasR, for coupling with already reduced vasodilator Ang-(1-7), and, on the other hand, more AT1R is available to be activated by a higher concentration of a vasoconstrictor, Ang II. We only measured the Ang peptides levels in plasma, which were significantly altered in AIA rats (Figure 4); however, considering the lowered ratio of the ACE/ACE expressions, it most likely is the case in all tissue (Figure 6).

The RAS exerts its physiological effects through AT1R, AT2R, and Mas receptors. Ang II has an affinity for AT1R and AT2R (with offsetting responses), but most Ang II effects are mediated through AT1R [46]. In the present study, we focused only on the AT1R and MasR; however, the AT2R receptor gene and protein expressions can also be impacted by inflammation and should be considered when the results are interpreted.

The depicted results in Figures 4–6 imply that treating AIA animals with plain Ang-(1-7) could compensate for the harmful impacts of activated RAS and its balance shift toward the classical arm. The restoration of the balance and exertion of anti-inflammatory effects was more significant and pronounced when Ang. Conj. was used. The improved efficacy can be attributed to the intermittently prolonged effect of Ang-(1-7) through MasR signaling. These observations are in concert with a previous study on the activation of MasR using its agonist, AVE 0991, in experimental models of arthritis [12]. The Ang. Conj. thrice-weekly application for two weeks significantly increased the reduced plasma Ang-(1-7), reduced the increased plasma level of Ang II, and restored the healthy control ratio of those peptides (Figure 4). The observed alteration of Ang peptides was in line with ACE and ACE2 gene and protein expression, which resulted in an overall significant improvement of the ACE2/ACE ratio in all tissues. The MasR/AT1R presents a similar trend in all tissues, although their expressions do not precisely follow suit. These findings indicate that the administration of Ang. Conj. is similar to treatment with ACEIs and ARBs and diminishes the anti-inflammatory effects of Ang II in different tissues to manage arthritis, renal, cardiovascular, pulmonary, and hepatic disease. For confirmation of these results, a head-to-head comparison study is warranted.

4. Materials and Methods

4.1. Animals

Adult male Sprague-Dawley rats weighing 200–250 g were obtained from Charles River (Wilmington, MA, USA) and were housed under ambient temperature and ventilation with 12-h day and night cycles. Rats were kept in standard cages with free access to drinking water and regular rat chow ad libitum. After 72 h of acclimatization, rats were randomly divided into healthy-control ($n = 6$), inflamed ($n = 6$), Ang-(1-7) ($n = 5$), and Ang. Conj. ($n = 5$) groups. The study protocol was approved by the Animal Care Facility Committee of Idaho State University (Protocol #772, 22 September 2021).

4.2. Arthritis Induction in Rats

AIA is associated with pain and discomfort for animals. To address that issue, we used a previously reported arthritis induction method that avoids unnecessary pain and discomfort while the inflammation effects are considerably noticeable [47,48]. This model is proven suitable for pharmacokinetics and pharmacodynamics studies. For induction of arthritis, animals were injected at the tail base on day 0 with a single dose of 200 µL of 50 mg/mL of Mycobacterium *butyricum* in squalene (Difco Laboratories, Detroit, MI, USA) (all groups except the healthy-control) or pyrogen-free sterile saline (healthy-control) to induce adjuvant arthritis. Subsequently, rats were monitored daily and assessed for the emergence of arthritis by assigning an arthritis index score for each rat. Arthritis index is a macroscopic scoring system [48]: for each hind paw on a 0–4 scale, 0 = no sign; 1 = single joint involved; 2 = more than one joint and ankle involved; 3 = several joints and ankle involved with moderate swelling; 4 = involvement of several joints and ankle with severe swelling. For each forepaw on a 0–3 scale, 0 = no sign; 1 = single joint involved; 2 = more than one joint and wrist affected; 3 = involvement of wrist and joints with moderate-to-severe swelling. The index was calculated by adding all the above scores to attain a maximum of 14. An arthritis index score of ≥5 was considered infliction of the disease, and its early signs and symptoms were evident typically in 8–10 days after adjuvant injection.

4.3. Body Weight, Paw, and Joint Diameter Measurement

Paw and joint diameters are indicatives of edema in rats and were measured three times per week, always at the same hour, using a micrometer caliper (Mitutoyo Canada Inc., Toronto, ON, Canada). The change in the rats' body weights was measured and recorded every other day using the animal balance.

4.4. Animal Dosage Regimens, Treatment, and Sampling

Ang-(1-7) (0.6 mg/kg) or Ang. Conj., containing an equivalent dose of Ang-(1-7), was dissolved in sterile normal saline and administered subcutaneously thrice per week for 14 days after the emergence of the signs and symptoms of inflammation at least in one hind on 8–10 days post adjuvant injection. The healthy-control and inflamed groups were injected with drug-free normal saline. At the end of the experiment, rats were anesthetized with isoflurane/oxygen, and blood samples were collected by cardiac puncture. Subsequently, the heart, kidney, lungs, and liver tissues were rapidly removed and washed with saline. Serum was separated from a portion of the collected blood that was not treated with an anti-coagulant for serum NO concentration assay. Fifty microliters of protease inhibitor cocktail was added per each 1 mL of blood. The cocktail was composed of 1 mM of p-hydroxymercury benzoate, 30 mM of 10-phenanthroline, 1 mM of phenylmethylsulfonyl fluoride, 1 mM of the pepstatin-A enzyme, and 7.5% of ethylenediaminetetraacetic acid. After centrifugation for 10 min at $2500 \times g$, the plasma was separated. All samples were stored at $-80\ ^\circ$C until further analysis.

4.5. Measurement of the Serum NO Concentration

According to the manufacturer's protocol, serum NO concentration was quantified using a Nitrate/Nitrite Colorimetric Assay Kit (#BCCB4059; Sigma, St. Luis, MO, USA). Briefly, blood samples were thawed, and each kit component was allowed to come to room temperature. Then, in a 96-well plate, 0, 20, 40, and 80 µL of 100 µM nitrite standard solution and 70 µL of each serum sample were added in triplicates. To measure the total nitrate and nitrite concentration, 10 µL of nitrate reductase and enzyme cofactor were added to the serum samples and shaken for 2 h. Then Griess reagents A and B were added to each well and incubated for 5 and 10 min, respectively. The absorbance intensity as an indicator of the sample NO concentration was measured at 540 nm in a microplate reader.

4.6. Ang Peptides Extraction and Quantification Using Liquid-Chromatography in Tandem with Mass Spectrometry (LC-MS/MS)

Ang peptides were extracted from plasma using solid-phase extraction (SPE) based on a previously published method with minor modifications [49]. Briefly, to 200 µL of plasma samples, 50 µL of the [Asn^1, Val^5] Ang II (IS) (100 ng/mL) was added and acidified with formic acid (final concentration of 0.5%). After a brief vortex mixing, the samples were loaded into preconditioned Waters C18 SPE cartridges (#WAT020805, Milford, MA, USA) with 2 mL of ethanol and water each by a wash step with 2 mL of deionized water. A positive nitrogen flow was applied to dry the cartridges. 2.5 mL of 5% formic acid in methanol was used to elute the Ang peptides. The eluted solution was collected and dried under the stream of nitrogen. The dried samples were reconstituted in 100 µL of 0.1% formic acid in acetonitrile: water (16:84), and 20 µL of it was injected into the LC-MS/MS system.

The plasma level of Ang-(1-7) and Ang II peptides, as one of the critical biomarkers of the protective and classical arms of the RAS, was measured using a validated LC-MS/MS method [49] in multiple reaction mode (MRM). The system was composed of liquid chromatography (Shimadzu, Columbia, MD, USA) with a controller (CBM-20A), two binary pumps (LC-30AD), an autosampler (SIL-30AC), and an AB Sciex (Foster City, CA, USA) QTRAP 5500 quadrupole mass spectrometer in positive electrospray ionization mode (ESI). The chromatograms were monitored and integrated by the Analyst 1.7 software (AB Sciex, Foster City, CA, USA).

The LC separation was performed on an analytical reversed-phase column Kinetex® 1.7 μm, C-18, 100 × 2.1 mm (Phenomenex, Torrance, CA, USA) by a combination of A: 0.1% formic acid in water and B: 0.1% formic acid in acetonitrile as mobile phases at a flow rate of 0.2 mL/min. The mobile phase gradient started at 5% B and increased to 30% B in 5 min, kept at 30% B for 5 min, returned to 5% B in 3 min, and held at 5% B for 2 min before the next injection for column re-equilibrium.

The positive electrospray ionization parameters were as follows: capillary voltage; 5.5 kV, temperature; 300 °C, declustering potential (DP); 100 V, and collision cell exit potential (CXP); 15 V. LC-MS/MS was performed with MRM transitions of m/z 300.6 → 371.2 (Ang-(1-7)), m/z 349.7 → 400.2 (Ang II), and m/z 516.5 → 769.4 (IS). Nitrogen was used as collision gas, and the collision energies were set at 20–30 eV. A calibration curve using peak height ratio (analyte over IS) was constructed over the concentration range of 500 pg/mL to 10 ng/mL in plasma and used to measure the Ang peptides' levels in plasma samples.

4.7. Quantitative Polymerase Chain Reaction (qPCR)

Total RNA was extracted from fifty mg tissue samples using the Quick-RNA™ Miniprep Plus kit (Zymo Research, Irvine, CA, USA) according to the manufacturer's protocol. Briefly, the samples were lysed into a yellow Spin-Away™ Filter in a collection tube and centrifuged to remove the genomic DNA. Then 95–100% ethanol was added to the flow-through (1:1) and mixed well. This mixture was transferred into a green Zymo-Spin™ IIICG Column in a collection tube and centrifuged. The flow-through was discarded. After DNase treatment, 400 μL of RNA Prep Buffer was added to the column, the solution was centrifuged, and the flow-through was discarded. This process was repeated with 700 μL of RNA Wash Buffer. To remove the wash buffer, 400 μL of RNA Wash Buffer was added, and the column was centrifuged for 1 min. The remaining solution was carefully transferred to a nuclease-free tube. A hundred μL of DNase/RNase-Free Water was added directly to the column matrix and centrifuged to elute the RNA.

Real-time qPCR mRNA analyses were performed using SYBR Green Supermix (Bio-Rad, Hercules, CA, USA). Relative expression of all genes was determined by the comparative threshold cycle method using $2^{-\Delta\Delta ct}$ normalized with GAPDH constitutive gene and expressed as fold change compared with control. All primers were designed based on the rat species, and a list of forward and reverse primers is shown in Table 2.

Table 2. List of the primer sequences used in a reverse transcription–quantitative polymerase chain reaction (qPCR) analysis of genes.

Gene	Primers	Sequences	Reference
ACE1	Forward (5′→3′) Reverse (5′→3′)	TTTGCTACACAAATGGCACTTGT CGGGACGTGGCCATTATATT	[50]
ACE2	Forward (5′→3′) Reverse (5′→3′)	ACCCTTCTTACATCAGCCCTACTG TGTCCAAAACCTACCCCACATAT	[51]
MasR	Forward (5′→3′) Reverse (5′→3′)	AGAAATCCCTTCACGGTCTACA GTCACCGATAATGTCACGATTGT	[52]
AT1R	Forward (5′→3′) Reverse (5′→3′)	CCTCTACAGCATCATCTTTGTGG CACACTGGCGTAGAGGTTGA	[53]
GAPDH	Forward (5′→3′) Reverse (5′→3′)	CCTGCACCACCAACTGCTTA AGTGATGGCATGGACTGTGG	[54]

4.8. Western Blot

The relative density of the proteins of interest (ACE, ACE2, AT1R, and MasR) was determined in the heart, lung, kidney, and liver tissues according to a previously reported western blotting method [55]. One hundred mg of each thawed tissue was sectioned, mixed in 1.5 mL of RIPA buffer containing a complete mini protease inhibitor tablet (Sigma

Aldrich, St. Louis, MO, USA), and mechanically homogenized. The samples were centrifuged, the supernatant was collected, and the protein concentrations were determined using a Qubit® protein reagent (Thermo Fisher Scientific, Waltham, MA, USA) based on the manufacturer's protocol. The same amount of protein was loaded onto each well and separated using an electrophoresis method by 4–12% tris-glycine gel. The proteins were then transferred to a polyvinylidene fluoride (PVDF) membrane, blocked in 5% skim milk in wash buffer, and incubated overnight at 4 °C with designated primary antibodies: ACE (ab25422; 1:1000), ACE2 (ab108252; 1:1000), MasR (ab66030; 1:1000), and AT1R (ab124734; 1:1000). As the housekeeping protein α-tubulin (ab 4074; 1:1000) was used. The next day, the membrane was washed in wash buffer containing 0.1% Tween 20 and incubated with secondary antibody (ab; 1:10,000) for 2 h at room temperature on a horizontal shaker. Visualization and density quantification of the images was carried out using Azure Biosystems Chemiluminescence Kit (190625-38; Azure Biosystems, Dublin, CA, USA). Results are presented as the ratio of densities of the band of interest over that of the housekeeping protein. The bands were quantified using ImageJ 1.53e (The National Institutes of Health and the Laboratory for Optical and Computational Instrumentation, LOCI, University of Wisconsin) software.

4.9. Statistical Analysis

Statistical analyses were performed by GraphPad Prism 8.0 statistical software (San Diego, CA, USA). Results are expressed as the mean ± SEM. One-way analysis of the variances (ANOVA) was used to evaluate the differences between groups after assessing the equality of means by the F-test, followed by Tukey multiple comparison post-hoc analysis. The level of significance was set at $p < 0.05$. In all cases, the p-value for F-test was lower than 0.05, except for right and left paw diameters.

5. Conclusions

In conclusion, the results of this study suggest that inflammation alters the balance of the RAS components, such as enzymes, peptides, and receptors, at the plasma and tissue levels. The exogenous administration of Ang-(1-7) and Ang. Conj. restores the imbalances in the activated RAS caused by inflammation. The observed superior Ang. Conj. protective effects in different tissue suggest that the bone-targeted delivery of Ang-(1-7) enhances its efficacy and can be a valuable therapeutic option for inflammatory diseases such as rheumatoid arthritis, renal, cardiovascular, pulmonary, and hepatic diseases.

Author Contributions: Conceptualization, A.A.-H.; methodology, A.A.-H.; validation, A.A.-H. and S.K.p.; formal analysis, A.A.-H and S.K.p.; investigation, A.A.-H. and S.K.p.; resources, A.A.-H.; data curation, S.K.p., A.R. and E.L.S.; writing—original draft preparation, S.K.p. and A.A.-H.; writing—review and editing, A.A.-H. and S.K.p.; visualization, A.A.-H. and S.K.p.; supervision, A.A.-H.; project administration, A.A.-H. and S.K.p.; funding acquisition, A.A.-H. and S.K.p. All authors have read and agreed to the published version of the manuscript.

Funding: This research received no external funding and was supported by the ISU startup fund.

Institutional Review Board Statement: Idaho State University's Institutional Animal Care Committee approved the animal study protocol under legal and ethical standards established by the National Research Council and published in the Guide for the Care and Use of Laboratory Animals (protocol #772, 22 September 2021).

Informed Consent Statement: Not applicable.

Data Availability Statement: Data is contained within the article.

Conflicts of Interest: The authors declare no conflict of interest.

References

1. Firestein, G.S. Evolving concepts of rheumatoid arthritis. *Nature* **2003**, *423*, 356. [CrossRef]
2. Choy, E. Understanding the dynamics: Pathways involved in the pathogenesis of rheumatoid arthritis. *Rheumatology* **2012**, *51*, v3–v11. [CrossRef] [PubMed]
3. Kaplan, M.J. Cardiovascular complications of rheumatoid arthritis: Assessment, prevention, and treatment. *Rheum. Dis. Clin.* **2010**, *36*, 405–426. [CrossRef] [PubMed]
4. Heinlen, L.; Humphrey, M. Skeletal complications of rheumatoid arthritis. *Osteoporos. Int.* **2017**, *28*, 2801–2812. [CrossRef]
5. Majithia, V.; Geraci, S.A. Rheumatoid arthritis: Diagnosis and management. *Am. J. Med.* **2007**, *120*, 936–939. [CrossRef] [PubMed]
6. Cobankara, V.; Öztürk, M.A.; Kiraz, S.; Ertenli, I.; Haznedaroglu, I.C.; Pay, S.; Çalgüneri, M. Renin and angiotensin-converting enzyme (ACE) as active components of the local synovial renin-angiotensin system in rheumatoid arthritis. *Rheumatol. Int.* **2005**, *25*, 285–291. [CrossRef] [PubMed]
7. Moreira, F.R.C.; de Oliveira, T.A.; Ramos, N.E.; Abreu, M.A.D.; Simões e Silva, A.C. The role of renin angiotensin system in the pathophysiology of rheumatoid arthritis. *Mol. Biol. Rep.* **2021**, *48*, 6619–6629. [CrossRef]
8. Gonzalez, A.; Icen, M.; Kremers, H.M.; Crowson, C.S.; Davis, J.M.; Therneau, T.M.; Roger, V.L.; Gabriel, S.E. Mortality trends in rheumatoid arthritis: The role of rheumatoid factor. *J. Rheumatol.* **2008**, *35*, 1009–1014.
9. Gabriel, S.E. Cardiovascular morbidity and mortality in rheumatoid arthritis. *Am. J. Med.* **2008**, *121*, S9–S14. [CrossRef]
10. Kaplan, M.J. Cardiovascular disease in rheumatoid arthritis. *Curr. Opin. Rheumatol.* **2006**, *18*, 289–297. [CrossRef]
11. Wei, C.-C.; Tian, B.; Perry, G.; Meng, Q.C.; Chen, Y.-F.; Oparil, S.; Dell'Italia, L.J. Differential ANG II generation in plasma and tissue of mice with decreased expression of the ACE gene. *Am. J. Physiol.-Heart Circ. Physiol.* **2002**, *282*, H2254–H2258. [CrossRef] [PubMed]
12. da Silveira, K.D.; Coelho, F.M.; Vieira, A.T.; Sachs, D.; Barroso, L.C.; Costa, V.V.; Bretas, T.L.B.; Bader, M.; de Sousa, L.P.; da Silva, T.A. Anti-inflammatory effects of the activation of the angiotensin-(1–7) receptor, MAS, in experimental models of arthritis. *J. Immunol.* **2010**, *185*, 5569–5576. [CrossRef] [PubMed]
13. Touyz, R.M.; Montezano, A.C. Angiotensin-(1–7) and vascular function: The clinical context. *Hypertension* **2018**, *71*, 68–69. [CrossRef] [PubMed]
14. Barroso, L.C.; Magalhaes, G.S.; Galvão, I.; Reis, A.C.; Souza, D.G.; Sousa, L.P.; Santos, R.A.; Campagnole-Santos, M.J.; Pinho, V.; Teixeira, M.M. Angiotensin-(1-7) promotes resolution of neutrophilic inflammation in a model of antigen-induced arthritis in mice. *Front. Immunol.* **2017**, *8*, 1596. [CrossRef]
15. McInnes, I.B.; Schett, G. The pathogenesis of rheumatoid arthritis. *New Engl. J. Med.* **2011**, *365*, 2205–2219. [CrossRef]
16. Liu, J.; Liu, Y.; Pan, W.; Li, Y. Angiotensin-(1–7) attenuates collagen-induced arthritis via inhibiting oxidative stress in rats. *Amino Acids* **2021**, *53*, 171–181. [CrossRef]
17. Connor, J.R.; Manning, P.T.; Settle, S.L.; Moore, W.M.; Jerome, G.M.; Webber, R.K.; Tjoeng, F.S.; Currie, M.G. Suppression of adjuvant-induced arthritis by selective inhibition of inducible nitric oxide synthase. *Eur. J. Pharmacol.* **1995**, *273*, 15–24. [CrossRef]
18. Fletcher, D.S.; Widmer, W.R.; Luell, S.; Christen, A.; Orevillo, C.; Shah, S.; Visco, D. Therapeutic administration of a selective inhibitor of nitric oxide synthase does not ameliorate the chronic inflammation and tissue damage associated with adjuvant-induced arthritis in rats. *J. Pharmacol. Exp. Ther.* **1998**, *284*, 714–721.
19. Stefanovic-Racic, M.; Meyers, K.; Meschter, C.; Coffey, J.; Hoffman, R.; Evans, C. Comparison of the nitric oxide synthase inhibitors methylarginine and aminoguanidine as prophylactic and therapeutic agents in rat adjuvant arthritis. *J. Rheumatol.* **1995**, *22*, 1922–1928.
20. McCartney-Francis, N.; Allen, J.B.; Mizel, D.E.; Albina, J.E.; Xie, Q.; Nathan, C.F.; Wahl, S.M. Suppression of arthritis by an inhibitor of nitric oxide synthase. *J. Exp. Med.* **1993**, *178*, 749–754. [CrossRef]
21. Yonekura, Y.; Koshiishi, I.; Yamada, K.-i.; Mori, A.; Uchida, S.; Nakamura, T.; Utsumi, H. Association between the expression of inducible nitric oxide synthase by chondrocytes and its nitric oxide-generating activity in adjuvant arthritis in rats. *Nitric Oxide* **2003**, *8*, 164–169. [CrossRef]
22. Villalobos, L.A.; San Hipólito-Luengo, Á.; Ramos-González, M.; Cercas, E.; Vallejo, S.; Romero, A.; Romacho, T.; Carraro, R.; Sánchez-Ferrer, C.F.; Peiró, C. The angiotensin-(1-7)/mas axis counteracts angiotensin II-dependent and -independent pro-inflammatory signaling in human vascular smooth muscle cells. *Front. Pharmacol.* **2016**, *7*, 482. [CrossRef] [PubMed]
23. Chappell, M.C.; Pirro, N.T.; Sykes, A.; Ferrario, C.M. Metabolism of angiotensin-(1-7) by angiotensin-converting enzyme. *Hypertension* **1998**, *31*, 362–367. [CrossRef]
24. Mordwinkin, N.M.; Russell, J.R.; Burke, A.S.; Dizerega, G.S.; Louie, S.G.; Rodgers, K.E. Toxicological and toxicokinetic analysis of angiotensin (1–7) in two species. *J. Pharm. Sci.* **2012**, *101*, 373–380. [CrossRef]
25. Aghazadeh-Habashi, A.; Khajehpour, S. Improved pharmacokinetics and bone tissue accumulation of Angiotensin-(1–7) peptide through bisphosphonate conjugation. *Amino Acids* **2021**, *53*, 653–664. [CrossRef] [PubMed]
26. Hirabayashi, H.; Fujisaki, J. Bone-specific drug delivery systems. *Clin. Pharmacokinet.* **2003**, *42*, 1319–1330. [CrossRef] [PubMed]
27. Cawthray, J.; Wasan, E.; Wasan, K. Bone-seeking agents for the treatment of bone disorders. *Drug delivery and translational research* **2017**, *7*, 466–481. [CrossRef]
28. Yang, Y.; Aghazadeh-Habashi, A.; Panahifar, A.; Wu, Y.; Bhandari, K.H.; Doschak, M.R. Bone-targeting parathyroid hormone conjugates outperform unmodified PTH in the anabolic treatment of osteoporosis in rats. *Drug Deliv. Transl. Res.* **2017**, *7*, 482–496. [CrossRef]

29. Bhandari, K.H.; Asghar, W.; Newa, M.; Jamali, F.; Doschak, M.R. Evaluation of bone targeting salmon calcitonin analogues in rats developing osteoporosis and adjuvant arthritis. *Curr. Drug Deliv.* **2015**, *12*, 98–107. [CrossRef]
30. Doschak, M.R.; Kucharski, C.M.; Wright, J.E.; Zernicke, R.F.; Uludag, H. Improved bone delivery of osteoprotegerin by bisphosphonate conjugation in a rat model of osteoarthritis. *Mol. Pharm.* **2009**, *6*, 634–640. [CrossRef]
31. Farrell, A.; Blake, D.; Palmer, R.; Moncada, S. Increased concentrations of nitrite in synovial fluid and serum samples suggest increased nitric oxide synthesis in rheumatic diseases. *Ann. Rheum. Dis.* **1992**, *51*, 1219–1222. [CrossRef] [PubMed]
32. McInnes, I.B.; Leung, B.P.; Field, M.; Wei, X.Q.; Huang, F.-P.; Sturrock, R.D.; Kinninmonth, A.; Weidner, J.; Mumford, R.; Liew, F.Y. Production of nitric oxide in the synovial membrane of rheumatoid and osteoarthritis patients. *J. Exp. Med.* **1996**, *184*, 1519–1524. [CrossRef] [PubMed]
33. Ialenti, A.; Ianaro, A.; Moncada, S.; Di Rosa, M. Modulation of acute inflammation by endogenous nitric oxide. *Eur. J. Pharmacol.* **1992**, *211*, 177–182. [CrossRef]
34. Asghar, W.; Aghazadeh-Habashi, A.; Jamali, F. Cardiovascular effect of inflammation and nonsteroidal anti-inflammatory drugs on renin–angiotensin system in experimental arthritis. *Inflammopharmacology* **2017**, *25*, 543–553. [CrossRef]
35. Maradit-Kremers, H.; Nicola, P.J.; Crowson, C.S.; Ballman, K.V.; Gabriel, S.E. Cardiovascular death in rheumatoid arthritis: A population-based study. *Arthritis Rheum.* **2005**, *52*, 722–732. [CrossRef]
36. Lemarié, C.A.; Schiffrin, E.L. The angiotensin II type 2 receptor in cardiovascular disease. *J. Renin-Angiotensin-Aldosterone Syst.* **2010**, *11*, 19–31. [CrossRef]
37. Ruiz-Ortega, M.; Lorenzo, O.; Suzuki, Y.; Rupérez, M.; Egido, J. Proinflammatory actions of angiotensins. *Curr. Opin. Nephrol. Hypertens.* **2001**, *10*, 321–329. [CrossRef]
38. Kortekaas, K.E.; Meijer, C.A.; Hinnen, J.W.; Dalman, R.L.; Xu, B.; Hamming, J.F.; Lindeman, J.H. ACE inhibitors potently reduce vascular inflammation, results of an open proof-of-concept study in the abdominal aortic aneurysm. *PLoS ONE* **2014**, *9*, e111952. [CrossRef]
39. Benicky, J.; Sánchez-Lemus, E.; Pavel, J.; Saavedra, J.M. Anti-inflammatory effects of angiotensin receptor blockers in the brain and the periphery. *Cell. Mol. Neurobiol.* **2009**, *29*, 781–792. [CrossRef]
40. Taguchi, I.; Toyoda, S.; Takano, K.; Arikawa, T.; Kikuchi, M.; Ogawa, M.; Abe, S.; Node, K.; Inoue, T. Irbesartan, an angiotensin receptor blocker, exhibits metabolic, anti-inflammatory and antioxidative effects in patients with high-risk hypertension. *Hypertens. Res.* **2013**, *36*, 608–613. [CrossRef]
41. Price, A.; Lockhart, J.; Ferrell, W.; Gsell, W.; McLean, S.; Sturrock, R. Angiotensin II type 1 receptor as a novel therapeutic target in rheumatoid arthritis: In vivo analyses in rodent models of arthritis and ex vivo analyses in human inflammatory synovitis. *Arthritis Rheum.* **2007**, *56*, 441–447. [CrossRef] [PubMed]
42. Mateo, T.; Nabah, Y.N.A.; Taha, M.A.; Mata, M.; Cerdá-Nicolás, M.; Proudfoot, A.E.; Stahl, R.A.; Issekutz, A.C.; Cortijo, J.; Morcillo, E.J. Angiotensin II-induced mononuclear leukocyte interactions with arteriolar and venular endothelium are mediated by the release of different CC chemokines. *J. Immunol.* **2006**, *176*, 5577–5586. [CrossRef] [PubMed]
43. Nabah, Y.N.A.; Mateo, T.; Estellés, R.; Mata, M.; Zagorski, J.; Sarau, H.; Cortijo, J.; Morcillo, E.J.; Jose, P.J.; Sanz, M.-J. Angiotensin II induces neutrophil accumulation in vivo through generation and release of CXC chemokines. *Circulation* **2004**, *110*, 3581–3586. [CrossRef] [PubMed]
44. Schlüter, K.-D.; Wenzel, S. Angiotensin II: A hormone involved in and contributing to pro-hypertrophic cardiac networks and target of anti-hypertrophic cross-talks. *Pharmacol. Ther.* **2008**, *119*, 311–325. [CrossRef] [PubMed]
45. Parajuli, N.; Ramprasath, T.; Patel, V.B.; Wang, W.; Putko, B.; Mori, J.; Oudit, G.Y. Targeting angiotensin-converting enzyme 2 as a new therapeutic target for cardiovascular diseases. *Can. J. Physiol. Pharmacol.* **2014**, *92*, 558–565. [CrossRef]
46. Xianwei, W.; Magomed, K.; Ding, Z.; Sona, M.; Jingjun, L.; Shijie, L.; Mehta, J.L. Cross-talk between inflammation and angiotensin II: Studies based on direct transfection of cardiomyocytes with AT1R and AT2R cDNA. *Exp. Biol. Med.* **2012**, *237*, 1394–1401. [CrossRef]
47. Ling, S.; Jamali, F. Effect of early phase adjuvant arthritis on hepatic P450 enzymes and pharmacokinetics of verapamil: An alternative approach to the use of an animal model of inflammation for pharmacokinetic studies. *Drug Metab. Dispos.* **2005**, *33*, 579–586. [CrossRef]
48. Aghazadeh-Habashi, A.; Kohan, M.G.; Asghar, W.; Jamali, F. Glucosamine dose/concentration-effect correlation in the rat with adjuvant arthritis. *J. Pharm. Sci.* **2014**, *103*, 760–767. [CrossRef]
49. Cui, L.; Nithipatikom, K.; Campbell, W.B. Simultaneous analysis of angiotensin peptides by LC–MS and LC–MS/MS: Metabolism by bovine adrenal endothelial cells. *Anal. Biochem.* **2007**, *369*, 27–33. [CrossRef]
50. Dai, S.-Y.; Peng, W.; Zhang, Y.-P.; Li, J.-D.; Shen, Y.; Sun, X.-F. Brain endogenous angiotensin II receptor type 2 (AT2-R) protects against DOCA/salt-induced hypertension in female rats. *J. Neuroinflammation* **2015**, *12*, 47. [CrossRef]
51. Liu, C.X.; Hu, Q.; Wang, Y.; Zhang, W.; Ma, Z.Y.; Feng, J.B.; Wang, R.; Wang, X.P.; Dong, B.; Gao, F. Angiotensin-converting enzyme (ACE) 2 overexpression ameliorates glomerular injury in a rat model of diabetic nephropathy: A comparison with ACE inhibition. *Mol. Med.* **2011**, *17*, 59–69. [CrossRef] [PubMed]
52. Liu, J.; Chen, Q.; Liu, S.; Yang, X.; Zhang, Y.; Huang, F. Sini decoction alleviates E. coli induced acute lung injury in mice via equilibrating ACE-AngII-AT1R and ACE2-Ang-(1-7)-Mas axis. *Life Sci.* **2018**, *208*, 139–148. [CrossRef] [PubMed]

53. Exner, E.C.; Geurts, A.M.; Hoffmann, B.R.; Casati, M.; Stodola, T.; Dsouza, N.R.; Zimmermann, M.; Lombard, J.H.; Greene, A.S. Interaction between Mas1 and AT1RA contributes to enhancement of skeletal muscle angiogenesis by angiotensin-(1-7) in Dahl salt-sensitive rats. *PLoS ONE* **2020**, *15*, e0232067. [CrossRef] [PubMed]
54. Wang, T.; Lian, G.; Cai, X.; Lin, Z.; Xie, L. Effect of prehypertensive losartan therapy on AT1R and ATRAP methylation of adipose tissue in the later life of high-fat-fed spontaneously hypertensive rats. *Mol. Med. Rep.* **2018**, *17*, 1753–1761. [CrossRef] [PubMed]
55. Hanafy, S.; Dagenais, N.; Dryden, W.; Jamali, F. Effects of angiotensin II blockade on inflammation-induced alterations of pharmacokinetics and pharmacodynamics of calcium channel blockers. *Br. J. Pharmacol.* **2008**, *153*, 90–99. [CrossRef]

Article

Synthesis, Neuroprotective Effect and Physicochemical Studies of Novel Peptide and Nootropic Analogues of Alzheimer Disease Drug

Radoslav Chayrov [1], Tatyana Volkova [2], German Perlovich [2], Li Zeng [3], Zhuorong Li [3], Martin Štícha [4], Rui Liu [3,*] and Ivanka Stankova [1,*]

1. Department of Chemistry, Faculty of Mathematics & Natural Sciences, South-West University "Neofit Rilski", 2700 Blagoevgrad, Bulgaria
2. G.A. Krestov Institute of Solution Chemistry, Russian Academy of Sciences, 153045 Ivanovo, Russia
3. Institute of Medicinal Biotechnology, Chinese Academy of Medical Sciences and Peking Union Medical College, Beijing 100050, China
4. Faculty of Science, Charles University in Prague, 128 43 Prague, Czech Republic
* Correspondence: liurui@imb.pumc.edu.cn (R.L.); ivastankova@abv.bg (I.S.); Tel.: +86-10-67087731 (R.L.); +359-897-295919 (I.S.)

Citation: Chayrov, R.; Volkova, T.; Perlovich, G.; Zeng, L.; Li, Z.; Štícha, M.; Liu, R.; Stankova, I. Synthesis, Neuroprotective Effect and Physicochemical Studies of Novel Peptide and Nootropic Analogues of Alzheimer Disease Drug. *Pharmaceuticals* 2022, 15, 1108. https://doi.org/10.3390/ph15091108

Academic Editors: Nuno Manuel Xavier and Gill Diamond

Received: 18 July 2022
Accepted: 30 August 2022
Published: 5 September 2022

Publisher's Note: MDPI stays neutral with regard to jurisdictional claims in published maps and institutional affiliations.

Copyright: © 2022 by the authors. Licensee MDPI, Basel, Switzerland. This article is an open access article distributed under the terms and conditions of the Creative Commons Attribution (CC BY) license (https://creativecommons.org/licenses/by/4.0/).

Abstract: Glutamate is an excitatory neurotransmitter in the nervous system. Excessive glutamate transmission can lead to increased calcium ion expression, related to increased neurotoxicity. Memantine is used for treating patients with Alzheimer's disease (AD) due to its protective action on the neurons against toxicity caused by over activation of N-methyl-D-aspartate receptors. Nootropics, also called "smart drugs", are used for the treatment of cognitive deficits. In this work, we evaluate the neuroprotective action of four memantine analogues of glycine derivatives, including glycyl-glycine, glycyl-glycyl-glycine, sarcosine, dimethylglycine and three conjugates with nootropics, modafinil, piracetam and picamilon. The new structural memantine derivatives improved cell viability against copper-induced neurotoxicity in APPswe cells and glutamate-induced neurotoxicity in SH-SY5Y cells. Among these novel compounds, modafinil-memantine, piracetam-memantine, sarcosine-memantine, dimethylglycine-memantine, and glycyl-glycine-memantine were demonstrated with good EC_{50} values of the protective effects on APPswe cells, accompanied with moderate amelioration from glutamate-induced neurotoxicity. In conclusion, our study demonstrated that novel structural derivatives of memantine might have the potential to develop promising lead compounds for the treatment of AD. The solubility of memantine analogues with nootropics and memantine analogues with glycine derivatives in buffer solutions at pH 2.0 and pH 7.4 simulating the biological media at 298.15 K was determined and the mutual influence of the structural fragments in the molecules on the solubility behavior was analyzed. The significative correlation equations relating the solubility and biological properties with the structural HYBOT (Hydrogen Bond Thermodynamics) descriptors were derived. These equations would greatly simplify the task of the directed design of the memantine analogues with improved solubility and enhanced bioavailability.

Keywords: memantine; glycine; nootropics; neuroprotective action; solubility

1. Introduction

Dementia has been known and is a significant cause of disability, dependence and mortality. Memantine 1-amino-3,5-dimethyladamantan (Figure 1). It is used on patients with Alzheimer's disease (AD). Progressive memory loss of and cognitive dysfunctions are the main symptoms. Currently this neurodegenerative disorder is irremediable. About 15 million people over the world are affected and are experiencing difficulties daily. The earlier symptoms come around the age of 65 and become more severe with age. Predominantly they are increasing from 0.5% at 65 to about 8% by the age of 85. Currently, AD is present in about 50–75% of all dementia patients [1–3].

Memantine action is based on the protection of neurons against toxicity caused by overactivation of N-methyl-D-aspartate receptors [4].

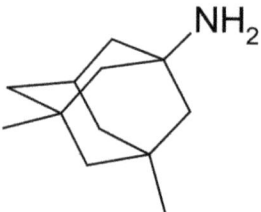

Figure 1. Chemical structure of memantine.

The concentrations of glutamate in the extracellular region should be low and is controlled by a multiple mechanism at the synapse. Disturbance to this system for regulation lead to harmful effects, for example, the excess releasing of glutamate. High levels of glutamate could cause increased excitability in post-synaptic neurons and excitotoxicity and cytotoxicity will probably appear. Glutamate-induced excitotoxicity in the hippocampus is related to less neuronal regeneration and dendritic branching, leading to weakened spatial learning. Any changes of glutamate uptake from the synapse are related to decreased sensitivity to reward, which is a symptom of depression. [5,6]. The main mechanism of excitotoxicity includes the excessive amounts of reactive oxygen radicals generated in overexcited neurons. On this basis, we can conclude that antioxidants could be effective in slowing down the development of neurological disorders by protecting the neurons from oxidative damage [7]. This state, called "glutamate neurotoxicity" or GNT, is explained by time-dependent cell injuries, which usually lead to cell death. In the biochemical mechanisms leading to the cell death the reactive oxygen species, known as ROS are hydroxyl radical ($^\bullet OH$), superoxide anion ($^\bullet O^{2-}$), and hydrogen peroxide (H_2O_2). They are all generated in different cell parts due to various reactions. However, in the presence of some molecules which have an antioxidant activity the cells could prevent ROS damage. [8]. According to our earlier research, glycyl-glycine has demonstrated scavenging potential [9]. Dimethylglycine, known as dietary supplement DMG improves the immune system and decreases the oxidative damage by scavenging the excess free radicals [10–12]. Similar to betaine and choline, the DMG molecule increases antioxidant capacity because it is a building material for the glutathione synthesis [13]. Another study showed that DMG decreases oxidative damage and keeps the growth and health in researches with animals [14]. Abnormal high glutamate delivery could lead to increased calcium ion flow which is related to neurotoxicity. On the other hand, insufficient delivery could significantly change the information stream in neurons causing the symptoms similar to schizophrenia [1]. The amino acid and dietary supplement Sarcosine has the biological activity leading to the increase of normal functioning of the glutamatergic N-methyl-D-aspartate receptors (NMDAR). This action can be considered as a rational treatment for schizophrenia. There are a few studies which provides evidence for its efficiency [15,16].

Nootropics (modafinil, piracetam, picamilon etc.) help the process of learning and memory, attention and motivation. They facilitate recovery processes and improve the metabolism of nerve cells, protecting them from hypoxia. [17]. Over time these substances become dopaminergic and serotonergic drugs. However, we should pay attention to cholinergically active nootropics because of the crucial role of acetylcholine in learning and memory.

Modafinil, known as Provigil® and Nuvigil®, is a widely preferred wake-promoting medication. It binds to the cell-membrane dopamine (DA) transporter competitively and is dependent on dopaminergic and adrenergic signaling for its wake-promoting action [18,19]. Additionally, Modafinil mediates the restoration of aerobic metabolism and hyper-glycolysis suppression, thereby resulting in an increase in pyruvate dehydrogenase and a decrease in lactate dehydrogenase activity, respectively, which ultimately reduced oxidative reperfusion injury [20].

The neuroprotective effect of piracetam has been well-known for a long time [21–23]. Piracetam has an influence on the excitatory neurotransmitters and inhibitors in the brain. It has been suggested that it has effects on increasing the availability of oxygen and permeability of the mitochondrial cell membrane in the Krebs cycle medium stages [24,25]. It has been reported that there are good effects of piracetam in the prevention of cognitive dysfunction related to the memory that usually occur after surgery under anesthesia [17].

Picamilon is a medication with proven cerebrovascular activity. From a chemical point of view, it is a nicotinoyl-gamma-aminobutyric acid and it is successfully used in neurological practice. Picamilon also exhibits antiplatelet activity [26–28].

Although there are various risk factors besides a genetic predisposition that are associated with the pathogenesis of such as oxidative stress, the formation of β-amyloid aggregates (Aβ), disorders in tau-protein, lowering of acetylcholine levels, dyshomeostasis of biometals, etc., the etiology of AD still remains a riddle. Acting as an open-channel blocker, the anti-AD drug memantine preferentially targets NMDAR overactivation, which has been proposed to trigger neurotoxic events mediated by amyloid β peptide (Aβ) and oxidative stress [29]. In this sense, the classic "drug-discovery" approach, based on the "one molecule-one target" paradigm, turned out to be ineffective. Based on the hypothesis that Alzheimer's disease is a multifactorial disease, in the last 10 years there has been a strong interest in developing the targeted therapy to affect different targets in AD. Currently, of the few accessible symptomatic therapies for AD, memantine is the only N-methyl-d-aspartate receptor (NMDAR) blocker. Turcu et al. explores a series of memantine analogs featuring a benzohomoadamantane scaffold. Most of the newly synthesized compounds block NMDARs in the micromolar range [30]. The aim of the present study is the design of hybrid molecules, including two pharmacophoric moieties—memantine (NMDA antagonist) and nootropics. The choice of this approach is that the NMDA antagonist can possibly stop or slow neurodegeneration, while the nootropics can improve memory and cognitive abilities by stimulating the surviving neurons.

Along with the tests on the biological activity and receptor affinity the investigations aimed at the characterization of the physicochemical properties of drugs and newly synthesized biologically active substances are highly demanded on the early stages of development [31]. This comes from the need for acceleration of the rational selection of the compounds with the best characteristics and, as follows, shortening the time period to the pharmaceutical market. Aqueous solubility and permeability across the biological membranes are the main properties for study in the case of new drug candidates. Moreover, since the vast array of the drugs exists in different ionization state according to the pH of the respective medium, the solubility should be estimated in the solutions with different acidity. Another important issue in the rational drug design is the creation of the approaches to the improvement of the physicochemical properties which is often possible if the structure-property correlations are disclosed. Along with various methods for increasing the solubility of the substances, the structural modification of molecules is of great interest to specialists. This is due to the fact that the solubility of organic compounds directly depends on the features of their structure. Changing the structural fragments and introducing various substituents lead to the structural and stereochemical changes in the molecule. The latter, in turn, affect the physicochemical properties, including the solubility. Since the hydrogen bonds are critical in the most of chemical and biological processes, it seems reasonable to predict the solubility using the correlations based on the hydrogen bonding physicochemical descriptors (HYBOT). Taking into consideration the objects of the present study and the structural analogues—the memantine derivatives reported previously [32], we tried to disclose the correlations of the biological activity and solubility with the structure of the compounds.

2. Results and Discussion

Herein we report on a design, synthesis and investigate the neuroprotective activity of newly memantine derivatives, QSAR and solubility study experiments.

2.1. Chemistry

Three hybrid memantine molecules with nootropics—modafinil, piracetam and picamilon and four memantine analogues with glycine—glycyl-glycine, glycyl-glycyl-glycine, methyl-glycine and dimethylglycine (Scheme 1) were formed using the TBTU coupling reagent [33]. The N-Boc protecting groups of glycine analogues were removed in CH_2Cl_2/TFA at 0 °C and the final TFA salt was removed with aqueous ammonia.

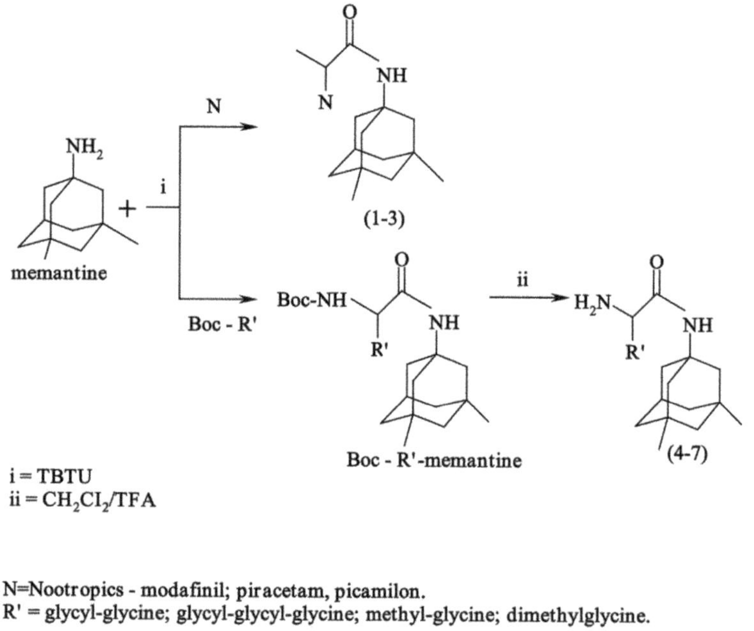

i = TBTU
ii = CH_2Cl_2/TFA

N = Nootropics - modafinil; piracetam, picamilon.
R' = glycyl-glycine; glycyl-glycyl-glycine; methyl-glycine; dimethylglycine.

Scheme 1. Synthesis of nootropics and glycine derivatives of memantine.

The structures of the synthesized memantine analogues are shown in Figure 2.

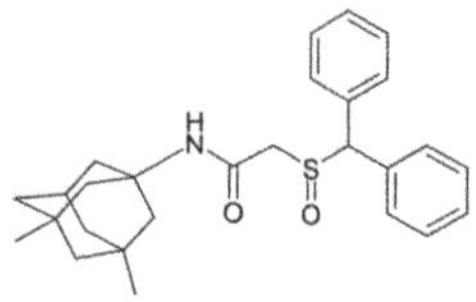

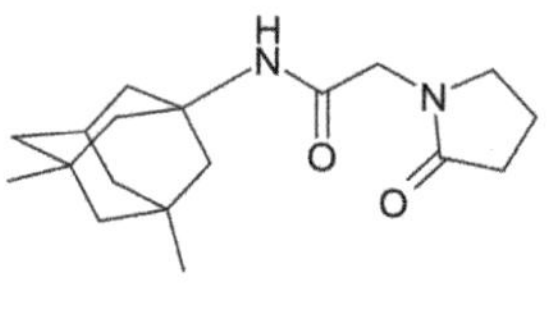

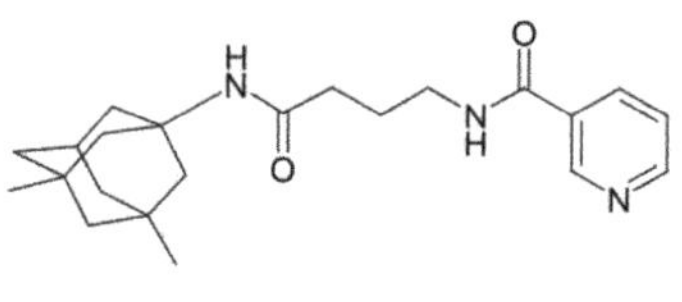

Figure 2. Cont.

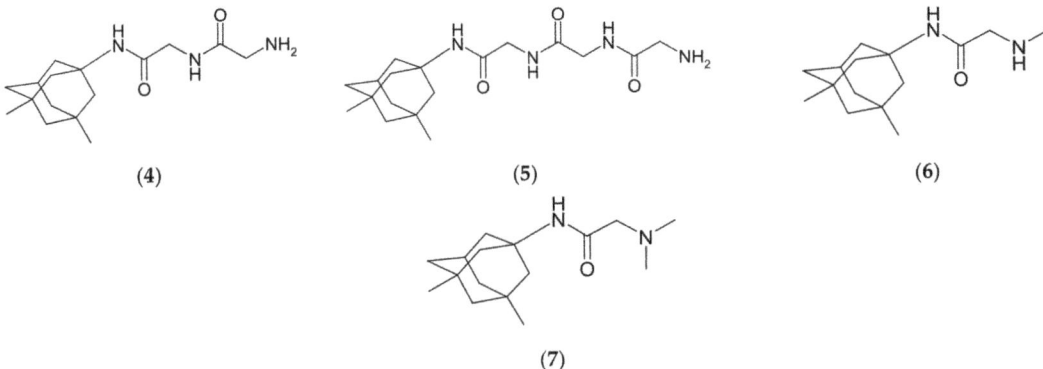

Figure 2. Chemical structures of the studied compounds. modafinil-memantine. (**1**); piracetam-memantine (**2**); picamilon-memantine (**3**); glycyl-glycyl-memantine (**4**); glycyl-glycyl-glycyl-memantine (**5**); sarcosine-memantine (**6**); dimethylglycyl-memantine (**7**).

2.2. Novel Structural Derivatives of Memantine Improve Cell Viability against Copper-Induced Neurotoxicity in APPswe Cells

As shown in Figure 3, 300 µM copper significantly reduced the cell viability of APPswe cells as compared to the control group (Figure 3A–H, all $p < 0.001$ vs. control group). Seven new compounds improved the cell viability of copper-injured APPswe cells with various effects (Figure 3A–H, $p < 0.05$–0.001). Among compounds **1** to **7**, modafinil-memantine, piracetam-memantine, picamilon-memantine, glycyl- glycine-memantine, glycyl-glycyl-glycine-memantine and dimethylglycine-memantine displayed neuroprotective effects comparable with the positive drug memantine in the AD cell model (Figure 3A–H, $p < 0.05$–0.001). Meanwhile, compounds **1–7** showed a good dose-dependent with EC_{50} values ranging from 1.120 µM to 94.88 µM (Figure 4A–H). Compounds modafinil-memantine ($EC_{50} = 1.120 \pm 0.398$ µM), piracetam-memantine ($EC_{50} = 3.217 \pm 0.139$ µM), picamilon-memantine ($EC_{50} = 4.905 \pm 1.267$ µM), sarcosine-memantine ($EC_{50} = 6.439 \pm 0.567$ µM), and dimethylglycine-memantine ($EC_{50} = 4.534 \pm 1.757$ µM) were potential anti-AD candidates with good EC_{50} value, especially compound modafinil-memantine. However, compounds modafinil-memantine and picamilon-memantine at 100 µM exhibited cell cytotoxicity, which was similar to the positive drug memantine at 100 µM (Figure 3B,D).

Taken together, these novel structural derivatives of memantine improved cell viability against copper-induced toxicity in APPswe cells, indicating that these active compounds may exert neuroprotective effects from Aβ toxicity in AD.

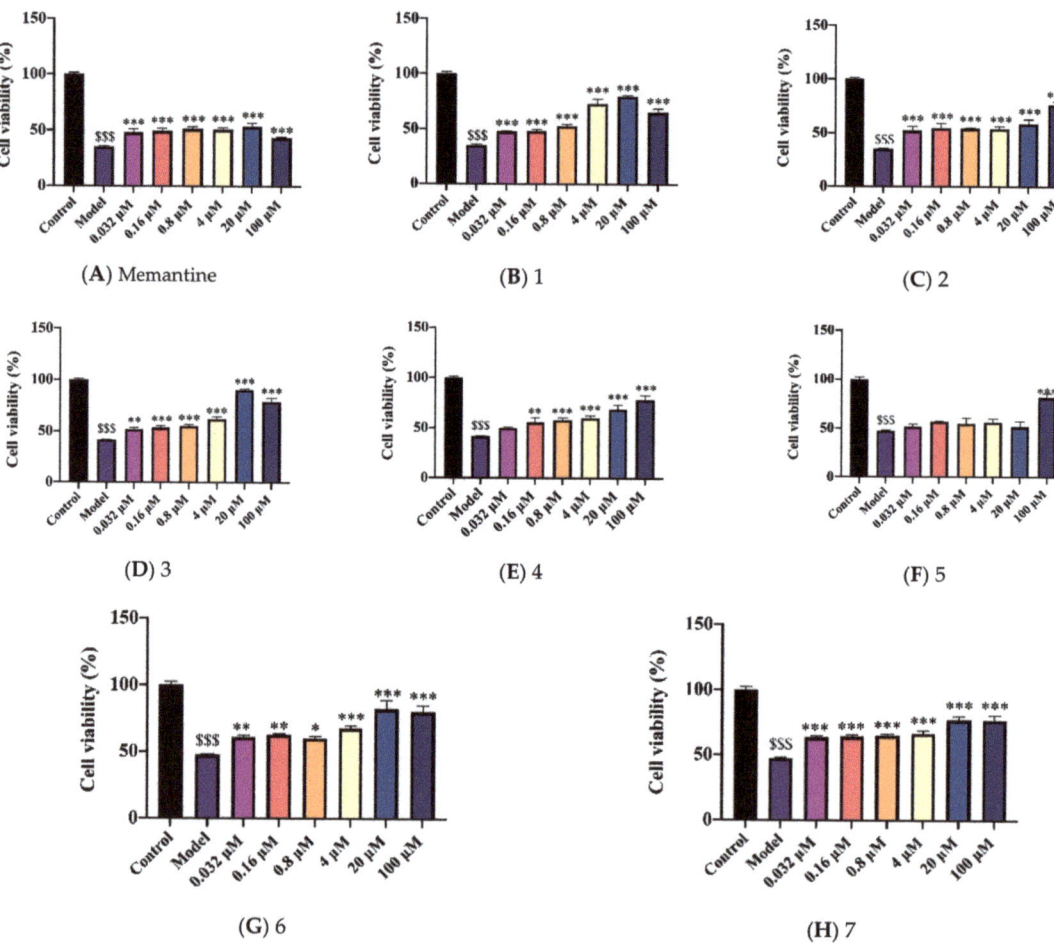

Figure 3. Cell viability of copper-injured APPswe cells treated with novel structural derivatives of memantine. (**A**–**H**) Cell viability of memantine (as a positive reference drug) (**A**) and memantine derivatives, including 1 (**B**) modafinil-memantine, 2 (**C**) piracetam-memantine, 3 (**D**) picamilon-memantine, 4 (**E**) glycyl-glycine-memantine, 5 (**F**) glycyl-glycyl-glycine-memantine, 6 (**G**) sarcosine-memantine, 7 (**H**) dimethylglycine-memantine, using MTS assay. Data are expressed as mean ± SEM, $n = 3$. $^{\$\$\$}$ $p < 0.001$ vs. the control group; * $p < 0.05$, ** $p < 0.01$, *** $p < 0.001$ vs. the model.

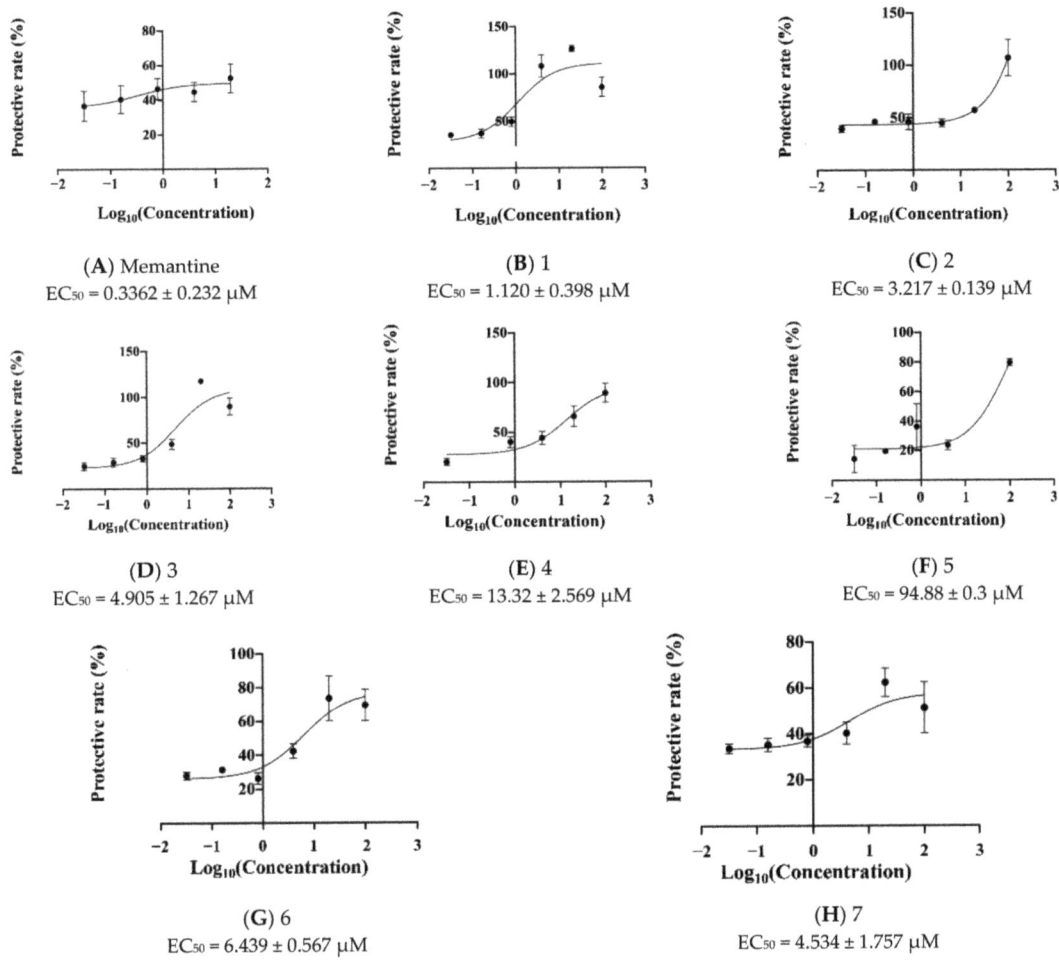

Figure 4. Value of EC_{50} of novel structural derivatives of memantine against copper-induced toxicity. (**A–H**) Dose-dependent curve of memantine (as a positive reference drug) (**A**) and memantine derivatives, including 1 (**B**) modafinil-memantine, 2 (**C**) piracetam-memantine, 3 (**D**) picamilon-memantine, 4 (**E**) glycyl-glycine-memantine, 5 (**F**) glycyl-glycyl-glycine-memantine 6 (**G**) sarcosine-memantine, 7 (**H**) dimethylglycine-memantine. Data are expressed as mean ± SEM, $n = 3$.

2.3. Novel Structural Derivatives of Memantine Improve Cell Viability against Glutamate-Induced Neurotoxicity in SH-SY5Y Cells

Glutamate is an excitatory neurotransmitter and participates in the plasticity of the central nervous system diseases affecting cognition, memory, and learning. Neuronal cell loss associated with glutamate neurotoxicity is a close pathological event in AD. The results showed that glutamate significantly decreased the cell viability of SH-SY5Y cells (Figure 5A–H, all $p < 0.001$). These compounds treated ranging from 0.032 μM to 4 μM showed different degrees of neuroprotection effects. Compound piracetam-memantine treated at 0.032 μM, 0.16 μM, 0.8 μM, and 4 μM significantly improved the cell viability of SH-SY5Y cells, and compound glycyl-glycine-memantine treated at 0.032 μM and 0.16 μM increased cell viability (Figure 5E, $p < 0.01$–0.001). These two compounds have a similar neuroprotective effect to the positive drug memantine. Compounds glycyl-glycyl-glycine-memantine, sarcosine-memantine, dimethylglycine-memantine, treated at 0.032 µM increased cell viability (Figure 5F–H, $p < 0.05$–0.01). Combined, these active compounds may have neuroprotective effects with a correlation with the glutamate receptors.

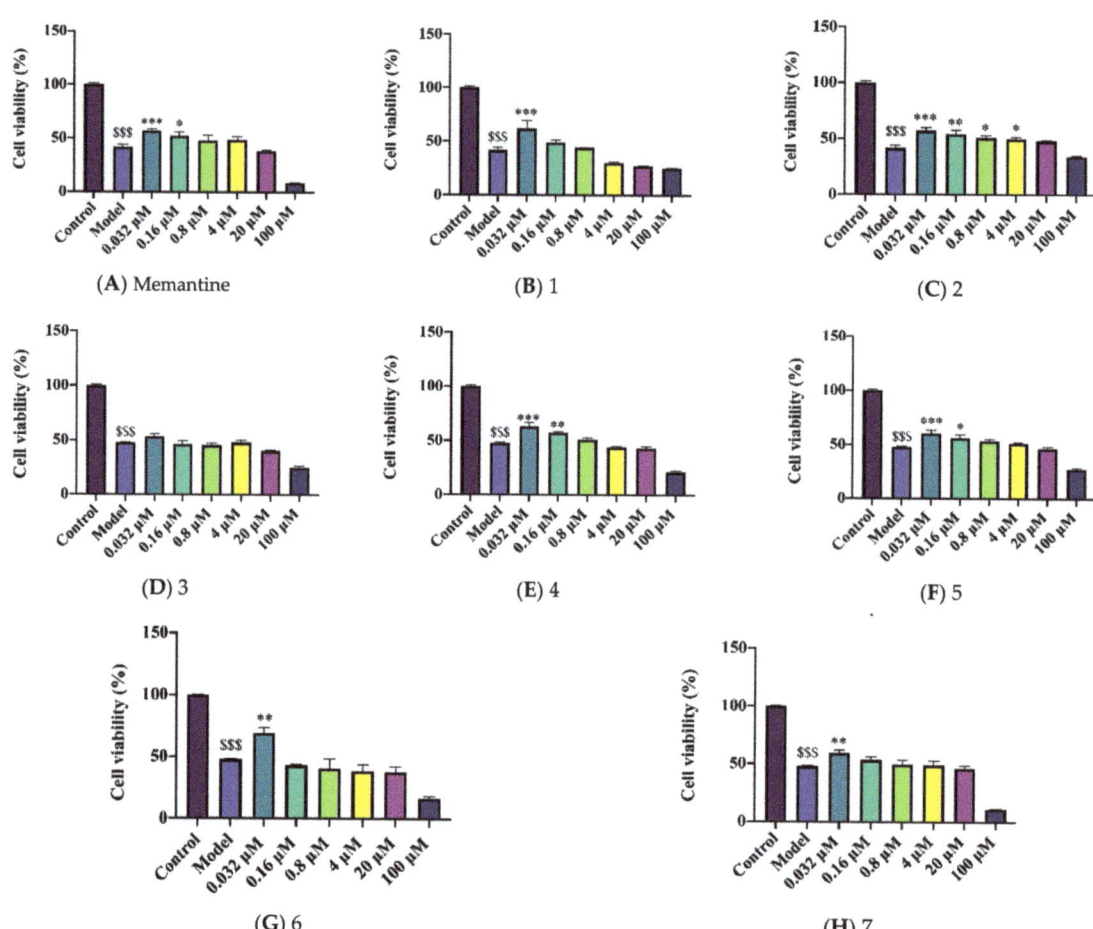

Figure 5. Cell viability of novel structural derivatives of memantine in glutamate-injured SH-SY5Y cells. (**A–H**) Cell viability of memantine (as a positive drug) (**A**) and memantine derivatives, including 1 (**B**) modafinil-memantine, 2 (**C**) piracetam-memantine, 3 (**D**) picamilon-memantine, 4 (**E**) glycyl-glycine-memantine, 5 (**F**) glycyl-glycyl-glycine-memantine, 6 (**G**) sarcosine-memantine, 7 (**H**) dimethylglycine-memantine, using MTS assay. Data are expressed as mean ± SEM, $n = 3$. $^{\$\$\$} p < 0.001$ vs. the control group; * $p < 0.05$, ** $p < 0.01$, *** $p < 0.001$ vs. the model.

2.4. Prediction of EC_{50} Using HYBOT Descriptors

One of the main tasks of the QSAR models is to predict the biological properties of the systems. In this work, we attempted to develop a correlation model based on the HYBOT (H-bond thermodynamics) descriptors in order to predict EC_{50}. To this end, all the physicochemical descriptors available in the software package were used. As a result of the fitting, the best correlation was obtained with the descriptor characterizing the sum of the donor and acceptor ability of atoms in a molecule to form hydrogen bonds normalized to molecular polarizability ($\Sigma(C_{ad}/\alpha)$) (Figure 6). The final correlation dependence can be described by the following equation:

$$\log(1/EC_{50}) = (0.677 \pm 0.236) - (4.762 \pm 0.703) \cdot \Sigma(C_{ad}/\alpha) \quad (1)$$

$$R = 0.9496;\ SD = 0.21;\ n = 7;\ F = 45.9$$

Evidently, the $\Sigma(C_{ad}/\alpha)$ descriptor allows for all types of the interactions of the studied molecules with the biological environment: specific (C_{ad}) and nonspecific (α). Thus, it is possible to predict the EC_{50} values for a given class of compounds with the help of Equation (1) based only on the structural formula without the expensive biological experiments.

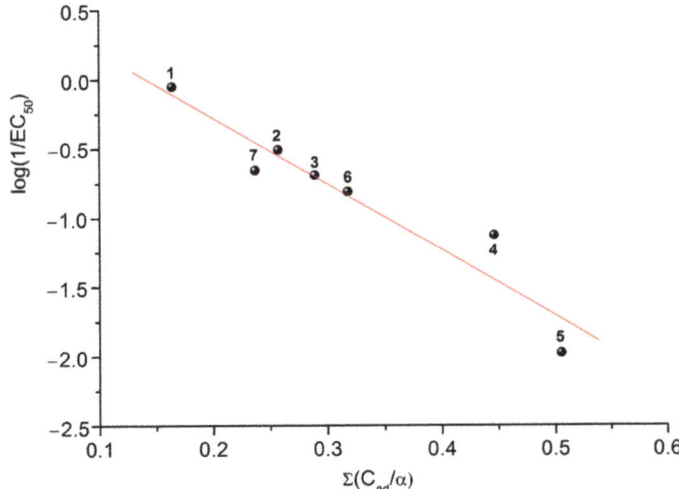

Figure 6. Dependence of $\log(1/EC_{50})$ versus $\Sigma(C_{ad}/\alpha)$ descriptor (compounds numbering corresponds to Figure 2).

2.5. Solubility Experiments

Both the objects of this investigation (**1–7**), represented in Figure 2, and the structural analogues from the previous study [32] (**8–12**), illustrated in Figure 7, were taken into consideration.

The solubility of compounds (**1**)–(**6**) was determined in buffer solutions at pH 2.0 and pH 7.4 simulating the gastric juice and the blood plasma/jejunum/ileum media at 298.15 K. The solid residuals after the dissolution experiments were isolated and analyzed. The comparison did not show any changes of the crystalline phases before and after the solubility tests (Figure 8). The solubility results for the studied derivatives are represented in Table 1 together with those of the compounds reported previously. For better visualization, the data are illustrated as a diagram in an ascending order of the solubility values (Figure 8).

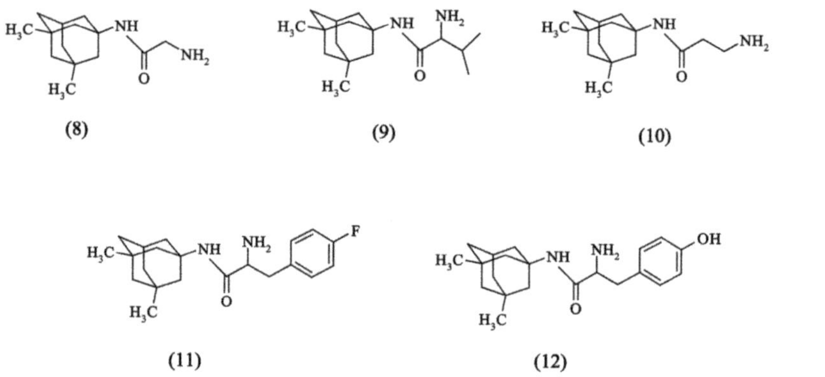

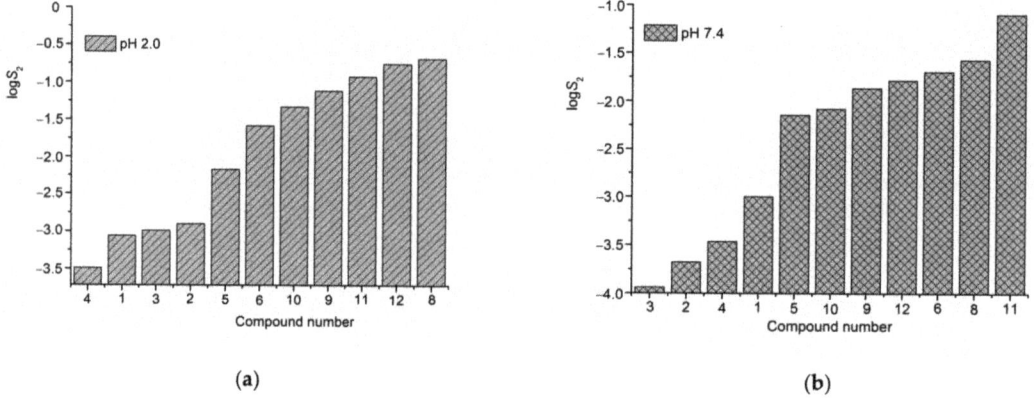

Figure 7. Formulas of the compounds studied in the previous study [30] structurally analogous to the derivatives investigated in the present work (see Figure 2 for the structures): **8**—glycine-memantine; **9**—valine-memantine; **10**—beta-alanine-memantine; **11**—4-F-phenylalanine-memantine; **12**—tyrosine-memantine.

Figure 8. Solubility of the considered compounds in buffer solutions pH 2.0 (a) and pH 7.4 (b) at 298.15 K (compounds numbering corresponds to Figures 2 and 7).

Table 1. Solubility of the compounds in buffer solutions pH 2.0 and pH 7.4 at 298.15 K.

№ Compound	pH 2.0 $S_2 \cdot 10^3$ (M)	pH 7.4 $S_2 \cdot 10^4$ (M)
(1)	0.873 ± 0.022	9.93 ± 0.25
(2)	1.25 ± 0.03	2.10 ± 0.05
(3)	1.02 ± 0.03	1.15 ± 0.03
(4)	0.321 ± 0.008	3.44 ± 0.09
(5)	6.67 ± 0.17	71.5 ± 1.8
(6)	25.88 ± 0.65	201.8 ± 5.1
(8) [a]	205.1 ± 5.1	269.2 ± 6.7
(9) [a]	75.9 ± 1.9	136.1 ± 3.4
(10) [a]	46.2 ± 1.1	83.0 ± 2.1
(11) [a]	118.0 ± 3.0	804 ± 20
(12) [a]	175.0 ± 4.4	164.1 ± 4.1

[a] solubility was reported in [30]. Compounds numbering corresponds to Figures 2 and 7.

As follows from the experiments, for the majority of the substances—(2), (3), (8), (9), (10), (11), (12)—the solubility in buffer solution pH 2.0 is essentially higher than in the medium of pH 7.4. The difference is the greatest for tyrosine-substituted compound (12). For piracetam—(2), valine—(9), and β-alanine—(10) derivatives these differences are very close (5.95 ÷ 5.57-fold). In their turn, compounds (1), (4), (5) and (6) have practically the same solubility values in both studied pHs. Appealing to the structures of the substances (Figures 2 and 7), it can be proposed that a higher solubility in muriatic buffer pH 2.0 is a result of the stronger ionization (protonation) of the molecules in the acidic solution. On the other hand, the compounds revealing the close solubilities in both investigated media are intended to be in the same ionization state. Some difference in the solubility (between pH 2.0 and pH 7.4) of compounds (1), (4), (5) and (6) can also be attributed to the impact of the buffer components in the aqueous solutions. The same order of the solubility of these substances in both pHs ($S_2(4) < S_2(1) < S_2(5) < S_2(6)$) proves the above proposal. It should be emphasized that if we consider all the compounds, the sequence of the solubility values depends on the pH. In buffer pH 2.0: $S_2(4) < S_2(1) < S_2(3) < S_2(2) < S_2(5) < S_2(6) < S_2(10) < S_2(9) < S_2(11) < S_2(12) < S_2(8)$, whereas at pH 7.4 the order is: $S_2(3) < S_2(2) < S_2(4) < S_2(1) < S_2(5) < S_2(10) < S_2(9) < S_2(12) < S_2(6) < S_2(8) < S_2(11)$. Obviously, the mutual influence of the structural fragments in the molecules impact greatly the solubility behavior. So, at the next step it was interesting to trace this influence. The analysis of the substituent impact on the solubility of the memantine derivatives studied in the present work at 298.15 K led to the following conclusions.

The addition of the hydrophilic NH-CO- fragment to compound (4) resulting in compound (5), expectedly, enhances the solubility in 20.8-fold in both pHs. The replacement of the NH_2- group in compound (10) by the fragment containing the CH_2-NH-CO-pyridine ring to obtain compound (3) results in 45.3-fold and 72.2-fold solubility decrease (in pH 2.0 and pH 7.4, respectively), probably, due to a higher hydrophilicity of the NH_2- group as compared to the NH- one, and addition of a bulky pyridine ring. Expectedly, the solubility reduction is less in pH 2.0 as compared to pH 7.4 due to the possibility of the nitrogen atom protonation. Taking into account the presence of the two bulky hydrophobic phenyl substituents in compound (1), very low solubility in aqueous medium can be expected. But, probably, the sulfinyl (SO) functional group in (1) molecule (capable of the hydrogen bonding) makes the solubility higher than for compound (4) in pH 2.0 and compounds (2), (3) and (4) in pH 7.4.

Since the solubility is an important physicochemical characteristic of a substance which defines the bioavailability of drugs, predicting the solubility values for this class of compounds was one of the goals of the study. To carry out the correlation analysis, we used the physicochemical descriptors HYBOT as the independent variables [34]. In turn, the logarithms of the solubility values were chosen as the dependent ones. Analysis of 32 independent descriptors from the HYBOT program showed that the solubility values in buffer pH 2.0 correlate well with the parameter describing the sum of the donor ability of atoms in a molecule to form hydrogen bonds normalized to polarizability ($\Sigma(C_d/\alpha)$). The results of the analysis are presented in Figure 9. If we neglect the experimental points (4) and (5) (which deviate significantly from the general dependence), then the correlation equation can be represented as follows:

$$\log(S_2^{pH2.0}) = (3.94 \pm 0.44) - (18.3 \pm 3.4) \cdot \Sigma(C_d/\alpha) \qquad (2)$$

$$R = 0.8991; SD = 0.47; n = 9; F = 29.5$$

In turn, if we analyze the solubility in buffer solution pH 7.4, we can disclose a relationship between these values and the descriptor characterizing the sum of the acceptor ability of the atoms in a molecule to form hydrogen bonds ($\Sigma(C_a)$). The results are given in Figure 10. As in the previous case, the experimental point for compound (5) deviates significantly from the observed dependence, which may be due to the amorphous state of the substance in the bottom phase (Figure 11), which leads to an overestimated solubility as compared to if it would be in the crystalline state. If we neglect of this value, then the correlation equation can be represented as follows:

$$\log(S_2^{pH7.4}) = (2.00 \pm 0.84) - (0.726 \pm 0.136) \cdot \Sigma(C_a) \qquad (3)$$

$$R = 0.8839;\ SD = 0.50;\ n = 10;\ F = 28.6$$

Analysis of Equations (2) and (3) shows that in buffer solution at pH 2.0, the solubility values are sensitive both to the donor ability of the substance to form hydrogen bonds (specific interactions) and to the polarizability of the molecule (nonspecific interactions). In contrast, in buffer pH 7.4 the dependence on the polarizability disappears, but the sensitivity to acceptor ability becomes evident. Apparently, this is due to different ionization states of the studied molecules in the solutions. The derived correlation dependences make it possible to predict the solubility of the studied class of compounds based only on their structural formulas. This fact greatly simplifies the task of designing the compounds with the improved solubility and enhanced bioavailability.

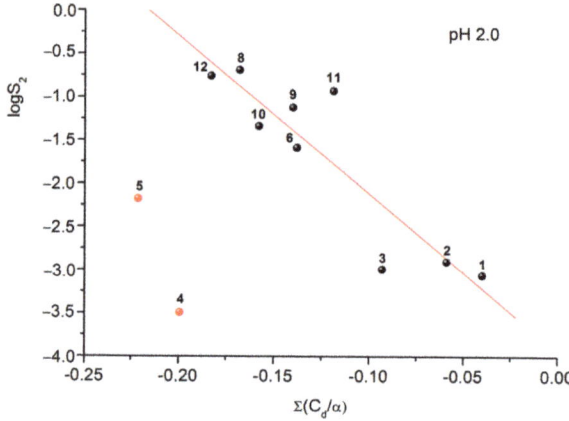

Figure 9. Dependence of $\log(S_2)$ versus $\Sigma(C_d/\alpha)$ descriptor. (compound numbering corresponds to Figures 2 and 7).

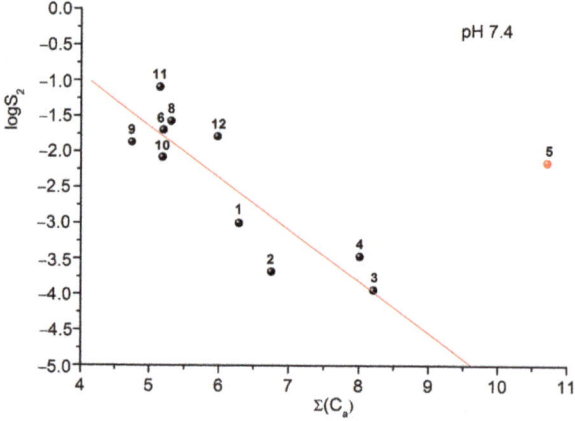

Figure 10. Dependence of $\log(S_2^{pH7.4})$ versus $\Sigma(C_d/\alpha)$ descriptor. (compound numbering corresponds to Figures 2 and 7).

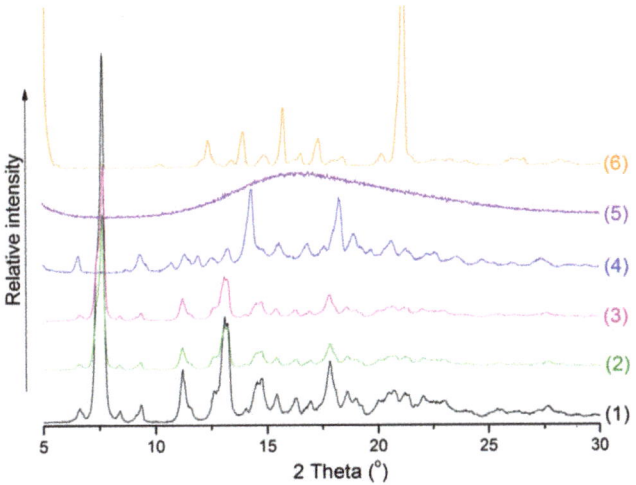

Figure 11. PXRD patterns of compounds (**1**–**6**). (compounds numbering corresponds to Figures 2 and 7).

3. Materials and Methods

3.1. Materials

3.1.1. Chemicals

Unless otherwise stated, the starting materials, reagents, and solvents were obtained from commercial purchase and used as supplied without further purification. Analytical thin-layer chromatography (TLC) was run on Merck silica gel 60 F-254, with detection by UV light (l j 254 nm). Memantine, amino acids, peptides, TBTU coupling regent, triethylamine and all necessary solvents for the synthesis were purchased from Sigma Aldrich. 2-Benzhydrylsulphinylacetic acid, (2-Oxopyrrolidin-1-yl)acetic Acid and 4-(Nicotinamido)butanoic acid were purchased from Shanghai Ruifu Chemical (Shanghai, China).

Identification of compounds

1H and 13C spectra were recorded on Bruker Avance II+ spectrometer (14.09 T magnet), operating at 600.11 MHz 1H frequencies, equipped with 5 mm BBO probe with z-gradient coil. The temperature is maintained at 293 K, using Bruker B-VT 3000 temperature unit with airflow of 535 L/h. All chemical shifts are reported in parts per million (ppm), referenced against teramethylsylane (TMS, 0.00 ppm) or using the residual solvent signal (7.27 ppm for CDCL3 of 2.5 ppm for DMSO). Electrospray mass spectrometry (ESI-MS) experiments were acquired on Bruker Compact QTOF-MS (Bruker Daltonics, Bremen, Germany) and controlled by the Compass 1.9 Control software. The data analysis was performed and the mono-isotopic mass values were calculated using Data analysis software v 4.4 (Bruker Daltonics, Bremen, Germany). The analyses were conducted in the positive ion mode at a scan range from m/z 50 to 1000, and nitrogen was used as nebulizer gas at a pressure of 4 psi and flow of 3 L/min^{-1} for the dry gas. The capillary voltage and temperature were set at 4500 V and 493 K, respectively. An external calibration for mass accuracy was carried out by using of sodium formate calibration solution. The precursor ion of each compound was selected, and ESI–MS/MS analysis was performed by collision-induced dissociation (CID); nitrogen was the collision gas, and the collision energy varied from 5 to 40 eV. MSn experiments were conducted on an ion trap instrument Esquire 3000 (Bruker Daltonics, Bremen, Germany) and controlled by the Esquire Control 5.3.11 software. ESI-MS data were collected in positive-ion mode at a scan range from m/z 50 to 500. In all ESI-MS measurements, the nebulizer gas pressure was 124.1 kPa at a flow rate of 5 L min^{-1}; the desolvation temperature was 573 K and capillary voltage was adjusted to 4000 V. The sample solutions were delivered to nebulizer by a syringe pump (Cole Parmer, Vernon Hills, IL, USA) at a flow rate 3 µL min^{-1}.

3.1.2. Solvents

Phosphate buffer pH 7.4 (I = 0.15 mol·L^{-1}) contained KHPO$_4$ (9.1 g in 1 L) and NaH$_2$PO$_4$·12H$_2$O (23.6 g in 1 L). Buffer solution pH 2.0 (I = 0.10 mol·L^{-1}) was prepared as follows: 6.57 g KCl was dissolved in water, and 119.0 mL of 0.1 mol·L^{-1} hydrochloric acid added. The volume was adjusted to 1 L with water. The pH value was controlled with a pH meter (Five GoTM F2, Mettler Toledo) using pH 4 and pH 7 standards. Bidistilled water (electrical conductivity equal to 2.1 µS cm^{-1}) was used throughout the experiments. The reagents were used as received without purification.

3.1.3. Cell Culture and Treatment

SH-SY5Y cells (ATCC, Manassas, VA, USA) were cultured in the Dulbecco's Modified Eagle Medium (DMEM) supplemented with 10% fetal bovine serum (FBS; Gibco/Invitrogen, Grand Island, NY, USA). SH-SY5Y cells transfected with the Swedish mutant form of human APP (referred to as "APPswe cells") is an established cell model of AD [35] in which copper triggers Aβ overproduction, cultured in Dulbecco's Modified Eagle Medium/Nutrient Mixture F-12 (DMEM/F12) supplemented with 2 mM L-glutamine, 10% fetal bovine serum (FBS; Gibco/Invitrogen, Grand Island, NY, USA), and 400 µM G418 (Sigma Chemical Company, St. Louis, MO, USA).

3.2. Chemical Synthesis

Amide bond formation for all seven compounds was realized according to method described by Knorr et al. [31]. Memantine (8 mmol) was dissolved in 5 mL DCM, TEA (8 mmol) was added to the solution. In separate flask the carboxyl component (8.8 mmol) was dissolved in 10 mL DCM. To the mixture was added TEA (8.8 mmol) and TBTU (12.3 mmol). Both solutions were mixed together after 30 min. After 24 h the reaction mixture was washed by 5% NaHCO$_3$ solution twice, then dried with anhydrous Na$_2$SO$_4$. Sarcosine and the peptides necessitated Boc-protection group removing (DCM/TFA 50:50 for 3 h) before any further purification. The obtained compound was purified by flash chromatography—system TCM/MeOH 97:3 or Hexan/EtOAc 5:4 ratio, depending of the polarity of the compounds.

Modafinil-memantine (1); ^{1}H NMR (500.17 MHz, DMSO-d6), δ, ppm: 0.58 (s, 6H, 2CH3), 0.88(s, 2H, AdmH), 1.00–1.15 (m, 4H, AdmH), 1.40 (m, 4H, AdmH), 1.59 (m, 2H, AdmH), 1.84 (q, 1H, AdmH), 3.02–3.27 (m, 2H, CH2, ModH), 5.15 (s, CH, ModH), 7.11–7.18 (m, 2H, ModH), 7.18–7.24 (m, 4H, ModH), 7.34–7.39 (m, 4H, ModH), 7.54 (s, NH, ModH).

^{13}C NMR (125.77 MHz, DMSO-d6), δ, ppm: 29.14 (2CH3), 29.45 (CH, AdmC), 31.78 (C, AdmC),34.39(C, AdmC), 39.42 (CH2, AdmC), 42.31 (2CH2, AdmC), 46.93 (2CH2, AdmC), 50.22 (C, AdmC), 53.17 (C, AdmC), 56.79 (CH2, ModH), 69.25 (CH, ModH), 128.32 (6C, para + orto, ModH), 129.89 (4C, meta, ModH), 134.19–138-42 (2C, ModH), 161.86 (C=O, ModH). Yield: 65%.

ESI-MS; *m/z* 436.63 [M + H]+.

Piracetam-memantine (2); ^{1}H NMR (500.17 MHz, DMSO-d6), δ, ppm: 0.58 (s, 6H, 2CH3), 0.88(s, 2H, AdmH), 1.00–1.15 (m, 4H, AdmH), 1.41 (m, 4H, AdmH), 1.59 (m, 2H, AdmH), 1.75 (q, 2H, ring, PiracH), 1.84 (q, 1H, AdmH), 2.02 (t, 2H, ring, PiracH), 3.21 (m, 2H, ring, PiracH), 3.62 (s, 2H, ring, PiracH), 7.17 (s, NH, PiracH).

^{13}C NMR (125.77 MHz, DMSO-d6), δ, ppm: 17.49 (CH2, ring, PiracC), 29.34 (2CH3, AdmC), 29.90 (CH, AdmC), 29.90 (CH2, ring, PiracC), 31.81 (C, AdmC), 34.39(C, AdmC), 39.60 (CH2, AdmC), 42.31 (2CH2, AdmC), 45.26 (CH2, PiracC), 47.09 (CH2, ring, PiracC), 47.09 (2CH2, AdmC), 50.28 (CH2, AdmC), 52.64 (C, AdmC), 161.86 (C=O, PiracC), 174.26 (C=O, ring, PiracC). Yield: 60%.

ESI-MS; *m/z* 305.44 [M + H]+.

Picamilon-memantine (3); ^{1}H NMR (500.17 MHz, DMSO-d6), δ, ppm: 0.57 (s, 6H, 2CH3), 0.87(s, 2H, AdmH), 1.00–1.15 (m, 4H, AdmH), 1.41 (m, 4H, AdmH), 1.59 (m, 2H, AdmH), 1.63 (m, 2H, CH2, PicaH), 1.83 (q, 1H, AdmH), 1.98 (t, 2H, CH2, PicaH), 3.17 (qt, 2H, CH2, PicaH), 7.11 (s, NH, MemH), 7.29 (qt, 1H, ring, PicaH), 7.79 (s, NH, MemH), 8.05 (qt, 1H, ring, PicaH), 8.50 (m, 1H, ring, PicaH), 8.89 (m, 1H, ring, PicaH).

^{13}C NMR (125.77 MHz, DMSO-d6), δ, ppm: 25.48 (CH2, PicaC), 29.27 (2CH3, AdC), 29.60 (CH, AdmC), 31.35 (C, AdmC), 34.35 (CH2, PicaC), 34.39(C, AdmC), 39.41 (CH2, AdmC), 39.41 (CH2, PicaC), 42.42 (2CH2, AdmC), 47.12 (2CH2, AdmC), 50.32 (CH2, AdmC), 52.30 (C, AdmC), 123.20 (CH, ring, PicaC), 130.26 (C, ring, PicaC), 134.59 (CH, ring, PicaC), 148.45 (CH, ring, PicaC), 151.67 (CH, ring, PicaC), 161.85 (C=O, PicaC), 171.44 (C=O, PicaC). Yield: 58%.

ESI-MS; m/z 370.51 [M + H]+.

Glycyl-glycine-memantine (4); ^{1}H NMR (500.17 MHz, DMSO-d6), δ, ppm: 0.81 (s, 6H, 2CH3), 1.10 (s, 2H, AdmH), 1.20–1.34 (m, 4H, AdmH), 1.56 (q, 4H, AdmH), 1.75 (m, 2H, AdmH), 2.07 (m, 1H, AdmH), 3.37 (2H, NH2), 3.71 (2H, CH2), 3.84 (2H, CH2), 7.48 (2H, NH2), 8.46 (1H, NH).

^{13}C NMR (125.77 MHz, DMSO-d6), δ, ppm: 29.96 (CH, AdmC), 30.58 (2CH3, AdmC), 32.30 (C, AdmC), 32.31 (C, AdmC), 40.06 (CH2, AdmC), 42.77 (CH2), 42.95 (CH2), 43.01 (2CH2, AdmC), 47.54 (2CH2, AdmC), 50.71 (CH2, AdmC), 52.91 (C, AdmC), 166.87 (C=O), 167.59 (C=O). Yield: 76%.

ESI-MS; m/z 294.41 [M + H]+.

Glycyl-glycyl-glycine-Memantine (5); ^{1}H NMR (500.17 MHz, DMSO-d6), δ, ppm: 0.81 (s, 6H, 2CH3), 1.10 (s, 2H, AdmH), 1.20–1.34 (m, 4H, AdmH), 1.36–1.48 (CH2), 1.56 (q, 4H, AdmH), 1.75 (m, 2H, AdmH), 2.07 (m, 1H, AdmH), 3.36 (1H, NH) 3.62 (2H, CH2), 3.82 (2H, CH2), 7.32 (1H, NH), 7.42 (2H, NH2), 8.66 (1H, NH).

^{13}C NMR (125.77 MHz, DMSO-d6), δ, ppm: 30.12 (CH, AdmC), 30.54 (2CH3, AdmC), 32.39 (C, AdmC), 32.41 (C, AdmC), 39.90 (CH2, AdmC), 42.00 (CH2), 42.70 (CH2), 43.00 (2CH2, AdmC), 46.50 (CH2), 47.60 (2CH2, AdmC), 50.68 (CH2, AdmC), 52.86 (C, AdmC), 166.84 (C=O), 167.97 (C=O), 168.86 (C=O). Yield: 61%.

ESI-MS; m/z 351.46 [M + H]+.

Methylglycine-memantine (6); ^{1}H NMR (500.17 MHz, DMSO-d6), δ, ppm: 0.83 (s, 6H, 2CH3), 1.12 (s, 2H, AdmH), 1.22–1.36 (m, 4H, AdmH), 1.58 (q, 4H, AdmH), 1.78 (m, 2H, AdmH), 2.09 (m, 1H, AdmH), 2.54 (3H, CH3), 3.35 (1H, NH), 3.59 (2H, CH2), 8.67 (1H, NH)

^{13}C NMR (125.77 MHz, DMSO-d6), δ, ppm: 29.94 (CH, AdmC), 30.48 (2CH3, AdmC), 32.31 (C, AdmC), 32.32 (C, AdmC), 33.06 (CH3), 39.58 (CH2, AdmC), 42.63 (2CH2, AdmC), 47.38 (2CH2, AdmC), 49.84 (CH2) 50.53 (CH2, AdmC), 53.43 (C, AdmC), 164.47 (C=O). Yield: 86%.

ESI-MS; m/z 251.39 [M + H]+.

Dimethylglycine-memantine (7); ^{1}H NMR (500.17 MHz, DMSO-d6), δ, ppm: 0.82 (s, 6H, 2CH3), (s, 2H, AdmH), 1.21–1.35 (m, 4H, AdmH), 1.63 (q, 4H, AdmH), 1.95 (m, 2H, AdmH)), 2.03 (m, 1H, AdmH), 2.78 (3H, CH3), 3.82 (2H, CH3), 8.27.

^{13}C NMR (125.77 MHz, DMSO-d6), δ, ppm: 29.20 (CH, AdmC), 30.49 (2CH3, AdmC), 32.33 (C, AdmC), 32.31 (C, AdmC), 36.24 (CH3), 41.26 (CH2, AdmC), 43.37 (2CH2, AdmC), 47.39 (2CH2, AdmC), 49.83 (CH2) 52.12 (CH2, AdmC), 58.25 (C, AdmC), 163.56 (C=O). Yield: 82%.

ESI-MS; m/z 265.41 [M + H]+.

3.3. Measurement of Neuroprotective Effects against Copper-Induced Toxicity in APPswe Cells

APPswe cells were seeded into 96-well plates at a density of 5×10^4 cells/well. APPswe cells were divided into three groups: control group (DMEM/F12 medium), copper-injured group (300 μM copper; model group), and copper-injured groups treated with compounds, in which cells were treated with 0.032 μM, 0.16 μM, 0.8 μM, 4 μM, 20 μM, and 100 μM tested compounds. The cell viability of APPswe cells was measured at 36 h incubation by MTS assay. The test was made and repeated four times.

3.4. Measurement of Neuroprotective Effects against Glutamate-Induced Toxicity in SH-SY5Y Cells

SH-SY5Y cells were seeded into 96-well plates at a density of 5×10^4 cells/well. Cells were divided into three groups: control group (DMEM medium), glutamate-injured group (7 mM glutamate; model group), and glutamate-injured groups treated with compounds, in which model groups were treated with 0.032 µM, 0.16 µM, 0.8 µM, 4 µM, 20 µM, and 100 µM tested compounds. The cell viability of SH-SY5Y cells was measured at 48 h incubation by MTS assay. The test was made and repeated four times.

3.5. Cell Viability Assay

Cell viability was assessed using an MTS assay and measured with a Spark 20M multimode microplate reader (Tecan Group Ltd., Mannedorf, Switzerland) at a wavelength of 490 nm. Cell viability = (OD compounds − OD blank)/(OD control − OD blank) × 100% (OD compounds: absorbance value of tested compounds group; OD control: absorbance value of control group; OD blank: absorbance value of culture medium without cells). Cell protection rate = (cell viability of compounds treatment groups—cell viability of model group)/cell viability of model group × 100%.

3.6. Statistical Analysis

Data represent means ± standard error of the mean (SEM). Data were analyzed using a one-way ANOVA. p-values < 0.05 were considered statistically significant. Analysis was performed using GraphPad Prism Version 8.0 software (GraphPad Prism Software; La Jolla, CA, USA).

3.7. Solubility Determination

The standard shake-flask isothermal saturation method [36] was applied to the solubility experiments. The saturated solutions of the studied compounds were prepared in the glass screw-capped vials using the respective buffer solution (pH 2.0 or pH 7.4) which was vigorously stirred at a constant temperature of 298.15 K. The time required for reaching the equilibrium between the solute and the solvent was estimated with the help of the solubility kinetic dependences and was stated as no less than 24 h. After this time period, the saturated solutions stayed in the thermostat for no less than 5 h and then the solution was separated from the solid residual via centrifugation (Biofuge pico, Thermo Electron LED GmbH, Germany) at 298.15 K for 20 min at 10,000 rpm. The concentrations of the compounds in the saturated solutions were measured with the help of a spectrophotometer (Cary-50, Varian, USA) and using the calibration curves for each compound in each buffer at the appropriate wavelength. The experimental results were reported as an average of at least three replicated experiments.

3.8. Powder X-ray Diffraction (PXRD) Were Recorded According to SUROV et al. [37]

The powder XRD data of the bulk materials were recorded under ambient conditions on a D2 Phaser (Bragg-Brentano) diffractometer (Bruker AXS, Germany) with a copper X-ray source ($\lambda CuK\alpha 1$ = 1.5406 Å) and a high-resolution position-sensitive LYNXEYE XE T detector. The samples were placed into the plate sample holders and rotated at a speed of 15 rpm during the data acquisition.

3.9. QSAR Modelling and Calculation of HYBOT Descriptors

QSAR models were constructed by partial least square regression (PLS) fitting (internal method of validation). Physicochemical HYBOT (Hydrogen Bond Thermodynamics) descriptors used as independent variables were calculated by the commercially available software HYBOT [32].

4. Conclusions

In this study, four memantine analogues of glycine derivatives, including glycyl-glycine, glycyl-glycyl-glycine, sarcosine, dimethylglycine and three conjugates with nootropics, modafinil, piracetam and picamilon which had favorable neuroprotective effects were synthesized for the first time.

The solubility of memantine analogues with nootropics and glycine derivatives in buffer solutions at pH 2.0 and pH 7.4 simulating the biological media at 298.15 K was determined. For some compounds the solubility was shown to be dependent on the pH of the aqueous solution as a result of the protonation processes in the acidic medium. Analysis of the mutual influence of the structural fragments in the molecules on the solubility behavior revealed that the addition of the hydrophilic fragments (substances (4)—glycyl-glycine-memantine and (5)—glycyl-glycyl-glycine-memantine) enhances the solubility, whereas the introduction of the bulky aromatic rings (compounds (2)—piracetam-memantine and (3)—picamilon-memantine) hampers the dissolution process in both media.

In order to disclose the structure-property correlations in the frame of the studied memantine analogues the physicochemical structural HYBOT descriptors were used. As a result, the significative correlation equations were derived. The equations demonstrated the sensitivity of the solubility in pH 2.0 to the donor ability of atoms in a molecule to form hydrogen bonds (specific interactions) and polarizability (nonspecific interactions), whereas, in pH 7.4 a determinative role of the sum of the acceptor ability to form hydrogen bonds on the solubility was shown. The derived correlation dependences make it possible to predict the solubility of the studied class of compounds based only on their structural formulas.

In the framework of the QSAR approach, the HYBOT descriptors were applied to predict the biological properties (EC_{50}) on the class of the memantine derivatives. The best correlation was obtained with the descriptor characterizing the sum of the donor and acceptor ability to form hydrogen bonds normalized to molecular polarizability allowing for both specific and nonspecific interactions of the studied molecules with the biological environment. The equation makes possible the prediction of the EC_{50} values based only on the structural formula, without the expensive biological experiments.

The correlation equations for the solubility and biological properties obtained in this study greatly simplifies the task of the directed design of the memantine analogues with improved solubility and enhanced bioavailability.

The new structural derivatives of memantine were investigated to improve cell viability against copper-induced neurotoxicity in APPswe cells and against glutamate-induced neurotoxicity in SH-SY5Y cells.

Seven novel compounds improved the cell viability of copper-damaged APPswe cells with different effects. Nootropics analogues of memantine (1–3), methyl-glycine (6) and dimethylglycine-memantine (7) are a good EC_{50} value, especially the modafinil-memantine (EC_{50} =1.120 ± 0.398 µM). The nootropics analogues of memantine—modafinil-memantine (1) and picamilon-memantine (3) showed cellular cytotoxicity that was similar to the positive drug memantine at 100 µM.

Studies against glutamate-induced neurotoxicity in SH-SY5Y cells showed that piracetam-memantine (2) and glycyl-glycine-memantine (4) treated at 0.032 µM significantly improved cell viability of SH-SY5Y cells. These two compounds have a similar neuroprotective effect to the positive drug memantine.

It was revealed that the highest activity combined with low cytotoxicity was demonstrated by the nootropic derivatives of memantine and derivatives of glycine with the simplest in structure—methyl-glycine and dimethylglycine-memantine.

Author Contributions: M.Š. analysis, M.Š.; neuroprotective effects, L.Z., Z.L. and R.L.; Synthesis, R.C.; Solubility experiments and QSAR, T.V. and G.P., writing—original draft, I.S., G.P. and R.L. All authors have read and agreed to the published version of the manuscript.

Funding: This research was funded by Bulgarian National Science Fund (BNSF), grant number KP-06-Russia/7-2019. This work was supported by the Russian Foundation for Basic Research for Russian-Bulgarian collaboration (project No. 19-53-18003).

Institutional Review Board Statement: Not applicable.

Informed Consent Statement: Not applicable.

Data Availability Statement: Data is contained within the article.

Conflicts of Interest: The authors declare no conflict of interest.

References

1. Czarnecka, K.; Chuchmacz, J.; Wójtowicz, P.; Szymański, P. Memantine in neurological disorders—Schizophrenia and depression. *J. Mol. Med.* **2021**, *99*, 327–334. [CrossRef] [PubMed]
2. Prince, M.; Albanese, E.; Guerchet, M.; Prina, M. *World Alzheimer Report 2014: Dementia and Risk Reduction: An Analysis of Protective and Modifiable Risk Factors*; Alzheimer's Disease International (ADI): London, UK, 2014.
3. Lane, C.A.; Hardy, J.; Schott, J.M. Alzheimer's disease. *Eur. J. Neurol.* **2018**, *25*, 59–70. [CrossRef] [PubMed]
4. Witt, A.; Macdonald, N.; Kirkpatrick, P. Memantine hydrochloride. *Nat. Rev. Drug Discov.* **2004**, *3*, 109–110. [CrossRef] [PubMed]
5. Altevogt, B.M.; Davis, M.; Pankevich, D.E. (Eds.) *Glutamate-Related Biomarkers in Drug Development for Disorders of the Nervous System: Workshop Summary*; National Academies Press: Washington, DC, USA, 2011.
6. Bechtholt-Gompf, A.J.; Walther, H.V.; Adams, M.A.; Carlezon, W.A.; Öngür, D.; Cohen, B.M. Blockade of astrocytic glutamate uptake in rats induces signs of anhedonia and impaired spatial memory. *Neuropsychopharmacology* **2010**, *35*, 2049–2059. [CrossRef] [PubMed]
7. Kang, H.S.; Kim, J.P. Butenolide derivatives from the fungus Aspergillus terreus and their radical scavenging activity and protective activity against glutamate-induced excitotoxicity. *Appl. Biol. Chem.* **2019**, *62*, 1–5. [CrossRef]
8. Atlante, A.; Calissano, P.; Bobba, A.; Giannattasio, S.; Marra, E.; Passarella, S. Glutamate neurotoxicity, oxidative stress and mitochondria. *FEBS Lett.* **2001**, *497*, 1–5. [CrossRef]
9. Stankova, I.; Stoilkova, A.; Chayrov, R.; Tsvetanova, E.; Georgieva, A.; Alexandrova, A. In Vitro Antioxidant Activity of Memantine Derivatives Containing Amino Acids. *Pharm. Chem. J.* **2020**, *54*, 268–272. [CrossRef]
10. Bai, K.; Jiang, L.; Zhu, S.; Feng, C.; Zhao, Y.; Zhang, L.; Wang, T. Dimethylglycine sodium salt protects against oxidative damage and mitochondrial dysfunction in the small intestines of mice. *Int. J. Mol. Med.* **2019**, *43*, 2199–2211. [CrossRef]
11. Hariganesh, K.; Prathiba, J. Effect of dimethylglycine on gastric ulcers in rats. *J. Pharm. Pharmacol.* **2000**, *52*, 1519–1522. [CrossRef] [PubMed]
12. Bai, K.; Xu, W.; Zhang, J.; Kou, T.; Niu, Y.; Wan, X.; Wang, T. Assessment of free radical scavenging activity of dimethylglycine sodium salt and its role in providing protection against lipopolysaccharide-induced oxidative stress in mice. *PLoS ONE* **2016**, *11*, e0155393. [CrossRef]
13. Friesen, R.W.; Novak, E.M.; Hasman, D.; Innis, S.M. Relationship of dimethylglycine, choline, and betaine with oxoproline in plasma of pregnant women and their newborn infants. *J. Nutr.* **2007**, *137*, 2641–2646. [CrossRef] [PubMed]
14. Clapes, P.; Rosa Infante, M. Amino acid-based surfactants: Enzymatic synthesis, properties and potential applications. *Biocatal. Biotransformation* **2002**, *20*, 215–233. [CrossRef]
15. Curtis, D. A possible role for sarcosine in the management of schizophrenia. *Br. J. Psychiatry* **2019**, *215*, 697–698. [CrossRef]
16. Chang, C.H.; Lin, C.H.; Liu, C.Y.; Chen, S.J.; Lane, H.Y. Efficacy and cognitive effect of sarcosine (N-methylglycine) in patients with schizophrenia: A systematic review and meta-analysis of double-blind randomised controlled trials. *J. Psychopharmacol.* **2020**, *34*, 495–505. [CrossRef] [PubMed]
17. Colucci, L.; Bosco, M.; Ziello, A.R.; Rea, R.; Amenta, F.; Fasanaro, A.M. Effectiveness of nootropic drugs with cholinergic activity in treatment of cognitive deficit: A review. *J. Exp. Pharmacol.* **2012**, *4*, 163. [CrossRef]
18. Wisor, J.P. Modafinil as a catecholaminergic agent: Empirical evidence and unanswered questions. *Front. Neurol.* **2013**, *4*, 139. [CrossRef]
19. Volkow, N.D.; Fowler, J.S.; Logan, J.; Alexoff, D.; Zhu, W.; Telang, F.; Apelskog-Torres, K. Effects of modafinil on dopamine and dopamine transporters in the male human brain: Clinical implications. *JAMA* **2009**, *301*, 1148–1154. [CrossRef]
20. Abbasi, Y.; Shabani, R.; Mousavizadeh, K.; Soleimani, M.; Mehdizadeh, M. Neuroprotective effect of ethanol and Modafinil on focal cerebral ischemia in rats. *Metab. Brain Dis.* **2019**, *34*, 805–819. [CrossRef]
21. Malykh, A.G.; Sadaie, M.R. Piracetam and piracetam-like drugs. *Drugs* **2010**, *70*, 287–312. [CrossRef] [PubMed]
22. Fesenko, U.A. Piracetam improves children's memory after general anaesthesia. *Anestezjol. Intensywna Ter.* **2009**, *41*, 16–21.
23. Holinski, S.; Claus, B.; Alaaraj, N.; Dohmen, P.M.; Kirilova, K.; Neumann, K.; Konertz, W. Cerebroprotective effect of piracetam in patients undergoing coronary bypass burgery. *Med. Sci. Monit.* **2008**, *14*, PI53–PI57. [PubMed]
24. Nickolson, V.J.; Wolthuis, O.L. Effect of the acquisition-enhancing drug ptracetam on rat cerebral energy metabolism. Comparison with naftidrofuryl and methamphetamine. *Biochem. Pharmacol.* **1976**, *25*, 2241–2244. [CrossRef]
25. Grau, M.; Montero, J.L.; Balasch, J. Effect of Piracetam on electrocorticogram and local cerebral glucose utilization in the rat. *Gen. Pharmacol. Vasc. Syst.* **1987**, *18*, 205–211. [CrossRef]

26. Mirzoian, R.S.; Gan'shina, T.S.; Kim, G.A.; Kurza, E.V.; Maslennikov, D.V.; Il'ya, N.K.; Gorbunov, A.A. The translational potential of experimental pharmacology for cerebrovascular disorders. *Ann. Clin. Exp. Neurol.* **2019**, *13*, 34–40.
27. Mishchenko, O.; Palagina, N. Experimental research of cerebroprotective activity of the new 4-aminobutatanoic acid derivative. *EUREKA Health Sci.* **2021**, *3*, 95–100. [CrossRef]
28. Tabassum, N.; Rasool, S.; Malik, Z.A.; Ahmad, F. Natural cognitive enhancers. *J. Pharm. Res.* **2012**, *5*, 153–160.
29. Rosini, M.; Simoni, E.; Caporaso, R.; Basagni, F.; Catanzaro, M.; Abu, I.F.; Fagiani, F.; Fusco, F.; Masuzzo, S.; Albani, D.; et al. Merging memantine and ferulic acid to probe connections between NMDA receptors, oxidative stress and amyloid-β peptide in Alzheimer's disease. *Eur. J. Med. Chem.* **2019**, *180*, 111–120. [CrossRef]
30. Turcu, A.; Companys-Alemany, J.; Phillips, M.B.; Patel, D.S.; Griñán-Ferré, C.; Loza, M.I.; Brea, J.M.; Pérez, B.; Soto, D.; Sureda, F.X.; et al. Design, synthesis, and in vitro and in vivo characterization of new memantine analogs for Alzheimer's disease. *Eur. J. Med. Chem.* **2022**, *236*, 114354–114382. [CrossRef]
31. Glomme, A.; März, J.; Dressman, J.B. Comparison of a miniaturized shake-flask solubility method with automated potentiometric acid/base titrations and calculated solubilities. *J. Pharm. Sci.* **2005**, *94*, 1–16. [CrossRef]
32. Tencheva, A.; Liu, R.; Volkova, T.V.; Chayrov, R.; Mitrev, Y.; Štícha, M.; Li, Y.; Jiang, H.; Li, Z.; Stankova, I.; et al. Synthetic analogues of memantine as neuroprotective and influenza viral inhibitors: In vitro and physicochemical studies. *Amino Acids* **2020**, *52*, 1559–1580. [CrossRef]
33. Knorr, R.; Trzeciak, A.; Bannwarth, W.; Gillessen, D. New coupling reagents in peptide chemistry. *Tetrahedron Lett.* **1989**, *30*, 1927–1930. [CrossRef]
34. Higuchi, T.; Connors, K. Phase-solubility techniques. *Adv. Anal. Chem. Instrum.* **1965**, *4*, 117–123.
35. Zeng, L.; Jiang, H.; Ashraf, G.M.; Liu, J.; Wang, L.; Zhao, K.; Liu, M.; Li, Z.; Liu, R. Implications of miR-148a-3p/p35/PTEN signaling in tau hyperphosphorylation and autoregulatory feedforward of Akt/CREB in Alzheimerés disease. *Mol. Ther. Nucleic Acids* **2021**, *27*, 256–275. [CrossRef]
36. Raevsky, O.A.; Grigor'ev, V.J.; Trepalin, S.V. HYBOT Program Package. Registration by Russian State Patent Agency. No. 990090 26 February 1999.
37. Surov, A.O.; Volkova, T.V. Solubility/distribution thermodynamics and permeability of two anthelmintics in biologically relevant solvents. *J. Mol. Liq.* **2022**, *354*, 118835–118862. [CrossRef]

Article

C16 Peptide and Ang-1 Improve Functional Disability and Pathological Changes in an Alzheimer's Disease Model Associated with Vascular Dysfunction

Xiaoxiao Fu [1,†], Jing Wang [2,†], Huaying Cai [2], Hong Jiang [2] and Shu Han [1,*]

[1] Institute of Anatomy, Medical College, Zhejiang University, Hangzhou 310058, China; 21818569@zju.edu.cn
[2] Department of Neurology, Sir Run Run Shaw Hospital, Medical College, Zhejiang University, Hangzhou 310058, China; wangjinjoy@zju.edu.cn (J.W.); caihuaying2004@zju.edu.cn (H.C.); jianghong1975@zju.edu.cn (H.J.)
* Correspondence: han00shu@zju.edu.cn; Tel.: +86-571-8820-8318
† These authors contributed equally to this work.

Citation: Fu, X.; Wang, J.; Cai, H.; Jiang, H.; Han, S. C16 Peptide and Ang-1 Improve Functional Disability and Pathological Changes in an Alzheimer's Disease Model Associated with Vascular Dysfunction. *Pharmaceuticals* 2022, 15, 471. https://doi.org/10.3390/ph15040471

Academic Editors: Giovanni N. Roviello and Rosanna Palumbo

Received: 19 February 2022
Accepted: 6 April 2022
Published: 13 April 2022

Publisher's Note: MDPI stays neutral with regard to jurisdictional claims in published maps and institutional affiliations.

Copyright: © 2022 by the authors. Licensee MDPI, Basel, Switzerland. This article is an open access article distributed under the terms and conditions of the Creative Commons Attribution (CC BY) license (https://creativecommons.org/licenses/by/4.0/).

Abstract: Alzheimer's disease (AD) is a neurological disorder characterized by neuronal cell death, tau pathology, and excessive inflammatory responses. Several vascular risk factors contribute to damage of the blood–brain barrier (BBB), secondary leak-out of blood vessels, and infiltration of inflammatory cells, which aggravate the functional disability and pathological changes in AD. Growth factor angiopoietin-1 (Ang-1) can stabilize the endothelium and reduce endothelial permeability by binding to receptor tyrosine kinase 2 (Tie2). C16 peptide (KAFDITYVRLKF) selectively binds to integrin $\alpha v \beta 3$ and competitively inhibits leukocyte transmigration into the central nervous system by interfering with leukocyte ligands. In the present study, 45 male Sprague-Dawley (SD) rats were randomly divided into three groups: vehicle group, C16 peptide + Ang1 (C + A) group, and sham control group. The vehicle and C + A groups were subjected to two-vessel occlusion (2-VO) with artery ligation followed by Aβ1-42 injection into the hippocampus. The sham control group underwent sham surgery and injection with an equal amount of phosphate-buffered saline (PBS) instead of Aβ1-42. The C + A group was administered 1 mL of drug containing 2 mg of C16 and 400 μg of Ang-1 daily for 2 weeks. The sham control and vehicle groups were administered 1 mL of PBS for 2 weeks. Our results showed that treatment with Ang-1 plus C16 improved functional disability and reduced neuronal death by inhibiting inflammatory cell infiltration, protecting vascular endothelial cells, and maintaining BBB permeability. The results suggest that these compounds may be potential therapeutic agents for AD and warrant further investigation.

Keywords: Alzheimer's disease; cerebrovascular dysregulation; inflammation; C16; Ang-1

1. Introduction

The structural/functional integrity of blood vessels and adequate blood supply are key to maintaining normal brain function [1]. The endothelial monolayer forms the blood–brain barrier (BBB), which maintains cerebrovascular integrity and regulates the transport of most blood cells (e.g., leukocytes), microbial pathogens, and blood-derived macromolecules between the blood and the brain [1]. Neurovascular dysfunction is becoming increasingly recognized in a chronic disorder, Alzheimer's disease (AD); the vascular risk factors have been shown to promote cognitive decline by acting synergistically with amyloid-β (Aβ) [2–5]. Brain vascular dysfunction is also implicated in other neuro-inflammatory disorders [4], including multiple sclerosis (MS), stroke, Parkinson's disease (PD), amyotrophic lateral sclerosis (ALS), traumatic brain injury (TBI), and epilepsy, all of which are associated with the downregulation of tight junction proteins and the activation of endothelial cells [6].

Neuronal damage, a hallmark of neurodegenerative diseases, is associated with chronic activation of innate immune responses in the central nervous system (CNS). The development of novel therapeutic strategies that target pathogenic mechanisms responsible for detrimental inflammation and harness endogenous protective pathways is the key to counteracting neuro-inflammatory diseases [6].

Increased BBB permeability induces edema and inflammatory cell recruitment into target tissues, resulting in nerve demyelination and dysfunction. Growth factor Angiopoietin-1 (Ang-1) binds to the receptor tyrosine kinase 2 (Tie2) to mediate angiogenesis. The binding of Ang-1 to Tie2 stabilizes the endothelium and reduces endothelial permeability, and therefore, is essential to the development of the cardiovascular system [7]. Moreover, a previous study showed that Ang-1 ameliorated inflammation-induced vessel leakage in the CNS and inhibited inflammatory cell infiltration into the spinal cord and the brain in a model of acute experimental allergic encephalomyelitis (EAE) [8].

The accumulation of surface cell adhesion molecules in peripheral blood cells during the course of CNS inflammation is a prerequisite for their crossing of the BBB and their interaction with activated endothelial cells. Integrins are heterodimeric membrane proteins primarily expressed in leukocytes, and serve as adhesion receptors for multiple extracellular matrix components. The integrin $\alpha v \beta 3$ is upregulated in autoimmune T cells and has been reported to mediate leukocyte adhesion to the intercellular adhesion molecule-1 (ICAM-1) and facilitate leukocyte migration across the BBB [9]. The synthetic C16 peptide (KAFDITYVRLKF) represents a functional laminin domain that selectively binds to integrin $\alpha v \beta 3$ and competitively inhibits leukocyte transmigration into the CNS by interfering with leukocyte ligands [10]. Importantly, C16 is not an immunosuppressant and has no effect on the amount of systemic leukocytes [10].

We investigated the efficacy of a combined treatment of C16 and Ang-1 in a rat model of AD associated with vascular dysfunction. In previous study, we analyzed the gene expression profiles of patients with AD and vascular dementia (VaD) in the Gene Expression Omnibus (GEO) database and found that the expressions of the Ras homolog gene family member A (RhoA), ICAM-1, angiotensinogen (AGT), phosphatidylinositol 3/kinase-protein kinase B (PI3K/AKT), and signal transducer and activator of transcription 3 (STAT3) were upregulated in both patient groups compared with healthy controls. The potential mechanisms underlying the upregulation of these genes were also investigated. Our data provide new insights into targeting the inflammatory microenvironment of the CNS to treat neurodegenerative diseases.

2. Results

2.1. Treatment with C16 Plus Ang-1 Alleviated Memory Impairment in AD Rats with Vascular Dysfunction

The average escape time of vehicle rats was markedly shorter compared with sham controls ($p < 0.05$; Table 1). Rats treated with C16 plus Ang-1 showed significantly improved escape performance compared with the vehicle group (Table 1). Vehicle rats also failed to find the hidden platform during the acquisition phase ($p < 0.05$, Figure 1A,B), suggesting impaired spatial memory in this group. In the hidden platform task, vehicle rats swam more slowly and crossed the target quadrant fewer times compared to the C + A group ($p < 0.05$, Figure 1A,B). However, the combined treatment with C16 plus Ang-1 increased the time that animals spent in the goal quadrant.

Table 1. The escape latency of three experimental groups in the positioning navigation test.

	Sham Control	Vehicle	C + A
1 day	13.11 ± 2.15	101 ± 7.74 *	30.11 ± 3.82 *&
2 days	13.22 ± 3.23	74.11 ± 3.51 *	27.88 ± 5.34 *&
3 days	5.0 ± 1.73	43.1 ± 8.28 *	14.2 ± 1.56 *&
4 days	4.3 ± 1.8	47.78 ± 5.1 *	15 ± 2.18 *&
5 days	4.4 ± 1.74	47.67 ± 3.67 *	15.1 ± 1.76 *&
6 days	3.89 ± 1.16	44.89 ± 5.64 *	10.3 ± 2 *&

* $p < 0.05$ vs. sham control group, & $p < 0.05$ vs. vehicle group.

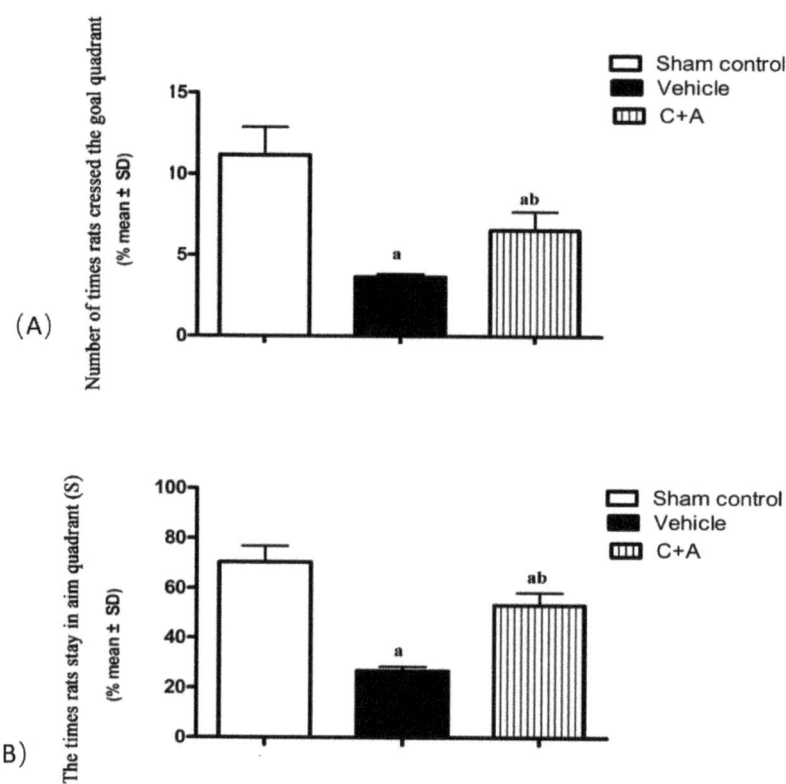

Figure 1. The results of the space exploration test showed that the number of times that rats crossed the platform quadrant (**A**) and the average time (s) that animals stayed in the platform quadrant (**B**) in the vehicle group were significantly decreased compared with the sham control. This phenomenon was effectively reversed by treatment with C16 plus Ang-1. a, $p < 0.05$ vs. sham control group; b, $p < 0.05$ vs. vehicle group.

2.2. Treatment with C16 Plus Ang-1 Suppressed Inflammation in the CNS

The expressions of macrophage-specific marker CD68 (Figure 2A,B), oxide stress-related factor Cox-2 (Figure 2C,D) and pro-inflammatory cytokine NF-κB (Figure 2E,F) were upregulated in the vehicle group, while treatment with C16 plus Ang-1 effectively suppressed inflammation-related gene expression (Figure 2A–F). ICAM-1, which enables leukocytes to effectively migrate across the endothelium, was upregulated in vehicle-treated animals, and this phenomenon was effectively reversed by treatment with C16 plus Ang-1 (Figure 2G,H). The results of immunofluorescence staining also confirmed the changes in the expression of CD68 (Supplementary Figure S1) and ICAM-1 (Figure 3). Moreover, compared to the sham control group, the concentration of pro-inflammatory cytokine IL-1β was significantly elevated in rats subjected to 2-VO operation and Aβ1-42 injection (Figure 4). C16 plus Ang-1, however, significantly inhibited the upregulation of IL-1β in AD rats (Figure 4).

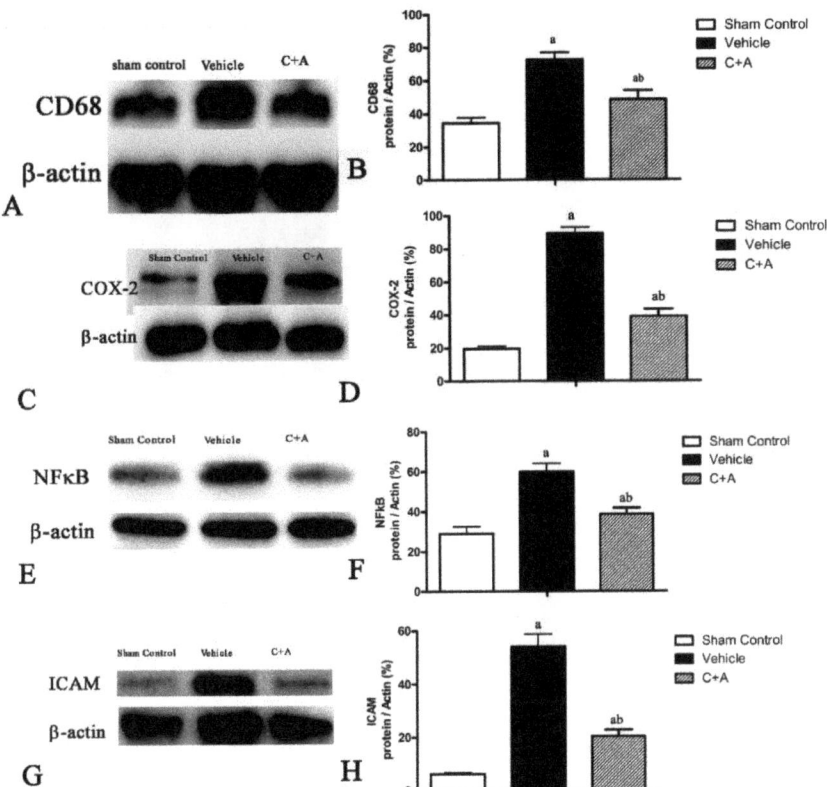

Figure 2. Western blot analysis of the expressions of (**A,B**) CD68/ED1, (**C,D**) cyclooxygenase-2 (COX-2), (**E,F**) nuclear factor kappa B (NF-κB), (**G,H**) intercellular cell adhesion molecule-1 (ICAM-1 in the sham control, vehicle, and C + A groups (**B,D,F,H**). Semi-quantitative profiles of protein bands showed that C16 plus Ang-1 suppressed the upregulation of CD68, COX-2, NF-κB, and ICAM-1 in the vehicle group. a, $p < 0.05$ vs. sham control group; b, $p < 0.05$ vs. vehicle group. The molecular weight (MW): CD68, 100 KD; COX-2, 75 KD; NF-κB, 65 KD; ICAM-1, 92 KD; β-actin, 42 KD.

2.3. Treatment with C16 Plus Ang-1 Reduced BBB Permeability and Blood Vessel Leakage

Evans blue dye extravasation is indicative of edema and compromised blood vessel integrity in tissues. The vehicle group showed severe vasculature leakage (Figure 5B,E), while the C + A group (Figure 5C,F) showed markedly less leakage from surrounding blood vessels (Figure 5J). Furthermore, ZO-1, a specific marker of tight junctions between endothelial cells in blood vessels, was downregulated in AD rats, but was restored by treatment with C16 plus Ang-1 (Figure 5G–I,K,L).

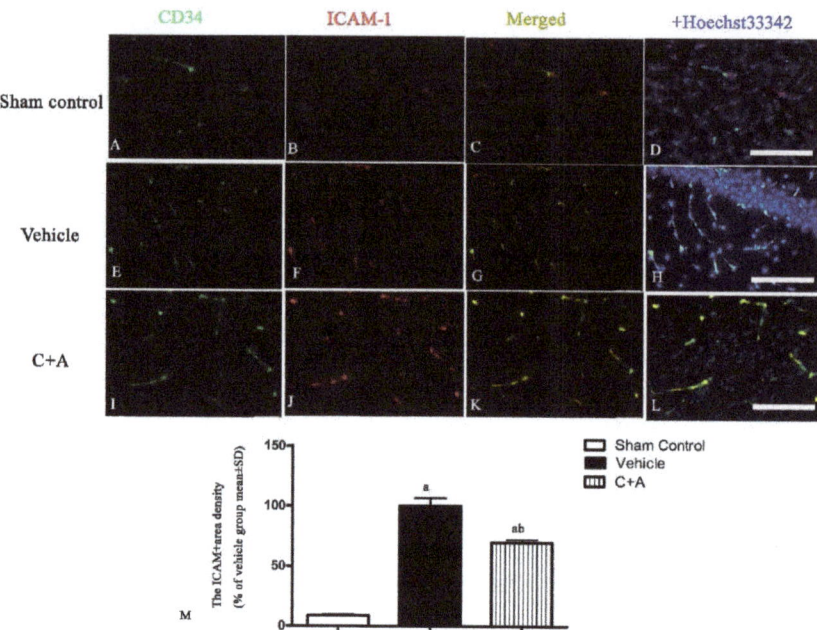

Figure 3. (**A–L**): Double immunofluorescence staining of microvessel marker CD34 (green) and ICAM-1 (red) in all groups. The nuclei were visualized by Hoechst 33342 staining (blue). (**M**) Quantitation of ICAM-1-labeled microvessels showed that treatment with C16 plus Ang-1 reduced the amount of ICAM-1-labeled vessels in AD rats. Scale bar = 100 µm. a, $p < 0.05$ vs. sham control group; b, $p < 0.05$ vs. vehicle group.

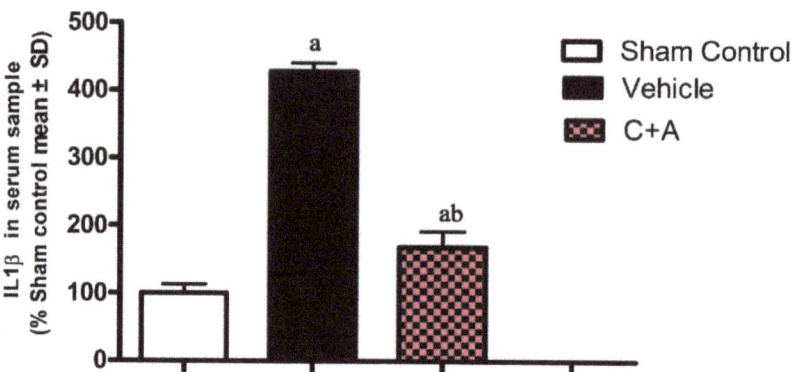

Figure 4. ELISA results showed that treatment with C16 plus Ang-1 decreased the serum level of pro-inflammatory factor IL-1β. a, $p < 0.05$ vs. sham control group; b, $p < 0.05$ vs. vehicle group.

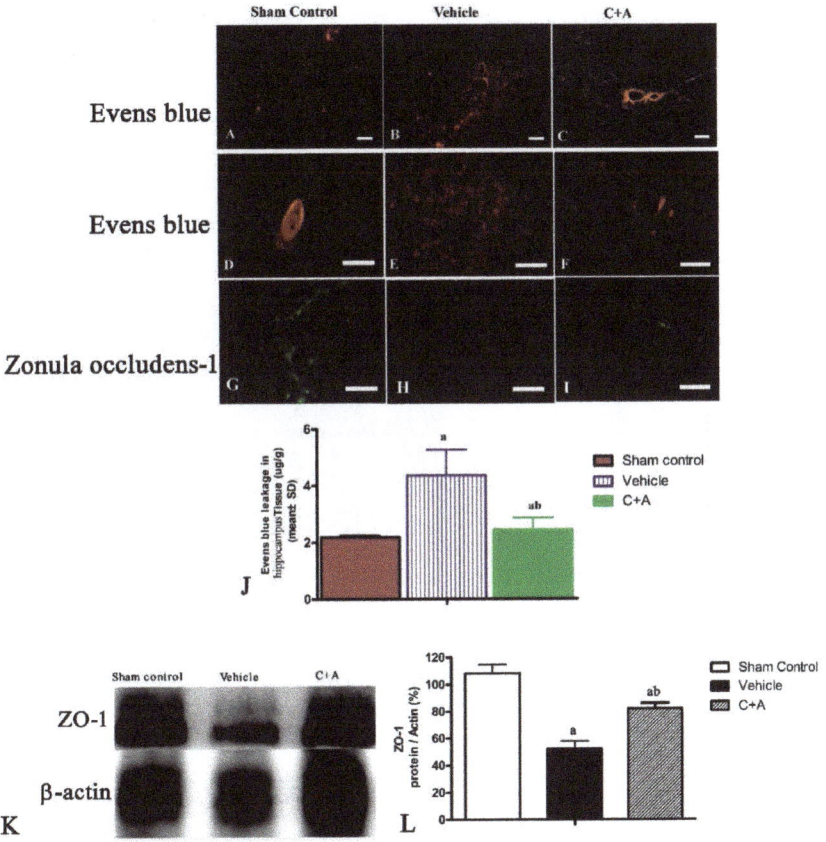

Figure 5. (**A–F**): Images of Evans blue (a dye to detect blood vessel leakage) staining (red) of the blood–brain barrier (BBB) in the (**A,D**) sham control, (**B,E**) vehicle, and (**C,F**) C + A groups. Increased blood vessel leakage in the vehicle group, as indicated by the amount of Evans blue dye that leaked out into surrounding blood vessels, was alleviated by the C + A treatment. (**G–I**): Immunofluorescence staining images (green) showed that ZO-1 (a marker of tight junctions) in micro-blood vessels was downregulated in the vehicle group, but was restored by C16 + Ang-1 treatment. Scale bar = 100 μm. (**J**) Semi-quantitative profiles of Evans blue staining. (**K**) Western blot analysis of the expression of ZO-1. (**L**) Semi-quantitative profiles of protein bands showed that C16 plus Ang-1 increased the expression of ZO-1 in the vehicle group. ZO-1, 240 KD; β-actin, 42 KD. a, $p < 0.05$ vs. sham control group; b, $p < 0.05$ vs. vehicle group. a, $p < 0.05$ vs. sham control group; b, $p < 0.05$ vs. vehicle group.

2.4. Treatment with C16 Plus Ang-1 Reduced Autophagy and Neuronal Apoptosis, Restored the Expression of ACH and CHAT, Alleviated Sn Loss in AD Rats

The injection of C16 plus Ang-1 decreased the cytoplasmic level of LC3BII, an autophagy marker, in both the cortex and hippocampus of AD rats (Figure 6). Moreover, the level of active caspase-3, an enzyme involved in mammalian cell apoptosis, as well as LC3BII, was increased in the vehicle group but was downregulated by the C16 + Ang-1 treatment (Figure 6A–L).

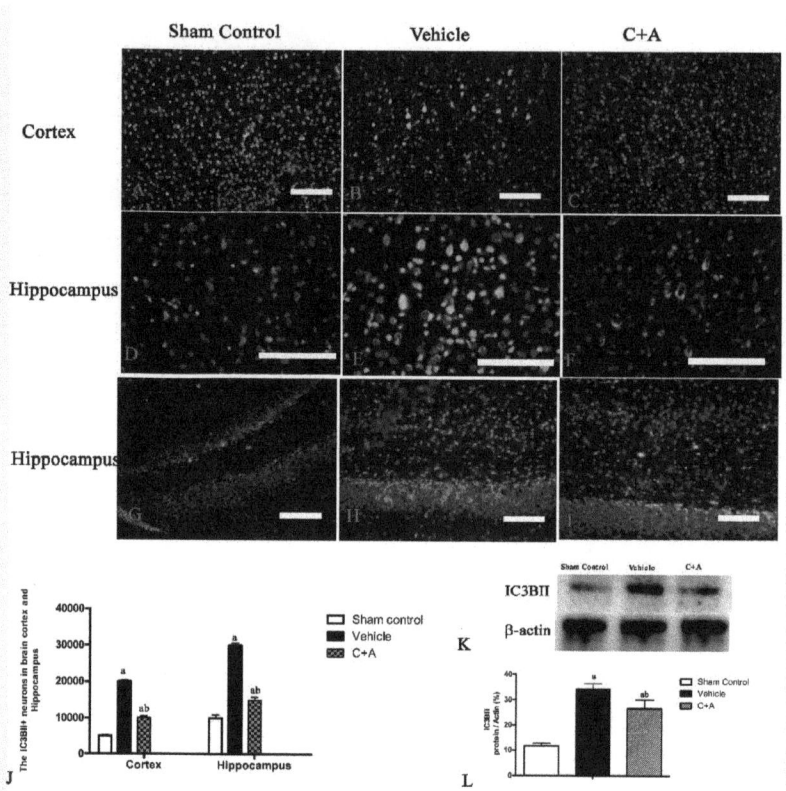

Figure 6. (A–I): Immunofluorescence staining (green) of LC3BII in the motor cortex and hippocampus. Scale bar = 100 μm. (J) The quantification of LC3BII-positive cells showed that treatment with C16 plus Ang-1 reduced the number of LC3BII-positive cells compared to the vehicle controls. (K) Western blot analysis of the expressions of autophagy marker LC3BII. (L) Semi-quantitative profiles of protein bands showed that C16 plus Ang-1 suppressed the upregulation of LC3BII in the vehicle group. LC3BII, 14.16 KD; β-actin, 42 KD. a, $p < 0.05$ vs. sham control group; b, $p < 0.05$ vs. vehicle group.

The ultrastructural morphology of the hippocampus and the cortex of the sham control (Figure 7A–D), vehicle (Figure 7E–H), and C + A (Figure 7I–L) groups were examined using transmission electron microscopy. The sham control group showed neuronal nuclei with uncondensed chromatin and intact tight junctions, and exhibited no blood vessel leakage or tissue edema. The myelinated axons were also surrounded by dark, ring-shaped myelin sheaths (Figure 7A–D). Neuronal apoptosis was found in vehicle animals, as indicated by shrunken nuclei with marginated, condensed, and fragmented nuclear chromatin, tissue edema in the extracellular space surrounding the vessels, blood vessel leakage, and the loosening of the tight junction between the endothelium. The loosening and splitting of myelin sheaths were also observed (Figure 7E–H). In the C + A group, the morphology of the nuclei was relatively normal. Destruction of tight junctions, perivascular edema, and myelin sheath splitting were also alleviated (Figure 7I–L).

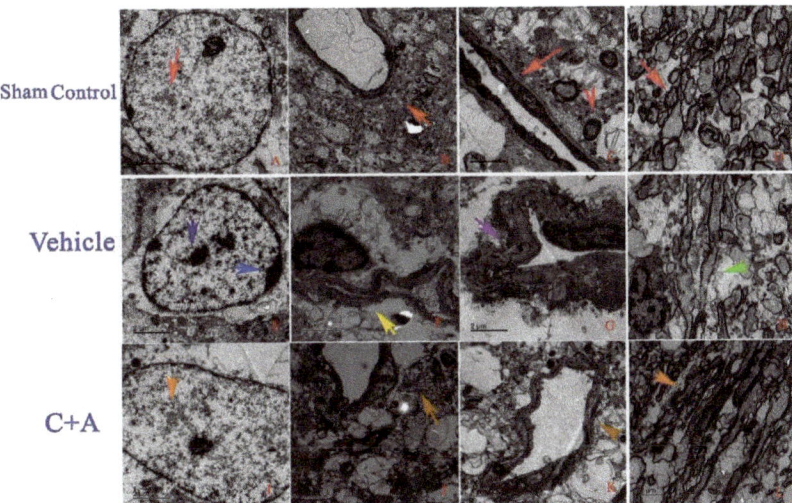

Figure 7. Electron micrographs show the ultrastructural morphology of the (**A–D**) sham control, (**E–H**) vehicle, and (**I–L**) C + A groups. The sham control rats showed (**A**) neuronal nuclei with uncondensed chromatin (red arrow in (**A**)). There was (**B**) no tissue edema or blood vessel leakage. Intact tight junctions (red arrow in (**C**)) and myelinated axons surrounded by dark, ring-shaped myelin sheaths were also observed (red arrows in (**C,D**)). In the vehicle group, (**E**) neuronal apoptosis was evidenced by shrunken nuclei with marginated, fragmented, and condensed nuclear chromatin (blue arrow in (**E**)). (**F**) Severe blood vessel leakage and tissue edema in the extracellular space surrounding the vessels (yellow arrow in (**F**)). (**G**) Loosening of tight junctions between the endothelium (purple arrow in (**G**)). (**H**) Myelin sheath loosening and splitting (green arrow in (**H**)). Treatment with C16 plus Ang-1 reduced morphological changes of the nuclei (orange arrow in (**I**)), alleviated perivascular edema (orange arrow in (**J**)), prevented destruction of tight junctions (orange arrow in (**K**)), and decreased myelin sheath splitting (orange arrow in (**L**)).

The levels of Syn (Figure 8A,B), CHAT (Figure 8C,D), and ACH (Figure 8E,F) in vehicle rats were decreased compared to those in the sham control group, while the C16 + Ang-1 treatment restored their expressions, suggesting that this treatment not only maintained the number of synapses, but also protected cholinergic neurons. Furthermore, the immunofluorescence staining also confirmed the changes in the expressions of Syn (Figure 9), CHAT (Figure 10), and ACH (Figure 11) in each group.

2.5. Treatment with C16 Plus Ang-1 Affected the Signaling Pathway Related to Inflammation

Western blot analysis and immunofluorescence staining showed that elevated expressions of p-tau (Figures 12A,B and 13), RhoA (Figure 12C,D; Supplementary Figure S3), AGT (Figure 12E,F; Supplementary Figure S4) and STAT3 (Figure 12G,H; Supplementary Figure S5) in AD rats were downregulated by treatment with C16 plus Ang-1.

2.6. The Activity of C16 Plus Ang1 in AD Is Mediated by the PI3K/Akt Pathway

Lastly, we measured the level of p85, a regulatory subunit of PI3K. Treatment with C16 plus Ang1 significantly induced the phosphorylation of AKT (Figure 12I,J), compared with the vehicle and sham control groups (Supplementary Figure S6A–G). Additionally, C16 plus Ang1 significantly upregulated the expression of PI3K/p85 (Figure 12K,L; Supplementary Figure S6H–N). These results indicate that the activity of C16 plus Ang1 was mediated by the PI3K/Akt pathway.

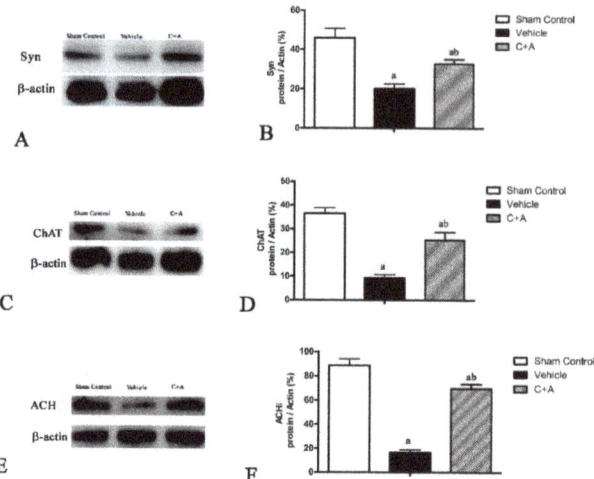

Figure 8. Western blot analysis of the expression of (**A**,**B**) synaptophysin (Syn), (**C**,**D**) choline acetyltransferase (CHAT), and (**E**,**F**) acetylcholine (ACH) in all three experimental groups. (**B**,**D**,**F**) Semi-quantitative profiles of protein bands showed that the C16 + Ang-1 treatment restored the downregulation of Syn, CHAT, and ACH in the vehicle group. a, $p < 0.05$ vs. sham control group; b, $p < 0.05$ vs. vehicle group. Molecular weight (MW): Syn, 38 KD; CHAT, 83 KD; ACH, 146KD; β-actin, 42 KD.

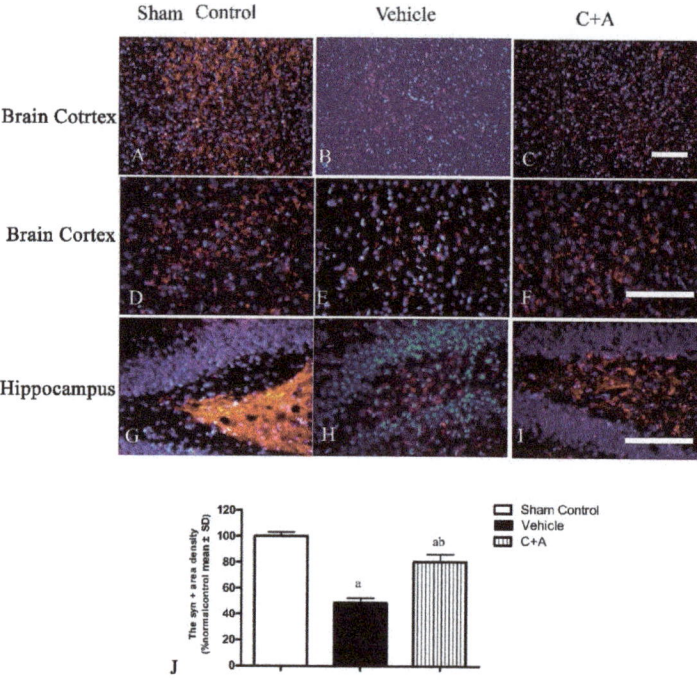

Figure 9. (**A–I**): Immunofluorescence staining (red) of Syn in the motor cortex (**A–F**) and hippocampus (**G–I**). The nuclei of neurons in the cortex and hippocampus were visualized by Hoechst 33342 staining (blue). (**J**): Quantification of Syn-positive areas showed that treatment with C16 plus Ang-1 induced the expression of Syn in AD rats. Scale bar = 100 μm. a, $p < 0.05$ vs. sham control group; b, $p < 0.05$ vs. vehicle group.

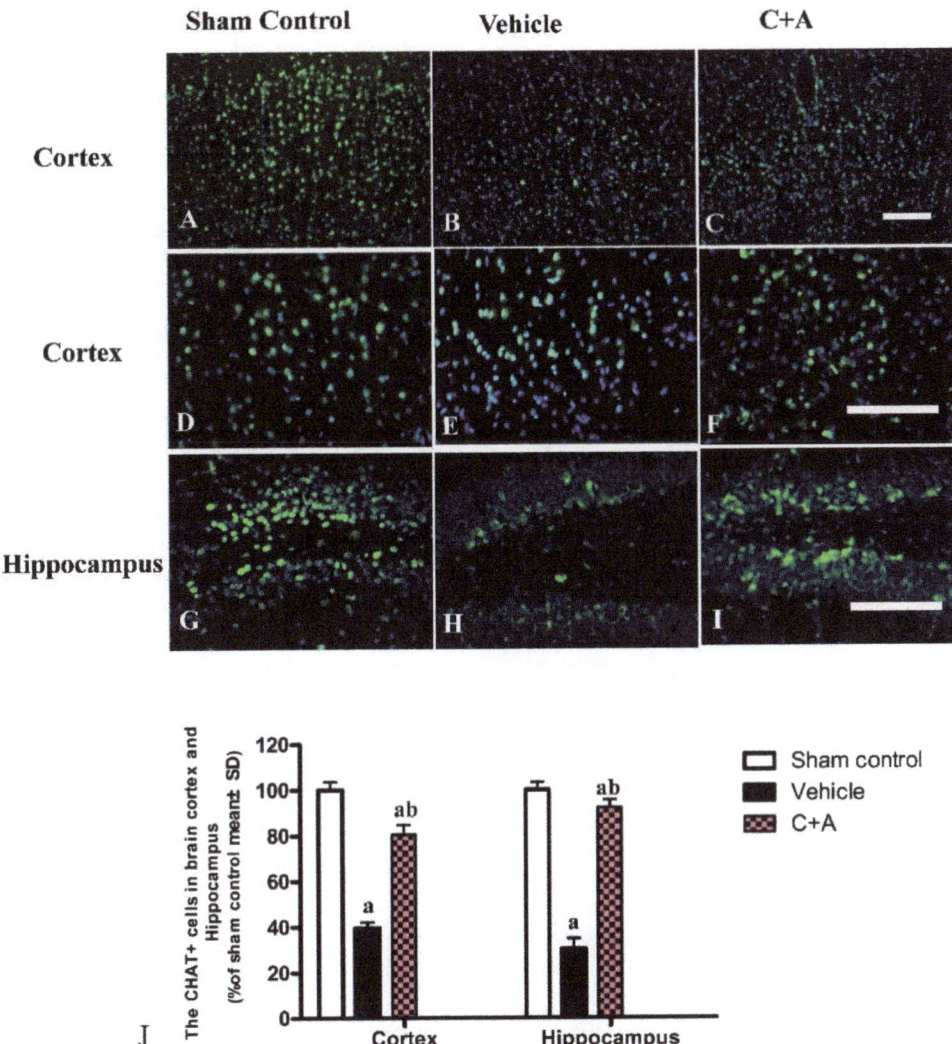

Figure 10. (**A–I**): Immunofluorescence staining (green) of CHAT in the motor cortex (**A–F**) and hippocampus (**G–I**). The nuclei of neurons in the cortex and hippocampus were visualized by Hoechst 33342 staining (blue). (**J**): Quantification of CHAT-positive cells indicated that treatment with C16 plus Ang-1 induced the expression of CHAT in AD rats. Scale bar = 100 µm. a, $p < 0.05$ vs. sham control group; b, $p < 0.05$ vs. vehicle group.

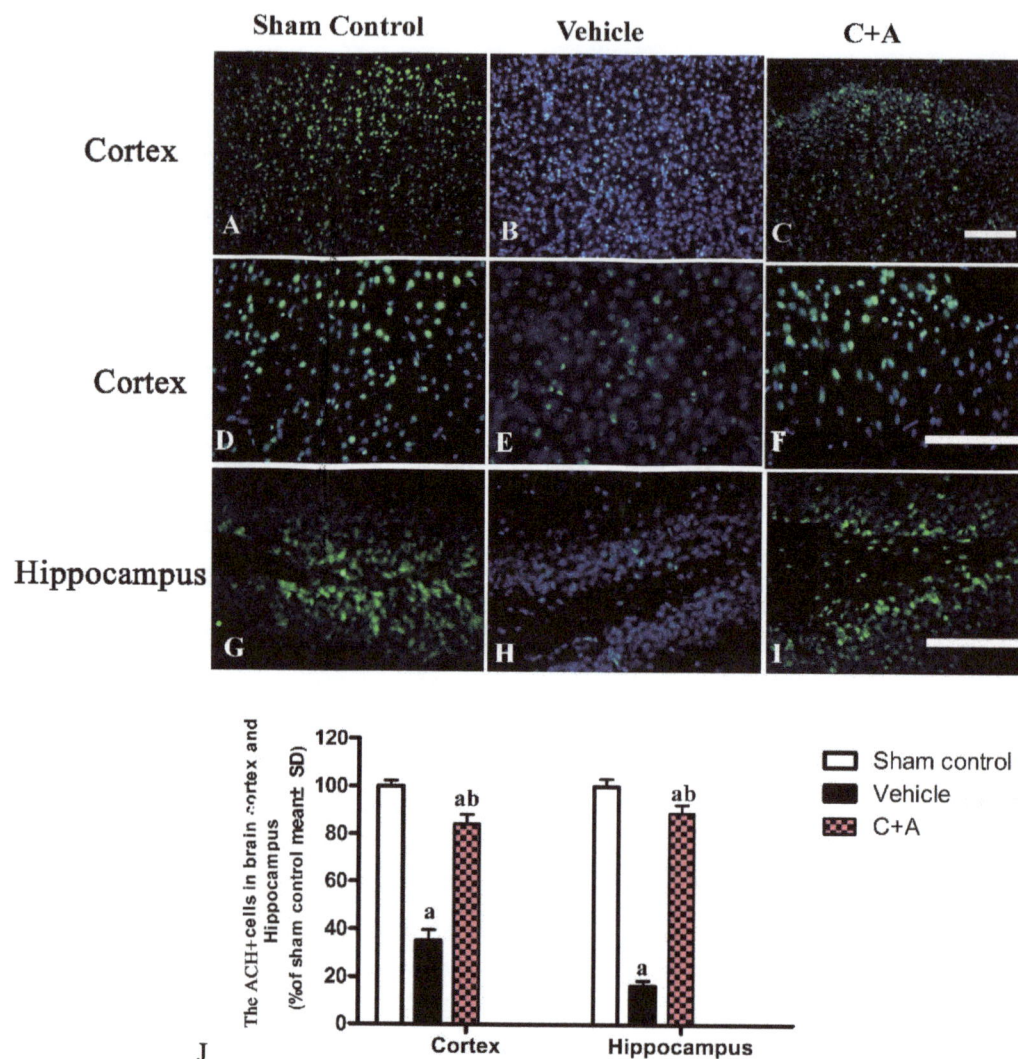

Figure 11. (**A–I**): Immunofluorescence staining (green) of ACH in the motor cortex (**A–F**) and hippocampus (**G–I**). The nuclei of neurons in the cortex and hippocampus were visualized by Hoechst 33342 staining (blue). (**J**): Quantification of ACH-positive cells showed that treatment with C16 plus Ang-1 induced the expression of ACH in AD rats. Scale bar = 100 μm. a, $p < 0.05$ vs. sham control group; b, $p < 0.5$ vs. vehicle group.

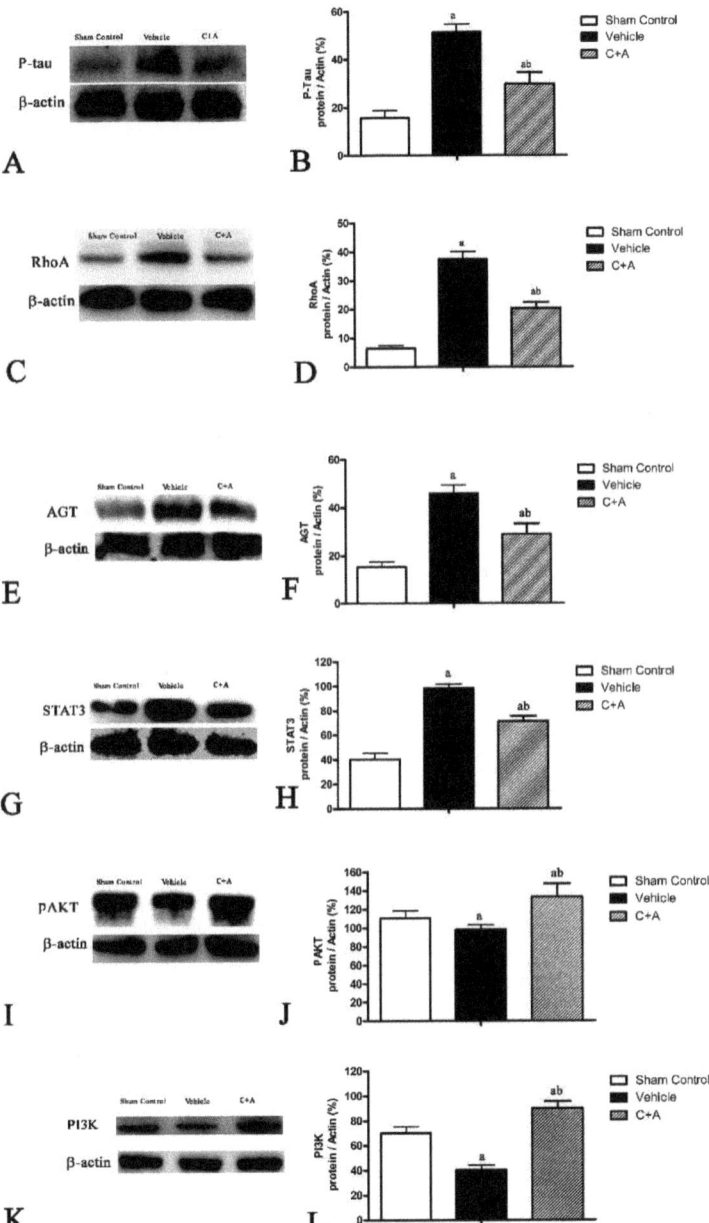

Figure 12. Western blot analysis of the level of (**A,B**) phosphorylated tau (p-tau), (**C,D**) RhoA/Rho kinase, (**E,F**) angiotensinogen (AGT), (**G,H**) signal transducers and activators of transcription 3 (STAT), (**I,J**) phosphorylated AKT (p-AKT), and (**K,L**) phospho-Tyr467 (PI3K/P85) in all groups. (**B,D,F,H,J,L**) Semi-quantitative profiles of protein bands showed that treatment with C16 plus Ang-1 suppressed the upregulation of p-tau, RhoA, AGT, and STAT3 in AD rats, but increased the expression of p-AKT and PI3K. a, $p < 0.05$ vs. sham control group; b, $p < 0.05$ vs. vehicle group. Molecular weight (MW): p-tau, 68 KD; RhoA, 22 KD; AGT, 52 KD; STAT3, 88 KD; p-AKT, 60 KD; PI3K, 85 KD; β-actin, 42 KD.

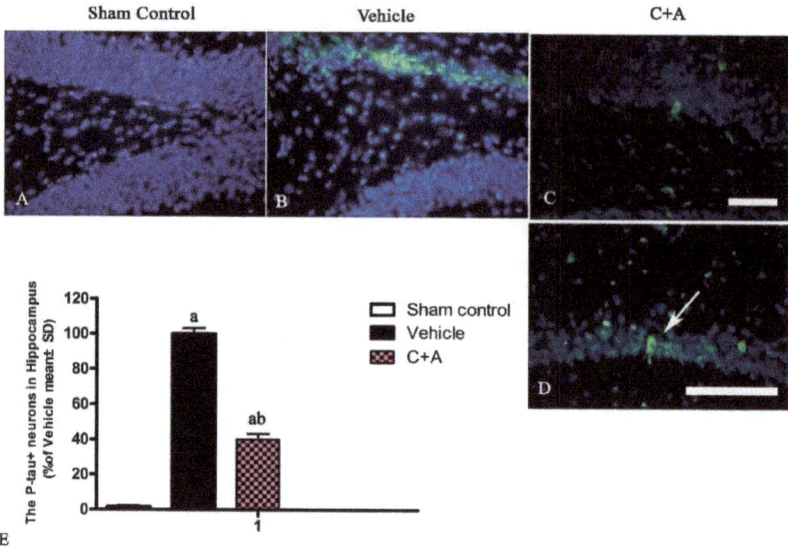

Figure 13. (**A–D**) Immunofluorescence staining (green) of p-tau in the hippocampus. The white arrow in (**D**) indicates a neuron expressing p-tau. The cell nuclei in the cortex and hippocampus were visualized by Hoechst 33342 staining (blue). (**E**) Quantification of p-tau-positive cells showed that treatment with C16 plus Ang-1 reduced the expression of p-tau in AD rats. Scale bar = 100 μm. a, $p < 0.05$ vs. sham control group; b, $p < 0.05$ vs. vehicle group.

3. Discussion

The CNS is an "immuno-privileged" organ that hosts defensive barriers, such as the choroid plexus, the perivascular space, and the meninges [11]. A coordinated response between microglia and other CNS cells is involved in neuro-inflammation, such as the infiltration of peripheral immune cells and astrocytes into the CNS. Acute neuro-inflammation, characterized by microgliosis and the release of inflammatory mediators, can be elicited by different types of stimuli, including ischemia, trauma, toxins, and infections [11]. Neuroinflammatory responses in AD are complex [6]. Previous studies have shown the strong contribution of inflammation to the pathological process of AD and have reported a reduced risk of developing AD for people who had been treated with anti-inflammatory drugs before the onset of the disease [6].

Neurodegenerative and neuro-inflammatory diseases are characterized by the recruitment of circulating innate and adaptive immune cells in the CNS. Neuro-inflammatory responses, ranging from glial cell activation to immune cell infiltration, are frequently associated with the pathogenesis of AD [12]. During neuro-inflammatory and neurodegenerative disorders, the protective function of the BBB may be severely impaired, resulting in detrimental neuro-inflammatory changes. Infiltrated leukocytes sustain detrimental responses in the CNS by releasing pro-inflammatory cytokines to aggravate BBB damage, glial cell activation, and neural cell death [13].

Our previous studies of C16 [9] and Ang-1 [10] showed that C16 peptide rescued blood vessels by acting as an agonist, and by binding only to αvβ3 and α5β1 integrin, two well-known factors that promote the survival of endothelial cells. Moreover, Ang-1, acting as a growth factor of vascular endothelial cells, protects micro-blood vessels, reduces blood–brain barrier (BBB) permeability, and prevents blood vessel leakage. Even though they may share the same mechanism, we assumed that Ang-1 and C16 might gain better synergistic effects through targeting different pathways.

Many factors are implicated in the changes of BBB permeability, such as matrix metalloproteinases, free radicals, and inflammatory cytokines/chemokines, which may lead to brain tissue damage during the course of cerebral ischemia [14]. Chemokines are a subgroup of cytokines initially described as "chemoattractant", and have been shown to induce leukocyte chemotaxis, thereby increasing BBB permeability and impairing the integrity thereof [15].

In our study, 2-VO operation followed by Aβ1-42 injection effectively destroyed the integrity of the BBB and induced neuro-inflammatory responses. The spatial memory decline, neuronal death in the bilateral hippocampus, increased BBB permeability and upregulated pro-inflammatory factors, as well as extensive inflammatory cell infiltration into the brain parenchyma confirmed that the AD model was successfully established.

Tauopathies are characterized by neuro-inflammation and intracellular accumulation of neurofibrillary tangles, in which hyper-phosphorylated tau (abnormal phosphorylation of tau) protein aggregates [15]. Our results showed that hyper-phosphorylated tau was upregulated in hippocampal neurons of the vehicle group compared with sham control rats, while this phenomenon was significantly attenuated by the C16 + Ang-1 treatment. Since accumulation of microtubule-associated protein tau within hippocampal neurons is related to multiple functional impairments, these findings further confirmed the protective effects of these compounds against memory impairment.

Cholinergic transmission is critical to high-order brain functions, such as memory, learning, and attention. Alzheimer's disease (AD) is characterized by cognitive decline associated with a specific degeneration of cholinergic neurons. A decrease in choline acetyltransferase (ChAT) enzymatic activity indicates the loss of cholinergic neurons [16]. Moreover, synapse-associated proteins, especially presynaptic Syn, can promote synaptic plasticity. Syn deficiency is correlated with cognitive decline in aging-related neurodegenerative disorders [17]. In the current study, the combination of C16 and Ang-1 alleviated AD-induced Syn reduction, indicating improved cognitive function in the drug-treated groups.

In the present study, we found that the combined use of C16 and Ang-1 exhibited potentially lasting effects. The intravenous injection of C16 and Ang-1 improved vascular-related outcomes (i.e., endothelial cell death and dysfunction, leukocyte extravasation, etc.). As Ang-1 only acts through Tie2, a tyrosine kinase receptor mainly expressed by endothelial cells, Ang-1 treatment mainly reduced the loss of blood vessels, and thus alleviated ischemia, prevented tissue loss, and improved memory, all of which were correlated with the amount of rescued blood vessels [10]. Moreover, integrins play a key role in the leukocyte accumulation and adhesion process by specifically binding to $\alpha v \beta 3$. C16 competitively inhibits the transmigration of leukocytes when they cross the endothelium. Inflammatory cell infiltration activates a number of noxious factors that may contribute to secondary injuries. Therefore, treatment with C16 plus Ang-1 suppressed secondary injuries and prevented further tissue damage in AD rats. Less abnormal ultrastructure observed in electron microscopy further confirmed the positive effects.

Our previous study also identified some upregulated genes in both AD and VaD patients, including RhoA, ICAM, AGT, and STAT3, all of which are related to the inflammatory microenvironment in the CNS and could be downregulated by anti-inflammatory therapeutic compounds.

The RhoA/Rho-associated protein kinase pathway plays an important role in regulating cell migration [18]. Angiotensinogen (AGT) is a unique precursor of all angiotensin peptides [19]. AGT is related to cardiovascular diseases, and the expression of AGT can be adjusted by the Rho-associated protein kinase pathway [20].

ICAM-1 is essential for the capture and migration of leucocytes [21]. Inflammation is a key driver of tissue destruction in ischemia reperfusion injury. IL-1β promotes inflammation in various autoimmune and inflammatory diseases [22]. The renin–angiotensin system also contributes to brain damage and cognitive decline caused by chronic cerebral ischemia [23]. A previous study revealed that chronic cerebral hypo-perfusion significantly increased renin activity in activated astrocytes and micro-vessels, and induced angiotensinogen expression in white matter. Thus, the inhibition of angiotensinogen might be a promising therapeutic option for subcortical vascular dementia [24].

Reactive astrogliosis is also a hallmark of AD. The transcription factor STAT3 is a canonical inducer of astrogliosis and has been reported to be activated in an AD mouse model. The STAT3-deficient APP/PS1 mice showed a decreased Aβ level and plaque burden, and were largely protected from neuronal network imbalance [25]. STAT3 deletion in astrocytes also greatly ameliorated memory decline and improved spatial learning in APP/PS1 mice, suggesting STAT3-mediated astrogliosis as a promising therapeutic target in AD [25].

The PI3K/Akt signaling pathway is involved in the antioxidant, neuroprotective, and anti-inflammatory effects of propofol, a widely used anesthetic agent that can attenuate subarachnoid hemorrhage-induced early brain injury via inhibiting oxidative reactions and inflammation [26]. The NF-κB/COX-2/tumor necrosis factor alpha (TNF-α)/IL-1β inflammatory pathway has been reported to positively mediate inflammation in the brain, while the PI3K/Akt signaling pathway negatively regulates inflammatory responses [27] and oxidative stress [28]. Here, treatment with C16 plus Ang-1 upregulated p85 and p-AKT and suppressed the NF-κB/COX-2/TNF-α/IL-1β inflammatory pathway compared with the vehicle animals, implying that brain inflammation in AD possibly acts through activation of the PI3K/Akt pathway.

Microglial activation (MA), a key feature of AD, has been considered a major contributor to progressive neuronal injury by releasing neurotoxic products. Microglial cells undergo two types of activation, each of which acquires a neurotoxic phenotype (M1-like) or a neuroprotective phenotype (M2-like). M1-like microglia build up a detrimental microenvironment for neurons by secreting reactive oxygen species (ROS) and inflammatory cytokines. By contrast, M2-like microglia produce anti-inflammatory mediators and neurotrophic factors, thereby generating a supportive microenvironment for neurons. The activation of the M1/M2 phenotype plays an essential role in AD, and the balance of the M1/M2 phenotype is regulated by the PI3K/AKT pathway. Akt downregulation and NF-κB upregulation have been observed in microglial cells after Aβ42 incubation. Moreover, LY294002, an Akt inhibitor, has been reported to enhance the expression of M1 marker [29]. The above evidence suggests a close relationship between MA and PI3K/AKT signaling. In a recent in vitro study, we explored the involvement of the PI3K/AKT pathway in the anti-MA activity of Ang1 plus C16. The results showed that C16 plus Ang1 significantly upregulated Tie2, which was markedly downregulated by PI3K/AKT inhibitors and the anti-Tie2 antibody [30]. Furthermore, treatment with Ang1 plus C16 significantly induced the expression of p85 and p-AKT, and this phenomenon was downregulated by the inhibition of α5β1 integrin, αvβ3 integrin, PI3K, AKT, and Tie2 [30]. These findings indicate that the anti-MA activity of C16 plus Ang1 was mediated by Tie2, αvβ3 and/or α5β1 integrins, as well as the PI3K/AKT pathway [30]. In the present study, the number of CD68/ED1 (a marker of activated microglia)-positive cells in the C + A group was notably lower compared to the vehicle group, which is consistent with our previous study. Thus, C16 plus Ang-1 not only reduced immune cell infiltration into the CNS but also inhibited the activation of glial cells, both of which were associated with the pathogenesis of AD and vascular dysfunction.

Although the therapeutic effects of C16 combined with Ang-1 have been confirmed, the side effects and toxicity at high doses remain unclear. The dose adjustment for this combined therapy is ongoing in our laboratory.

4. Materials and Methods

4.1. Animals

Forty-five male SPF SD rats (age: 10–12 weeks; weight: 250 ± 30 g) were purchased from the Shanghai SIPPR-BK Laboratory Animal Center of the Chinese Academy of Sciences. All rats were housed at a constant room temperature (20 ± 2 °C), supplied with sterilized water and food, and acclimated for a week prior to experiments. The experimental protocols were approved by the Ethics Committee of Zhejiang University Medical College (SRRSH202102016).

Rats were randomly assigned into three groups (n = 15 per group): the vehicle (2-VO + Aβ1-42 + PBS) group, the C + A (2-VO + Aβ1-42 + C16 + Ang-1) group, and the sham control group. The vehicle and the C + A groups were subjected to two-vessel occlusion (2-VO) followed by Aβ1-42 injection into the hippocampus. Rats in the sham control group underwent the same procedure of 2-VO without artery ligation. Further, instead of Aβ1-42, sham control rats were injected with an equal amount of PBS. The C + A group was administered 1 mL of drug containing 2 mg of C16 and 400 μg of Ang-1 daily for 2 weeks via intravenous tail vein injection. The vehicle and sham control groups were administered 1 mL of PBS for 2 weeks. The flowchart of the study is shown in Supplementary Figure S7.

For each group, 11 out of 15 rats were intracardially perfused with cold saline, followed by 4% paraformaldehyde. Then, 5 (one was infused with EB dye) of the 11 brains were fixed in 4% paraformaldehyde and subsequently soaked in 30% sucrose to make frozen sections; 3 of the 11 brains were fixed in 2.5% glutaraldehyde solution and immersed in 1% osmium tetroxide to make transmission electron sections; another 3 of the 11 brains were dehydrated with alcohol, isopropanol, and n-butanol to make paraffin-embedded sections for the TUNEL assay. Five sections (three visual fields per slide) of each brain were randomly selected for staining and analysis. The remaining 4 rats were sacrificed by decapitation; their blood samples were collected for ELISA and the fresh tissues were used for Western blot analysis.

4.2. 2-VO

Rats were anesthetized with 3% pentobarbital sodium (1.5 mL/kg body weight; 922L038, Shanghai Curie biological science and Technology), and then, the hair on the anterior neck was shaved. The target animal was placed in the supine position and disinfected with iodophor (131101 A, ShanDong LIRCON Medical, De Zhou, Shandong, china). Next, a midline cervical incision was made; the fascia was cut, the sternocleidomastoid muscles were separated, and the bilateral common carotid arteries were exposed. A double ligation was performed between the distal and proximal ends using a 5-0 silk thread. Then, the incision was sprayed with penicillin (F4046203, North China Pharmaceutical Group), and the skin was sutured. The body temperature of the rat was maintained between 36.5 and 37.5 °C [30]. The mortality rate following 2-VO operation was 20%. Most animals died 1 day after the operation. The deceased animals were added up immediately. If the animal survived the 2-VO operation for 1 week, it was subjected to Aβ1-42 injection and further analysis.

4.3. Aβ1-42 Injection

Aβ1-42 (Millipore, Billerica, MA, USA) was dissolved in 1% $NH_3 \cdot H_2O$ to prepare a solution with a concentration of 1 μg/μL and incubated for 5 days at 37 °C. At 1 week after the 2-VO operation, rats in the vehicle and C + A groups were anesthetized via an intraperitoneal injection of sodium pentobarbital (60 mg/kg) and then placed in a stereotaxic frame (Ambala, Haryana, India). An incision was made in the dorso-ventral region at 3.6 mm (a site located at 4 mm posterior and 3 mm medio-lateral to bregma) [30]. After 72 h, 1.0 μL of mature Aβ1-42 was injected bilaterally into the hippocampus using a 1-μL Hamilton syringe at a rate of 0.1 μL/min. The sham control group was injected with an equal amount of PBS. The needle was left in place for 5 min to allow complete diffusion of the solution. A small amount of bone wax was used to seal the bone. After being sprayed with penicillin, the skin was sutured. The body temperature of the rat was maintained between 36.5 and 37 °C throughout the operation.

4.4. Behavioral Test

The Morris water maze (Smart-Mass, Panlab, Barcelona, Spain) test was performed, including the place navigation test and the space exploration experiment.

The place navigation test assesses memory performance and spatial learning. One day before grouping, animals were placed in water for 120 s to become familiar with the environment. The formal experiment began 3 days after modeling and lasted for 6 days.

Rats were trained twice a day (morning and afternoon). The target animal was placed in a pool and released facing the sidewall in each trial. The starting points were the same for all rats in the same training session. The average time that each group spent was measured to determine the escape latency, with an upper limit of 120 s. If the animal failed to find the platform within 120 s, it was guided to the platform and allowed to stay there for 10 s to strengthen the memory. The escape latency of these animals was recorded as 120 s.

On day 7, the space exploration experiment was performed to assess long-term spatial memory. The platform was first removed from the water environment. The number of times that the rat crossed the area where the platform was located and the average time that the animal stayed in the target quadrant were recorded. The test was performed by investigators who were blinded to the treatments of the animals.

4.5. Perfusion and Tissue Processing

At 4 weeks after Aβ1-42 injection, animals were anesthetized with sodium pentobarbital and perfused intracardially with cold saline, followed by 4% paraformaldehyde. The brain was harvested from each animal, fixed in 4% paraformaldehyde for 4 h and then soaked in 30% sucrose. Coronal sections (20 um) were obtained using a Leica cryostat and a freezing microtome (Leica, Buffalo Grove, IL, USA). The sections were mounted onto 0.02% poly-L-lysine-coated slides and subjected to immunofluorescence staining [9].

4.6. Transmission Electron Microscopy

A proportion of hippocampus and cortex was fixed in 2.5% glutaraldehyde solution, immersed in 1% osmium tetroxide at 4 °C, and then washed 3 times with 0.1 M PB. After fixation, the following steps were performed using an EM processor with agitation at room temperature. The tissues were dehydrated in graded ethanol (30%, 50%, 70%, 80%, 90%, 95% ethanol) for 5 min, each followed by three changes of absolute ethanol (each 10 min). After two changes in 1, 2-propylene oxide (PO) (15 min each), tissues were immersed in a 1:1 PO:Epon mixture for 1 h. These tissues were then incubated overnight in pure Epon and embedded in pure Epon at 60 C for 3 days. The Epon-embedded tissues were cut into 90 nm sections with a diamond knife on an ultracut microtome and collected on a 200 mesh copper grid. Lead citrate (approximately 3%) and 8% uranyl acetate were filtered before use. The grids were stained with lead citrate droplets for 20 min in a Petri dish, washed 3 times with distilled water, and were then ready for electron microscopic analysis [7,9].

4.7. Evans Blue (EB) Assay

The vascular permeability of the BBB was analyzed using the modified EB extravasation method (n = 5 per group) [9]. Rats were anesthetized with 60 mg/kg sodium pentobarbital and infused with EB dye (4 mL/kg; 2% in 0.9% normal saline) via the right femoral vein at 37 °C for 5 min. Two hours later, blood vessels were perfused with 300 mL of saline, and the brain was harvested. The EB-stained tissue sections (20 μm) were observed under an ultraviolet light filter using red laser excitation. Red staining indicated high vascular permeability in tissues. The staining intensity was quantified using ImageJ (NIH, Bethesda, MD, USA). One-half of the tissues were homogenized with 750 μL of N, N-dimethylformamide (Sigma, St. Louis, MO, USA). The tissue homogenates were maintained in the dark for 72 h at room temperature, centrifuged at $10,000 \times g$ for 25 min, and then analyzed using a spectrophotometer (Molecular Devices OptiMax, San Jose, CA, USA) at 610 nm. The dye concentration was expressed as μg/g of tissue weight [7].

4.8. Immunofluorescence Staining

Slides were warmed for 20 min on a slide warmer. A ring of wax was applied around the sections with a PAP pen (Invitrogen, Carlsbad, CA, USA). After being rinsed with 0.01 M Tris-buffered saline (TBS) for 10 min, the sections were permeabilized and blocked with 0.3% Triton X-100/10% normal goat serum in 0.01 M PBS for 30 min. Tissue sections were stained with primary polyclonal rabbit antibodies against phosphorylated tau (1:200;

Thermo Fisher Scientific, Waltham, MA, USA), nuclear factor kappa B (NF-κB; 1:500; R&D Systems, Millipore, Billerica, MA, USA), CD68/ED1 (1:200; Santa Cruz Biotechnology, Santa Cruz, CA, USA), zonula occludens 1 (ZO-1,1:500; Santa Cruz Biotechnology, Santa Cruz, CA, USA), cyclooxygenase-2 (COX-2; 1:1000; Neuromics, Edina, MN, USA), synaptophysin (Syn; 1:1000; Thermo Fisher Scientific, Waltham, MA, USA), RhoA/Rho-kinase (1:300; Upstate Biotechnology, Lake Placid, NY, USA), intercellular cell adhesion molecule-1 (ICAM-1,1:200; Novus Biologicals, Centennial, CO, USA), angiotensinogen (AGT, 1:1000; Chemicon, Temecula, CA, USA), phospho-Tyr467 (PI3K/p85,1:200; Thermo Fisher Scientific, Waltham, MA, USA), Microtubule-associated protein 1A/1B-light chain 3 beta II (LC3BII,1:400; Novus Biologicals, Centennial, CO, USA), signal transducers and activators of transcription 3 (STAT3,1:500; Cayman Chemical, Ann Arbor, MI, USA), phosphorylated AKT (p-AKT; 1:500; Abcam, Cambridge, MA, USA), acetylcholine (ACH; 1:100; St. Louis, MO, USA), anti-caspase 3 (1:500; Cayman Chemical, Ann Arbor, MI, USA) and choline acetyltransferase (CHAT; 1:500; Abcam, Waltham, MA, USA) overnight at 4 °C. After being rinsed with PBS three times, sections were stained with a goat anti-rabbit/mouse IgG secondary antibody (1:200; Invitrogen, Carlsbad, CA, USA) at 37 °C for 1 h and mounted with Gel Mount antifade aqueous mounting medium (Southern Biotech, Birmingham, AL, USA). Primary antibody omission controls were used in order to further confirm the specificity of the immunohistochemical labeling. For all histological and immunohistochemical analyses, five transverse sections from each animal were randomly selected, and digital photomicrographs were obtained from three visual fields per section. Micrographs of three visual fields on each section were taken at 200× magnification. The quantification of the histological results was performed by an investigator that was blinded to the treatments. The Zo-1, CD34 syn, and ICAM-1-immunoreactive areas were calculated using the NIH ImageJ software. Cells expressing NF-κB, CD68, COX-2, RhoA, LC3BII, AGT, CHAT, ACH, STAT3, tau, p-AKT, caspase-3, and PI3K/p85 in each image were counted.

4.9. Enzyme-Linked Immunosorbent Assay (ELISA)

Rats were sacrificed by decapitation at 4 weeks post-induction. Peripheral blood samples were collected (n = 3 per group per time point) at 4 °C. The blood samples were centrifuged at 1000× g for 20 min and then at 10,000× g at 4 °C for 10 min before storage. To measure the expression levels of cytokines, plasma samples were added to 96-well plates pre-coated with anti-interleukin-1 (IL-1β; BioLegend Inc. San Diego, CA, USA) primary antibodies for 1 h at 37 °C. Then, samples were incubated with a goat anti-rabbit IgG secondary antibody (1:2000; Bio-Rad, Hercules, CA, USA) for 60 min at 37 °C. The optical density of protein bands at 450 nm was measured using a Model 680 Microplate Reader (Bio-Rad), and the results were analyzed using GraphPad Prism (version 4; GraphPad, San Diego, CA, USA).

4.10. Western Blot

Rats were sacrificed by decapitation at 4 weeks post-induction (n = 4 per group per time point). The whole brain was collected from each animal. Proteins were separated by 12% SDS-PAGE electrophoresis and blotted onto polyvinylidene difluoride membranes. The membranes were then incubated with primary polyclonal rabbit antibodies against phosphorylated tau (p-tau; 1:400; Thermo Fisher Scientific), LC3BII (1:800; Novus Biologicals), NF-κB (1:800; R&D Systems), CD68/ED1 (1:500; Santa Cruz Biotechnology), ZO-1 (1:800; Santa Cruz Biotechnology), COX-2 (1:1500; Neuromics, Edina, MN, USA), Syn (1:1200; Thermo Fisher Scientific), RhoA (1:500; Upstate Biotechnology), ICAM-1 (1:500; Novus Biologicals), AGT (1:1200; Chemicon), PI3K/p85 (1:500; Thermo Fisher Scientific), STAT3 (1:800; Cayman Chemical), p-AKT (1:800; Abcam), ACH (1:500; Sigma), caspase 3 (1:1000; Cayman Chemical), and CHAT (1:500; Abcam) at room temperature for 12 h. The membranes were then re-probed with a rabbit anti-β-actin antibody (1:5000; Abcam). After incubation with a goat anti-rabbit secondary antibody (1:5000; Santa Cruz), the protein bands were detected using an enhanced chemiluminescence reagent.

4.11. Statistical Analysis

Data are presented as mean ± standard deviation (SD) unless otherwise indicated. All data were analyzed using the SPSS 13.0 software. Differences between protein levels were analyzed by a two-way analysis of variance (ANOVA) followed by Tukey's post hoc test. Differences between histological scores were analyzed using a Mann–Whitney U test. The Kruskal–Wallis nonparametric one-way ANOVA was used to compare data presented as percentages. A p-value of less than 0.05 was considered statistically significant. All graphs were plotted using GraphPad Prism Version 4.0. Statistical analysis was also performed by a statistician that was blinded to the study design.

5. Conclusions

In conclusion, treatment with C16 plus Ang-1 alleviated functional disability and reduced neuronal death in rat brain cortex and hippocampus by ameliorating inflammatory cell infiltration into the CNS, protecting endothelial cells of blood vessels, and maintaining BBB permeability. This treatment might exert an anti-MA effect by activating the PI3K/AKT pathway. The combined treatment with C16 plus Ang-1 can be considered as a potential therapeutic option for AD and warrants further investigation.

Supplementary Materials: The following are available online at https://www.mdpi.com/article/10.3390/ph15040471/s1, Supplementary Figure S1: (A–F): Immunofluorescence staining (green) of macrophage-specific marker CD68 in the hippocampus of all three experimental groups. (G) Quantification of CD68-positive cells showed that treatment with C16 plus Ang-1 downregulated CD68 in AD rats. Scale bar = 100 μm. a, $p < 0.05$ vs. sham control group; b, $p < 0.05$ vs. vehicle group. Supplementary Figure S2: (A–H): Immunofluorescence staining of caspase-3 showed apoptotic cells (red) in all three experimental groups. The arrow in the high magnification images in C and G demonstrated the expression of caspase-3 in the cytoplasm. The cell nuclei in the motor cortex and hippocampus were visualized by Hoechst 33342 staining (blue). (I) Quantification of caspase-3-labeled nuclei showed that treatment with C16 plus Ang-1 reduced apoptotic cell death in AD rats. Scale bar = 100 μm. (J) Western blot analysis of the expression of apoptotic marker caspase-3. (K) Semi-quantitative profiles of protein bands showed that C16 plus Ang-1 suppressed the upregulation of caspase-3 in the vehicle group. caspase-3, 31 KD; β-actin, 42 KD. a, $p < 0.05$ vs. sham control group; b, $p < 0.05$ vs. vehicle group. Supplementary Figure S3: (A–I): Immunofluorescence staining (green) of RhoA/Rho kinase (RHOA) in the motor cortex and hippocampus. (J) Quantification of RHO-positive cells showed that treatment with C16 plus Ang-1 reduced the expression of RHO in AD rats. Scale bar = 100 μm. a, $p < 0.05$ vs. sham control group; b, $p < 0.05$ vs. vehicle group. Supplementary Figure S4: (A–I): Immunofluorescence staining (red) of angiotensinogen (AGT) in the motor cortex and hippocampus. The cell nuclei were visualized by Hoechst 33342 staining (blue). (J) Quantification of AGT-positive cells showed that treatment with C16 plus Ang-1 reduced the expression of AGT in AD rats. Scale bar = 100 μm. a, $p < 0.05$ vs. sham control group; b, $p < 0.05$ vs. vehicle group. Supplementary Figure S5: (A–I): Immunofluorescence staining (green) of activators of transcription 3 (STAT3) in the motor cortex and hippocampus. (J) Quantification of STAT3-positive cells showed that treatment with C16 plus Ang-1 reduced the expression of STAT3 in AD rats. Scale bar = 100 μm. a, $p < 0.05$ vs. sham control group; b, $p < 0.05$ vs. vehicle group. Supplementary Figure S6: (A–F): Immunofluorescence staining (green) of phosphorylated AKT (p-AKT) in the motor cortex and hippocampus. (H-M): Immunofluorescence staining (green) of PI3K in the motor cortex and hippocampus. (G, N) Quantification of p-AKT- and PI3K-positive cells showed that treatment with C16 plus Ang-1 increased the expressions of p-AKT (G) and PI3K (N) in AD rats. Scale bar = 100 μm. a, $p < 0.05$ vs. sham control group; b, $p < 0.05$ vs. vehicle group. Supplementary Figure S7: The flowchart o drug administration in the different groups.

Author Contributions: S.H. conceived and designed research; X.F. and J.W. collected data and conducted research; H.C. and H.J. analyzed and interpreted data; X.F. and J.W. wrote the initial paper; S.H. revised the paper; H.C. and S.H. had primary responsibility for final content. All authors have read and agreed to the published version of the manuscript.

Funding: We thank the National Natural Science Foundation of China (project No. 81971069) and Zhejiang Provincial Natural Science Foundation of China (project No. LY22H090017) for the funding support.

Institutional Review Board Statement: The experimental protocols were approved by the Ethics Committee of Zhejiang University Medical College (SRRSH20212016).

Informed Consent Statement: Not applicable.

Data Availability Statement: Data is contained within the article and Supplementary Material.

Conflicts of Interest: The authors declare no conflict of interest.

References

1. Sweeney, M.D.; Kisler, K.; Montagne, A.; Toga, A.W.; Zlokovic, B.V. The role of brain vasculature in neurodegenerative disorders. *Nat. Neurosci.* **2018**, *21*, 1318–1331. [CrossRef] [PubMed]
2. Kisler, K.; Nelson, A.R.; Montagne, A.; Zlokovic, B.V. Cerebral blood flow regulation and neurovascular dysfunction in Alzheimer disease. *Nat. Rev. Neurosci.* **2017**, *18*, 419–434. [CrossRef] [PubMed]
3. Iturria-Medina, Y.; Sotero, R.C.; Toussaint, P.J.; Mateos-Pérez, J.M.; Evans, A.C. Early role of vascular dysregulation on late-onset Alzheimer's disease based on multifactorial data-driven analysis. *Nat. Commun.* **2016**, *7*, 11934. [CrossRef] [PubMed]
4. Sweeney, M.D.; Sagare, A.P.; Zlokovic, B.V. Blood-brain barrier breakdown in Alzheimer disease and other neurodegenerative disorders. *Nat. Rev. Neurol.* **2018**, *14*, 133–150. [CrossRef]
5. Toledo, J.B.; Arnold, S.E.; Raible, K.; Brettschneider, J.; Xie, S.X.; Grossman, M.; Monsell, S.E.; Kukull, W.A.; Trojanowski, J.Q. Contribution of cerebrovascular disease in autopsy confirmed neurodegenerative disease cases in the National Alzheimer's Coordinating Centre. *Brain* **2013**, *136*, 2697–2706. [CrossRef] [PubMed]
6. Stephenson, J.; Nutma, E.; van der Valk, P.; Amor, S. Inflammation in CNS neurodegenerative diseases. *Immunology* **2018**, *154*, 204–219. [CrossRef]
7. Wang, B.; Tian, K.W.; Zhang, F.; Jiang, H.; Han, S. Angiopoietin-1 and C16 Peptide Attenuate Vascular and Inflammatory Responses in Experimental Allergic Encephalomyelitis. *CNS Neurol. Disord. Drug Targets* **2016**, *15*, 496–513. [CrossRef]
8. David, S.; Park, J.K.; Meurs, M.; Zijlstra, J.G.; Koenecke, C.; Schrimpf, C.; Shushakova, N.; Gueler, F.; Haller, H.; Kümpers, P. Acute administration of recombinant Angiopoietin-1 ameliorates multiple-organ dysfunction syndrome and improves survival in murine sepsis. *Cytokine* **2011**, *55*, 251–259. [CrossRef]
9. Zhang, F.; Yang, J.; Jiang, H.; Han, S. An $\alpha v \beta 3$ integrin-binding peptide ameliorates symptoms of chronic progressive experimental autoimmune encephalomyelitis by alleviating neuroinflammatory responses in mice. *J. Neuroimmune. Pharmacol.* **2014**, *9*, 399–412. [CrossRef]
10. Han, S.; Arnold, S.A.; Sithu, S.D.; Mahoney, E.T.; Geralds, J.T.; Tran, P.; Benton, R.L.; Maddie, M.A.; D'Souza, S.E.; Whittemore, S.R.; et al. Rescuing vasculature with intravenous angiopoietin-1 and alpha v beta 3 integrin peptide is protective after spinal cord injury. *Brain* **2010**, *133*, 1026–1042. [CrossRef]
11. Mammana, S.; Fagone, P.; Cavalli, E.; Basile, M.S.; Petralia, M.C.; Nicoletti, F.; Bramanti, P.; Mazzon, E. The Role of Macrophages in Neuroinflammatory and Neurodegenerative Pathways of Alzheimer's Disease, Amyotrophic Lateral Sclerosis, and Multiple Sclerosis: Pathogenetic Cellular Effectors and Potential Therapeutic Targets. *Int. J. Mol. Sci.* **2018**, *19*, 831. [CrossRef] [PubMed]
12. Schenk, G.J.; de Vries, H.E. Altered blood-brain barrier transport in neuro-inflammatory disorders. *Drug Discov. Today Technol.* **2016**, *20*, 5–11. [CrossRef] [PubMed]
13. Runtsch, M.C.; Ferrara, G.; Angiari, S. Metabolic determinants of leukocyte pathogenicity in neurological diseases. *J. Neurochem.* **2021**, *158*, 36–58. [CrossRef] [PubMed]
14. Michalicova, A.; Majerova, P.; Kovac, A. Tau Protein and Its Role in Blood-Brain Barrier Dysfunction. *Front. Mol. Neurosci.* **2020**, *13*, 570045. [CrossRef] [PubMed]
15. Kong, L.L.; Wang, Z.Y.; Hu, J.F.; Yuan, Y.H.; Li, H.; Chen, N.H. Inhibition of chemokine-like factor 1 improves blood-brain barrier dysfunction in rats following focal cerebral ischemia. *Neurosci. Lett.* **2016**, *627*, 192–198. [CrossRef] [PubMed]
16. Kim, M.S.; Yu, J.M.; Kim, H.J.; Kim, H.B.; Kim, S.T.; Jang, S.K.; Choi, Y.W.; Lee, D.I.; Joo, S.S. Ginsenoside Re and Rd enhance the expression of cholinergic markers and neuronal differentiation in Neuro-2a cells. *Biol. Pharm. Bull.* **2014**, *37*, 826–833. [CrossRef]
17. Liu, Y.; Zhang, Y.; Zheng, X.; Fang, T.; Yang, X.; Luo, X.; Guo, A.; Newell, K.A.; Huang, X.F.; Yu, Y. Galantamine improves cognition, hippocampal inflammation, and synaptic plasticity impairments induced by lipopolysaccharide in mice. *J. Neuroinflammation* **2018**, *15*, 112. [CrossRef]
18. Rao, J.; Ye, Z.; Tang, H.; Wang, C.; Peng, H.; Lai, W.; Li, Y.; Huang, W.; Lou, T. The RhoA/ROCK Pathway Ameliorates Adhesion and Inflammatory Infiltration Induced by AGEs in Glomerular Endothelial Cells. *Sci. Rep.* **2017**, *7*, 39727. [CrossRef]
19. Xu, Y.; Rong, J.; Zhang, Z. The emerging role of angiotensinogen in cardiovascular diseases. *J Cell Physiol.* **2021**, *236*, 68–78. [CrossRef]
20. Verma, S.K.; Lal, H.; Golden, H.B.; Gerilechaogetu, F.; Smith, M.; Guleria, R.S.; Foster, D.M.; Lu, G.; Dostal, D.E. Rac1 and RhoA differentially regulate angiotensinogen gene expression in stretched cardiac fibroblasts. *Cardiovasc. Res.* **2011**, *90*, 88–96. [CrossRef]
21. Vainer, B.; Nielsen, O.H. Changed colonic profile of P-selectin, platelet-endothelial cell adhesion molecule-1 (PECAM-1), intercellular adhesion molecule-1 (ICAM-1), ICAM-2, and ICAM-3 in inflammatory bowel disease. *Clin. Exp. Immunol.* **2000**, *121*, 242–247. [CrossRef] [PubMed]

22. Furuichi, K.; Wada, T.; Iwata, Y.; Kokubo, S.; Hara, A.; Yamahana, J.; Sugaya, T.; Iwakura, Y.; Matsushima, K.; Asano, M.; et al. Interleukin-1-dependent sequential chemokine expression and inflammatory cell infiltration in ischemia-reperfusion injury. *Crit. Care Med.* **2006**, *34*, 2447–2455. [CrossRef]
23. Dong, Y.F.; Kataoka, K.; Toyama, K.; Sueta, D.; Koibuchi, N.; Yamamoto, E.; Yata, K.; Tomimoto, H.; Ogawa, H.; Kim-Mitsuyama, S. Attenuation of brain damage and cognitive impairment by direct renin inhibition in mice with chronic cerebral hypoperfusion. *Hypertension* **2011**, *58*, 635–642. [CrossRef]
24. Reichenbach, N.; Delekate, A.; Plescher, M.; Schmitt, F.; Krauss, S.; Blank, N.; Halle, A.; Petzold, G.C. Inhibition of Stat3-mediated astrogliosis ameliorates pathology in an Alzheimer's disease model. *EMBO Mol. Med.* **2019**, *11*, e9665. [CrossRef] [PubMed]
25. Zhang, H.B.; Tu, X.K.; Chen, Q.; Shi, S.S. Propofol Reduces Inflammatory Brain Injury after Subarachnoid Hemorrhage: Involvement of PI3K/Akt Pathway. *J. Stroke Cerebrovasc. Dis.* **2019**, *28*, 104375. [CrossRef] [PubMed]
26. Tu, X.K.; Zhang, H.B.; Shi, S.S.; Liang, R.S.; Wang, C.H.; Chen, C.M.; Yang, W.-Z. 5-LOX Inhibitor Zileuton Reduces Inflammatory Reaction and Ischemic Brain Damage Through the Activation of PI3K/Akt Signaling Pathway. *Neurochem. Res.* **2016**, *41*, 2779–2787. [CrossRef]
27. Ayadi, A.E.; Zigmond, M.J.; Smith, A.D. IGF-1 protects dopamine neurons against oxidative stress: Association with changes in phosphokinases. *Exp. Brain Res.* **2016**, *234*, 1863–1873. [CrossRef]
28. Shi, X.; Cai, X.; Di, W.; Li, J.; Xu, X.; Zhang, A.; Qi, W.; Zhou, Z.; Fang, Y. MFG-E8 Selectively Inhibited Aβ-Induced Microglial M1 Polarization via NF-κB and PI3K-Akt Pathways. *Mol. Neurobiol.* **2017**, *54*, 7777–7788. [CrossRef]
29. Dai, S.J.; Zhang, J.Y.; Bao, Y.T.; Zhou, X.J.; Lin, L.N.; Fu, Y.B.; Zhang, Y.J.; Li, C.Y.; Yang, Y.X. Intracerebroventricular injection of Aβ(1-42) combined with two-vessel occlusion accelerate Alzheimer's disease development in rats. *Pathol. Res. Pract.* **2018**, *214*, 1583–1595. [CrossRef]
30. Fu, X.; Chen, H.; Han, S. C16 peptide and angiopoietin-1 protect against LPS-induced BV-2 microglial cell inflammation. *Life Sci.* **2020**, *256*, 117894. [CrossRef]

Article

Structure-Activity Relationship Investigations of Novel Constrained Chimeric Peptidomimetics of SOCS3 Protein Targeting JAK2

Sara La Manna [1], Marilisa Leone [2], Flavia Anna Mercurio [2], Daniele Florio [1] and Daniela Marasco [1,*]

1 Department of Pharmacy, Research Center on Bioactive Peptides (CIRPEB), University of Naples "Federico II", 80131 Naples, Italy; sara.lamanna@unina.it (S.L.M.); daniele.florio@unina.it (D.F.)
2 Institute of Biostructures and Bioimaging (CNR), 80145 Naples, Italy; marilisa.leone@cnr.it (M.L.); flaviaanna.mercurio@cnr.it (F.A.M.)
* Correspondence: daniela.marasco@unina.it; Tel.: +39-0812534607

Abstract: SOCS3 (suppressor of cytokine signaling 3) protein suppresses cytokine-induced inflammation and its deletion in neurons or immune cells increases the pathological growth of blood vessels. Recently, we designed several SOCS3 peptidomimetics by assuming as template structures the interfacing regions of the ternary complex constituted by SOCS3, JAK2 (Janus Kinase 2) and gp130 (glycoprotein 130) proteins. A chimeric peptide named KIRCONG chim, including non-contiguous regions demonstrated able to bind to JAK2 and anti-inflammatory and antioxidant properties in VSMCs (vascular smooth muscle cells). With the aim to improve drug-like features of KIRCONG, herein we reported novel cyclic analogues bearing different linkages. In detail, in two of them hydrocarbon cycles of different lengths were inserted at positions i/i+5 and i/i+7 to improve helical conformations of mimetics. Structural features of cyclic compounds were investigated by CD (Circular Dichroism) and NMR (Nuclear Magnetic Resonance) spectroscopies while their ability to bind to catalytic domain of JAK2 was assessed through MST (MicroScale Thermophoresis) assay as well as their stability in biological serum. Overall data indicate a crucial role exerted by the length and the position of the cycle within the chimeric structure and could pave the way to the miniaturization of SOCS3 protein for therapeutic aims.

Keywords: mimetic peptides; cytokine signaling; JAK/STAT; SOCS3; stapled peptides

Citation: La Manna, S.; Leone, M.; Mercurio, F.A.; Florio, D.; Marasco, D. Structure-Activity Relationship Investigations of Novel Constrained Chimeric Peptidomimetics of SOCS3 Protein Targeting JAK2. *Pharmaceuticals* **2022**, *15*, 458. https://doi.org/10.3390/ph15040458

Academic Editor: Gill Diamond

Received: 18 March 2022
Accepted: 7 April 2022
Published: 9 April 2022

Publisher's Note: MDPI stays neutral with regard to jurisdictional claims in published maps and institutional affiliations.

Copyright: © 2022 by the authors. Licensee MDPI, Basel, Switzerland. This article is an open access article distributed under the terms and conditions of the Creative Commons Attribution (CC BY) license (https://creativecommons.org/licenses/by/4.0/).

1. Introduction

Suppressor of cytokine signaling (SOCSs) are a family of cytokine-inducible proteins able to inhibit cytokine signaling mainly through the negative regulation of the Janus kinase/signal transducer and activator of transcription (JAK/STAT) pathway [1]. JAK/STAT is a major contributor of chronic inflammatory diseases and is largely involved in the regulation of the expression of many genes involved in cellular activation, differentiation, migration, apoptosis, and proliferation [2].

This protein family includes eight members: SOCS1-SOCS7 and cytokine-inducible SH2-containing protein (CIS). Structurally, these proteins contain a Src homology 2 (SH2) domain, a variable N-terminal domain and a C-terminal SOCS box [3]. Two members of this family, SOCS1 and SOCS3 are the unique to have a motif in the N-terminal region, called kinase inhibitory region (KIR) [4] that is crucially involved in the inhibition of JAKs [5]. Biochemical studies have highlighted different mechanisms of action (MOA) of these proteins in the inhibition of JAKs: while the SH2 domain of SOCS1 directly binds to the activation loop of JAK [6], the SH2 domains of CIS, SOCS2 and SOCS3 bind to phosphorylated tyrosine residues on activated cytokine receptors (glycoprotein (gp) 130 in IL-6 signaling) [7]. Dysfunctions of and genetic alterations in JAK/STAT/SOCS axis are linked with a wide range of cardiovascular [8], inflammatory and autoimmune disorders [9]. SOCS1 is a critical regulators of IFNs signaling [10,11] and interleukins (IL) as −2 [12], −12, −23 [13], −6 [14]. Its deficiency leads to inflammation [15], while its overexpression represses pro-inflammatory genes [16] and apoptosis caused by reactive oxygen species (ROS) [17,18]. In atherosclerotic models, the inhibition of JAK2, STAT1 and STAT3 prevented lesion formation [19,20]. Otherwise, mice lacking SOCS1 (Socs1$^{-/-}$) resulted protected from viral infection [21,22].

SOCS3 has a key role in controlling IL-6 signaling in the hepatocyte priming stage and it attenuates hepatocyte proliferation [23]. SOCS3 overexpression revealed favorable effects in colorectal [24], ovarian cancer lines [25] in MCF7 BC (Breast Cancer) cells and several solid tumors [26]. It represses cell proliferation through the reduction of STAT3 expression [27] and tumor growth and metastasis formation in mouse xenograft models [28]. In non-small cell lung cancers (NSCLC) it inhibits many tumor cell functions [29,30]; while a downregulation of SOCS3 was found in gastric adenocarcinoma tissues [31] and linked with enhanced risks of recurrent disease in BC patients [32]. In the chronic constriction injury of the sciatic nerve (CCI), SOCS3 prevented the abnormal expression of IL-6 and C–C motif chemokine ligand 2 (CCL2) and attenuated allodynia in rats [33]. A direct link with cardiovascular diseases was often observed: in diabetic cardiac fibrosis tissue, diabetic cardiomyopathy (DCM) patients' heart tissue and cardiac fibroblasts (after long term high-glucose treatment), SOCS3 resulted downregulated while its overexpression inhibited cardiac fibroblast activation and collagen production [34].

Interestingly, in neovascular age-related macular degeneration (nAMD), SOCS3 overexpression in myeloid lineage cells suppressed laser-induced choroidal neovascularization (CNV) through the block of myeloid lineage-derived macrophage/microglia recruitment and proinflammatory factors [35].

In this *scenario*, the identification of SOCS 1, 3 proteomimetics endowed with anti-inflammatory/antioxidant properties, is considered a valid strategy to employ peptide-based compounds as novel therapeutics [36]. For SOCS1, the miniaturization process is well advanced since our [37–40] and other research groups [41–47], starting from KIR domain (52–67), developed promising peptidomimetics with important anti-inflammatory properties. Very recently, in turn, several JAK1 peptidomimetics able to interact tightly with the SOCS1-SH2 domain and to block its activity have been designed as potential antiviral drugs [48].

Conversely, given the structural complexity of the ternary complex formed by SOCS3 with JAK2 and gp130 [7], reports on SOCS3 mimetics have been developed only from our research activity, at the best of our knowledge. Few years ago, we identified, a linear peptide spanning 22–45 residues of SOCS3, including KIR and ESS (Extended SH2 Subdomain) regions, called KIRESS. Intratumoral administration of this peptide significantly reduced growth of squamous cell carcinoma [49] and in triple-negative breast cancer (TNBC) it prevented the formation of pulmonary metastasis [50] and it was applied as an in vivo mimetic of the whole SOCS3 protein [35].

Recently, we designed another SOCS3 mimetic, named KIRCONG chim, that includes non-contiguous protein regions. It demonstrated able to bind to the catalytic domain of JAK2 and to act as a potent anti-inflammatory and antioxidant agent in VSMCs (vascular smooth muscle cells) [51].

Herein, to improve the stability to proteolytic degradation and to rigidify the structure, we introduced conformational constraints into the linear KIRCONG chim sequence: in detail, four cyclic compounds were designed and their conformational features analyzed through CD (Circular Dichroism) and NMR (Nuclear Magnetic Resonance) spectroscopies while their binding abilities were assessed by MST (MicroScale Thermophoresis) technique. To evaluate if introduced chemical modifications could provide major stability with respect to linear peptide in cellular contexts, serum stability assays were carried out.

2. Materials and Methods

2.1. Peptide Synthesis

Peptides were synthesized as carboxyl C-termini on Wang resin, using 9-fluorenylmethoxycarbonyl/*tert*-butyl (Fmoc/*t*Bu) strategy. For KIRCONG *amide*, the formation of a lactam bridge was obtained on solid support, by employing the super-acid labile protecting groups of side chains in Fmoc-Lys(Mtt)-OH and Fmoc-Glu(O-2-PhiPr)OH derivatives [37]. For KIRCONG *disulfide*, the peptide was dissolved in sodium carbonate 100 mM (0.1 mg/mL) and the mixture was left open to atmosphere under magnetic stirring, until the intramolecular oxidation was complete (confirmed by LC-MS analysis). Stapled analogues, KIRCONG i/i+5 and i/i+7, were obtained via ruthenium-based ring-closing metathesis (RCM) of olefin-derivatized amino acid residues (2-(4′-pentenyl)alanine and 2-(7′-octenyl)alanine) at the (i) and (i+5) or (i+7) positions in the peptide backbone. Peptides were purified through RP-HPLC and identified as already reported [52]. Purified peptides were lyophilized and stored at −20 °C until use.

2.2. Shake Flask Procedure for Determination of Log P

The logarithmic partition coefficient Log P between 1-octanol and water phases was determined for KIRCONG analogues using the shake flask method [53]. All peptides were dissolved in an equal volume of water (pre-saturated with 1-octanol) and 1-octanol (pre-saturated with ultrapure water) to reach a final concentration of each peptide of 300 µM. The mixtures were shaken mechanically for 120 min at 25 °C. The samples were centrifuged to assist with bilayer formation. The experiments were performed at least in duplicates.

The concentrations in water phase were determined by UV/Vis absorption (BioDrop-DUO-spectrophotometer, Biochrom, Waterbeach Cambridge, UK) employing as $\varepsilon_{275\,nm} = 8450\ M^{-1}\ cm^{-1}$, due to the presence of two Phe, two Tyr and one Trp residues in the sequences. The Log P was calculated according to the following equation, by assuming $C_{octanol} = C_{total} - C_{water}$ (C: peptide concentration):

$$\log P = \log \frac{C_{octanol}}{C_{water}}$$

2.3. Circular Dichroism (CD) Spectroscopy

CD spectroscopy experiments were carried out by employing a Jasco J-815 spectropolarimeter (JASCO, Tokyo, Japan), at room temperature in the spectral range 190–260 nm and spectra are averaged over two scans, to which blanks were subtracted. CD signals were converted to mean residue ellipticity with deg*cm^2*$dmol^{-1}$*res^{-1} as units. Scan speed value was 20 nm/min, band width 2.0 nm, resolution 0.2 nm, sensitivity 50 mdeg and response 4 s. Samples were prepared by dilution of freshly prepared stock solutions in 100% TFE (2,2,2-Trifluoroethanol) (1 mM on average). In the CD samples compound concentrations were 100 µM. Spectra were acquired in a quartz cuvette with a path-length of 0.1 cm in mixtures TFE/phosphate buffer, 10 mM at pH 7.4 [52].

2.4. NMR Studies

NMR spectra of KIRCONG i/i+5 and KIRCONG i/i+7 stapled peptides were registered at 25 °C on a Varian Unity Inova 600 MHz spectrometer provided with a cold probe. KIRCONG i/i+5 NMR experiments were acquired in H_2O/TFE (2,2,2-trifluoroethanol-D3 −99.5% isotopic purity, Sigma-Aldrich, Milan, Italy) 60/40 v/v and 85/15 v/v, peptide concentration equal to 411 µM, pH 4.52 and 6.84 at the highest and lowest TFE percentages, respectively. NMR spectra for KIRCONG i/i+7 were recorded in H_2O/TFE 60/40 v/v (peptide concentration 411 µM, pH 4.65). The volume of all NMR samples was equal to 500 µL. To conduct NMR structural analyses the following 2D [^{1}H, ^{1}H] experiments were recorded: TOCSY (Total Correlation Spectroscopy) [54], NOESY (Nuclear Overhauser Enhancement Spectroscopy) [55], ROESY (Rotating Frame Overhauser Enhancement Spectroscopy) [56] and DQFCOSY (Double Quantum-Filtered Correlated Spectroscopy) [57].

Typical acquisition parameters included 16–64 scans, 128–256 FIDs in t1, 1024 or 2048 data points in t2. Mixing times for TOCSY experiments were set to 70 ms whereas, mixing times equal to 200 ms and 300 ms were used to record NOESY spectra; ROESY experiments were acquired with a mixing time equal to 250 ms. Residual water signal was suppressed through excitation sculpting [58]. A standard strategy was followed to gain proton resonance assignments [59]. Trimethylsilyl-3-propionic acid sodium salt-D4 (TSP) (99% D, Armar Scientific, Döttingen, Switzerland) was used as internal standard for chemical shifts referencing. The Varian software VNMRJ 1.1D (Varian/Agilent Technologies, Milan, Italy) was implemented to process NMR spectra that were next analyzed with the program NEASY [60] included in CARA (Computer Aided Resonance Assignment) (http://www.nmr.ch/, accessed on 2 March 2022).

Chemical shift deviations from random coil values for Hα protons (CSD) for KIRCONG i/i+7 in H_2O/TFE 60/40 v/v were evaluated with the method by Kjaergaard and collaborators [61]. Random-coil chemical shift reference values refer to T = 25 °C and pH 4.65 in H_2O/TFE 60/40 v/v (https://www1.bio.ku.dk/english/research/bms/sbinlab/randomchemicalshifts1, accessed on 2 March 2022).

2.5. NMR Structure Calculations and Analysis

The NMR solution structure of KIRCONG i/i+7 in H_2O/TFE 60/40 v/v was calculated with the software CYANA (version 2.1) [62]. CYANA library entries for the non-standard amino acids β-Alanine, (R)-N-2-(7'-octenyl) alanine and (S)-2-(4'-pentenyl) alanine were generated with the CLYB software [63]. To simulate the olefinic linker, the distance between CZ1 atom of (R)-N-2-(7'-octenyl) alanine and CE atom of (S)-2-(4'-pentenyl) alanine was imposed equal to 1.34 Å. Distance constraints (i.e., upper distance limits) were generated from manual integration of peaks in 2D [^{1}H,^{1}H] NOESY spectrum (300 ms mixing time); the GRIDSEARCH module of CYANA software [62] was implemented to obtain angular constraints. 100 random conformers were initially generated and in the end the 20 structures provided with the lowest CYANA target functions and better obeying to experimental constraints, were selected as representative NMR conformers [62,64]. The software MOLMOL [65] was used to additionally analyze the NMR peptide structure and to generate images.

2.6. Serum Stability

KIRCONG analogues (~1 mg/mL, 500 μM on average), were incubated with fetal bovine serum (FBS) at 25% (w/v) at 37 °C as previously described [39]. At the following time points: 0, 3, 17, 20, 23 and 42 h, 50 μL aliquots of the solutions were mixed with 50 μL of 15% trichloroacetic acid (TCA) to allow the precipitation of serum proteins, they were stored at −20 °C for at least 15 min. After centrifugation (13,000 rpm for 15 min) the supernatants were recovered. Samples were analyzed by RP-HPLC on a HPLC LC-4000 series (Jasco) equipped with UV detector using a C18-Kinetek column from Phenomenex (Milan, Italy), by employing a gradient from 5 to 70% of B (acetonitrile 0.1% TFA) versus A (water 0.1% TFA) in 20 min. Peptide compounds were detected by recording the absorbance at 210 nm and percentages were quantified by assuming 100% their peak areas at t = 0. All stability tests were performed at least in triplicates and reported data are averaged values.

2.7. Microscale Thermophoresis

MST experiments were carried out with a Monolith NT 115 system (NanoTemper Technologies, München, Germania) equipped with 40% LED and 40% IR-laser power. Labeling of His-tagged Catalytic Domain of JAK2 (residues 826–1132) (Carna Biosciences, Kobe, Japan) was achieved with the His-Tag labeling Kit RED-tris-NTA. The protein concentration was adjusted to 200 nM in labeling buffer (Nano Temper Technologies), while the dye concentration was set to 100 nM. Equal volumes (100 μL) of protein and fluorescent dye solutions were mixed and incubated at room temperature in the dark for 30 min. KIRCONG chim analogues were used in the following concentrations: KIRCONG amide

459 µM, KIRCONG disulfide 470 µM, KIRCONG i/i+5 525 µM, and KIRCONG i/i+7 585 µM in labeling buffer. Standard capillaries were employed for analysis, at 25 °C in 50 mM Tris-HCl, 150 mM NaCl, 0.05% Brij35, 1 mM DTT, 10% glycerol, 15% TFE buffer at pH 7.5, as already reported [51]. The equation implemented by the software MO-S002 MO Affinity Analysis [66], used for fitting data at different concentrations, is based on Langmuir binding isotherm.

3. Results

3.1. Design of Constrained KIRCONG Chim Mimetics

Our previous investigations on mimetics of SOCS3 pointed out that different protein regions provided hot spots for JAK2 recognition. In detail, KIRCONG chim peptide is a chimeric peptide mostly centered on KIR domain, since it includes the stretch 25–33 (while KIR spans 22–35 residues) covalently conjugated to a small aromatic stretch, named CONG (46–52 residues) (Table 1) that provides specific aromatic interactions with the catalytic domain of the kinase. This compound exhibited a low micromolar value of K_D through MST assay, but NMR studies performed in H_2O revealed a flexible conformation lacking regular secondary structure elements. The presence of 15% TFE induced a slight decrease of flexibility of the compound that remained prevalently disordered, even if a certain helical content was encountered [50]. On this basis, herein we report the design and analysis of KIRCONG chim analogues in which structural modifications were inserted (Table 1) and their chemical structures are reported in Figure 1. These constraints were introduced with the aim to ameliorate drug-like features of mimetics indeed cyclization is a powerful approach to improve selectivity, metabolic stability, and bioavailability of bioactive peptides [39,67–69].

Table 1. Sequences and names of compounds investigated in this study. Residues belonging to different human SOCS3 regions are colored in blue (KIR) and orange (CONG). X: (S)-2-(4′-pentenyl) alanine Z: (R)-2-(7′-octenyl) alanine).

Name	Sequence	K_D (µM)	LogP
KIRCONGchim [51]	AcNH-KβAlaF^{25}SSKSEYQL^{33}βAlaβAlaF^{46}YWSAVT^{52}-CONH$_2$	$(1.1 \pm 0.3) \times 10$	−1.32
KIRCONG amide	H-CβAlaF^{25}SSKSEYQL^{33}βAlaβAlaF^{46}YWSAVT^{52}CG-OH	$(2.0 \pm 0.9) \times 10^2$	−0.92
KIRCONG disulfide	H-CβAlaF^{25}SSKSEYQL^{33}βAlaβAlaF^{46}YWSAVT^{52}CG-OH	$(6 \pm 3) \times 10^2$	−1.46
KIRCONG i/i+5	H-F^{25}SSKSEYQL^{33}βAla XF^{46}YWSXVT^{52}G-OH	No binding	−0.27
KIRCONG i/i+7	H-F^{25}SSKSEYZL^{33}βAlaF^{46}YWSXVT^{52}G-OH	No binding	0.29

Different cyclic compounds were conceived: the first two are macrocycles containing disulfide and lactam bridges, respectively, between residues located in the C- and N-terminal extremities. In detail: KIRCONG amide was obtained through the formation of a peptide bond between the side chains of a Glu residue at the C-termini (native position 53 of CONG domain) and a non-native Lys upstream of KIR; in KIRCONG *disulfide* a cysteine bond was obtained between thiol groups of two non-native Cys (Table 1). After a detailed analysis of point mutations of SOCS3 residues that contact JAK2 protein, reported in [7], we introduced stapled constraints [70]. Bridges were generated by substituting residues not crucial for the interaction: in KIRCONG i/i+5, βAla and Ser^{50} and in KIRCONG i/i+7, Gln^{32} and Ser^{50} of reference KIRCONG chim (Table 1).

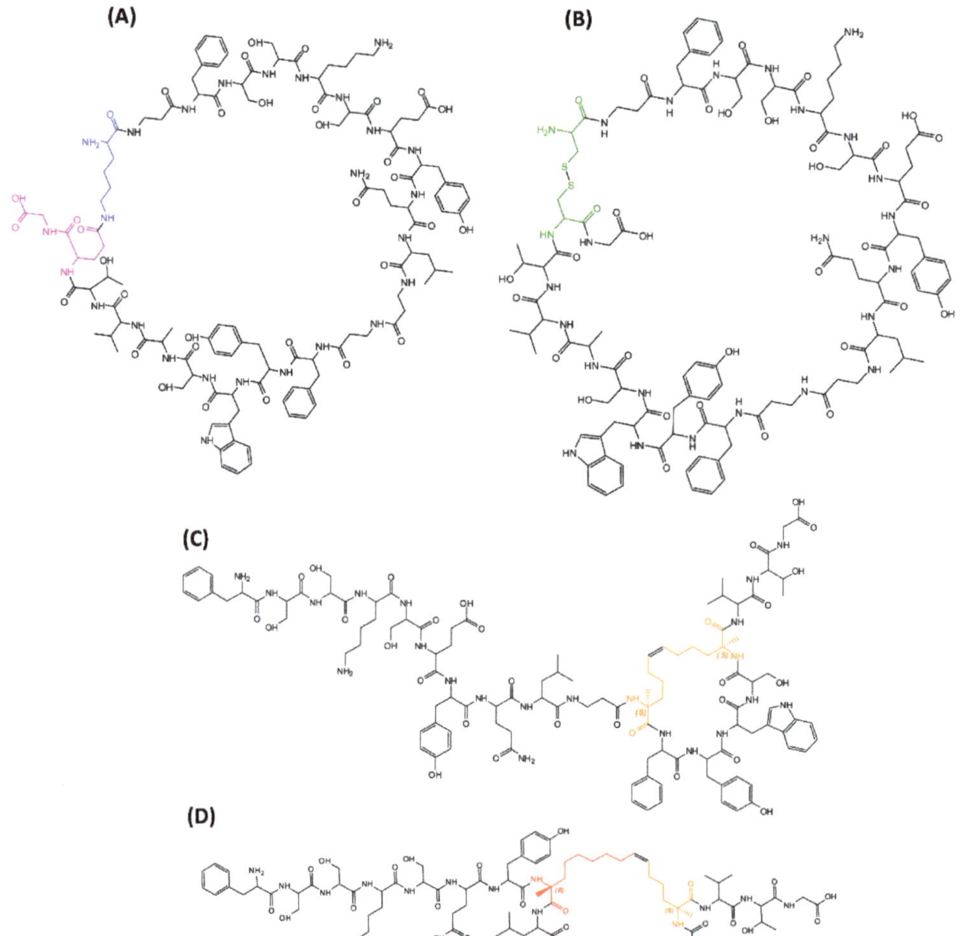

Figure 1. Chemical structures of KIRCONG analogues with residues involved in cycle formation colored: (**A**) KIRCONG amide (blue: Lys, purple: Asp), (**B**) KIRCONG disulfide (green: Cys), (**C**) KIRCONG i/i+5 (orange: (S)-N- 2-(4′-pentenyl) alanine) and (**D**) KIRCONG i/i+7 (orange: (S)-N-2-(4′-pentenyl) alanine, red: (R)-2-(7′-octenyl) alanine).

The α-helix is the most common secondary structure present in nature but different helical structures can be found in proteins, as 3_{10}- and π-helix, even if less frequently [71]. To evaluate the effects of different helical structures into KIRCONG chim we introduced stapling residues at different positions: (i) i and i+7 to generate a α-helix and (ii) i and i+5 to *f*orm of the so-called π-helix [71,72]. As consequence, in KIRCONG i/i+7 the bridge encompasses both KIR and CONG stretches, while in KIRCONG i/i+5 the cycle is located only in CONG region leaving the KIR region more flexible.

To study the lipophilicity of designed compounds, we evaluated LogP values through the classical shake flask method [53] and reported them in Table 1. All KIRCONG analogues exhibited negative values except KIRCONG i/i+7, indicating its minor water solubility with respect to others.

3.2. Conformational Studies of Constrained KIRCONG Analogues

The conformational features of KIRCONG analogues were investigated through CD and NMR spectroscopies.

3.2.1. Circular Dichroism

CD spectra were recorded in the far UV region, for the low water solubility of all analogues, their stock solutions were prepared in 100% TFE and then diluted in aqueous buffer, till 15% v/v (TFE/aqueous buffer). The overlays of CD spectra are reported in Figure 2. As expected, for their non-native construction, KIRCONG analogues did not present canonical CD profiles. By comparing their spectral features with that of the lead compound [73] at the lowest TFE percentage (15%) only KIRCONG amide presents the slight positive band at ~235 nm due to an aromatic contribution to the peptide conformation. In this solvent system, also the deconvolution of CD spectra (Table S1) indicates prevalent random + beta contents, particularly evident for the minimum at ~218 nm for KIRCONG disulfide and KIRCONG i/i+7 (Figure 2B,D). For both stapled compounds (Figure 2C,D) higher helical contents were already present at 15% TFE (8.2 and 3.8%, respectively) with respect to the other two analogues, for which helix percentage was 0 (Table S1). Increasing amounts of TFE induced more ordered conformations especially for stapled structures: for them a clear transition toward helical conformations was detected, as testified by the progressive appearance of a secondary minimum at 220 nm.

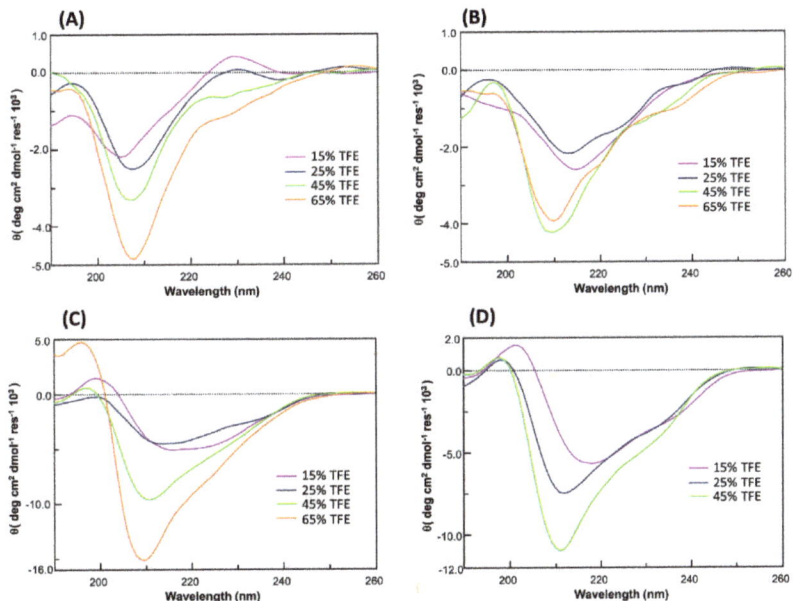

Figure 2. CD spectra of: (**A**) KIRCONG amide; (**B**) KIRCONG disulfide; (**C**) KIRCONG i/i+5; (**D**) KIRCONG i/i+7 in TFE/buffer 15-65% v/v. KIRCONG i/i+7 resulted not soluble in 65/25 v/v, TFE/buffer and the related spectrum is absent.

3.2.2. NMR Studies

Since stapled analogues are generally characterized by an increase of helical conformations with respect to linear counterparts, NMR studies coupled to CD analyses are useful to reveal such tendencies [74].

NMR characterization of stapled analogues was conducted under similar experimental conditions to those employed for KIRCONG chim [51]. Indeed, NMR spectra of KIRCONG i/i+5 were recorded in H_2O/TFE mixtures containing increasing amounts of TFE (e.g., 15 and 40%) (Figure S1). By increasing TFE amount, NMR spectra showed a certain improvement of dispersion (Figure S1A) likely indicating an increase of folded peptide population. However, a better comparison of 2D NMR experiments (Figure S1B) showed a similar number of NOE peaks at both TFE concentrations highlighting only a minor improvement of structuration. By analyzing TOCSY and NOESY spectra many protons chemical shifts were identified although the assignments resulted ambiguous, particularly in the peptide region encompassing residues 10-16 (KIRCONG i/i+5, peptide numeration), where the non-natural residues (β-Alanine and (S)-N-2-(4′-pentenyl) alanine) are located (Tables S2 and S3). Within this region many duplicated spin systems are detected; this variability could be due to a CIS-TRANS equilibrium around the double bond of the staple coupled to the larger backbone flexibility induced by the β-Alanine. The presence of multiple conformers and the ambiguity of certain signals hampered to assign all proton residues and achieve a reliable structural model.

Better results were obtained with KIRCONG i/i+7, for which NMR spectra were recorded in solution at 40% TFE (Figure S2). Indeed, this compound resulted poorly soluble at NMR concentration in 15% TFE (Table 1): this is likely due to the presence of the longer alkyl chain of the i/i+7 stapled analogue. The comparison of TOCSY and NOESY spectra allowed to clearly assign almost completely the resonances of peptide protons (Table S4) [59]. The presence of a strong contact between the $H\zeta 1$ and $H\varepsilon$ protons in the spectral region around 5 ppm (Table S4), indicated a trans arrangement of the double bond. The analysis of chemical shifts for $H\alpha$ protons with respect to the random coil reference values showed a negative trend (Figure 3A), mainly, [75,76] that is characteristic of helical/turn conformations, more evident in the region between residues Y^7 and V^{16}, including the cyclic arrangement of the stapled peptide. The NOE pattern (Figure 3B) even if not canonically defined in helical conformation, presented a few signals of the type $H\alpha$-$H\beta$ i/i+3 and $H\alpha$-H_N i/i+4 pointing out some helical content [59] in the peptide fragment 5-12. The NMR structure of KIRCONG i/i+7 (Figure 3C,D, Table S5) is represented by a distorted helical arrangement extending through the whole sequence and appearing more regular in the C-terminal region between Y^{12} and (S)-N-2-(4′-pentenyl) alanine^{15} (KIRCONG i/i+7, peptide numeration) due to the presence of the hydrocarbon stapling (Figure 3D). Analysis of the NMR ensemble with the software MOLMOL [65] revealed the presence of a few backbone H-bonds characteristic of α-helix (i.e., (S)-N-2-(4′-pentenyl) alanine^{15} H_N-F^{11} $_CO$ in 12/20 conformers; V^{16} H_N-Y^{12} $_CO$ in 20/20 conformers; G^{18} H_N-S^{14} $_CO$ in 5/20 conformers).

3.3. Serum Stability

To evaluate if introduced chemical modifications in KIRCONG analogues influence their stabilities to proteolytic degradation, pure compounds were incubated with fetal bovine serum (FBS) and the decrease of chromatographic peaks was followed during time In Figure 4, the area percentages of new derivatives of KIRCONG chim versus time are reported. After 3 h of incubation, KIRCONG chim is degraded by 30% while KIRCONG disulfide of 20% and others less than 10%. Interestingly, in both stapled peptides, the presence of unnatural amino acids used for staples formation, greatly increased their stability: after 42 h both peptides still showed a residual concentration of 85% and 75%, respectively while KIRCONG chim peptide appeared completely degraded.

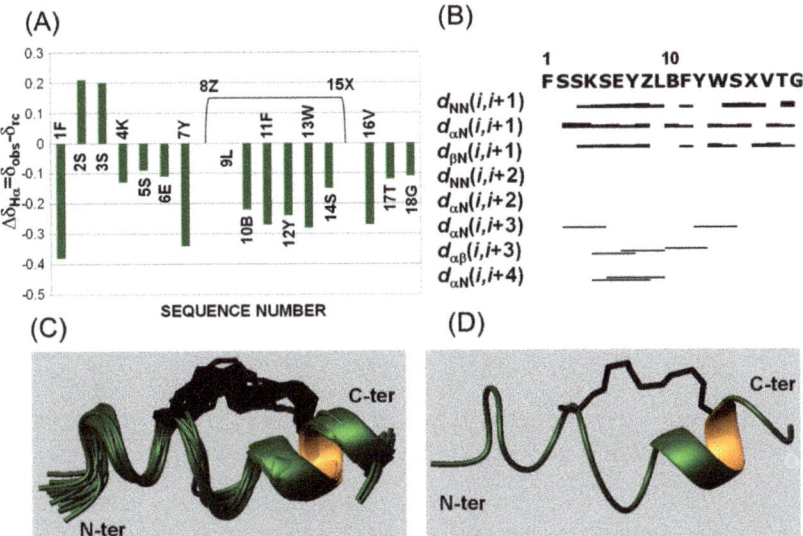

Figure 3. NMR analysis of KIRCONG i/i+7 in H$_2$O/TFE 60/40 v/v. (**A**) Chemical shift deviations of Hα protons from random coil values ($\Delta\delta_{H\alpha}$). Standard amino acids are reported with the one letter code whereas, B stands for β-Alanine, Z stands for (R)-N-2-(7′-octenyl) alanine and X stands for (S)-2-(4′-pentenyl) alanine. $\Delta\delta_{H\alpha}$ for Z and X is set equal to 0 as the reference random coil value is missing. For $\Delta\delta_{H\alpha}$ evaluation β-Alanine was assimilated to Glycine. (**B**) NOEs pattern. (**C**) Ribbon representation of KIRCONG *i/i+7* NMR structures: 20 conformers are superimposed on the backbone atoms of residues 3–17, and (**D**) ribbon representation of the first NMR conformer. The aliphatic linker between (R)-N-2-(7′-octenyl) alanine (residue n.8) and (S)-2-(4′-pentenyl) alanine (residue n.15) is shown in black. The NMR structure was generated from 261 upper distance limits (170 intraresidue, 58 short-range, 27 medium-range and 6 long-range) and 73 angular constraints.

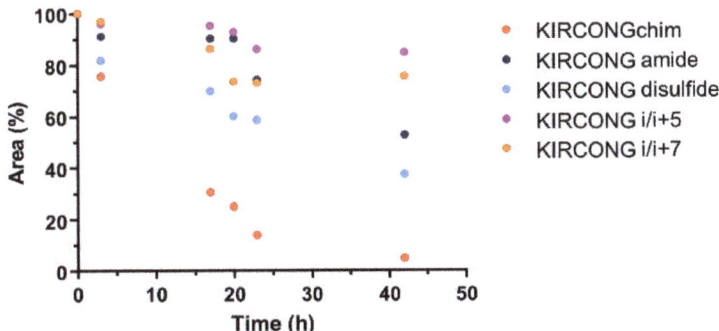

Figure 4. Serum stability assay of KIRCONG chim derivatives. The serum stability was evaluated by incubation in 25% FBS for 42 h. Residual peptide amount is expressed as the percentage of the initial amount versus time.

3.4. MST Investigations

To evaluate the ability of KIRCONG analogues to recognize JAK2 catalytic domain, in vitro MST experiments were carried out (Figure 5), keeping JAK2 concentration constant and increasing ligands' concentrations.

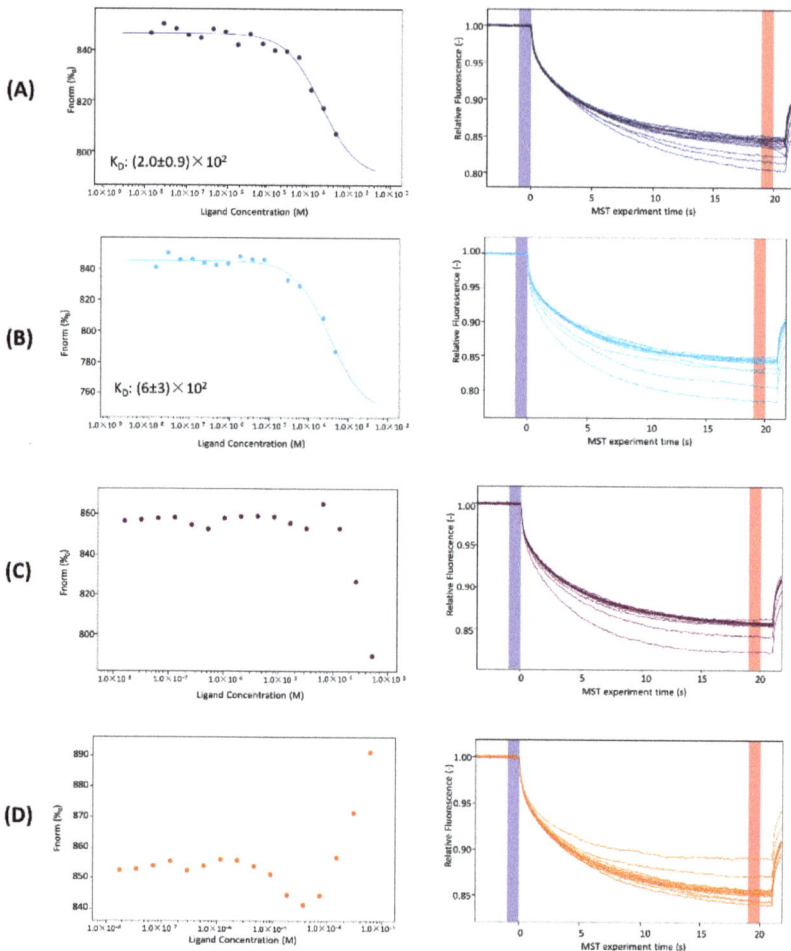

Figure 5. Left column: binding isotherms for MST signals versus peptide concentrations. Right column: thermophoretic traces of MST assays for the binding to JAK2 of (**A**) KIRCONG amide; (**B**) KIRCONG disulfide; (**C**) KIRCONG i/i+5; (**D**) KIRCONG i/i+7.

Even if all signals exhibit a dose–response curve for all four cyclic peptides (capillary shape and scan are reported in Figure S4), for KIRCONG i/i+5 and KIRCONG i/i+7 the signal variation was not meaningful and did not allow to obtain K_D values [77,78]. Concerning the other two analogues, KIRCONG amide and KIRCONG disulfide, the signal did not reach saturation and the fitting of data provided very high micromolar K_D values (Figure 5 and Table 1).

4. Discussion

Given the key role that the JAK/STAT pathway plays in many disorders, the identification of molecules capable of modulating its activity, is a powerful strategy for the treatment of several serious diseases. In last years, many researchers have focused their attention on the identification of various small molecules capable to interfere with this pathway by inhibiting the activity of the JAK proteins and many of them have already been approved by U.S. Food and Drug Administration (FDA) (e.g., two JAK inhibitors (JAKins), Ruxolitinib for myeloproliferative disorders [79] and Afatinib in non-small-cell lung carcinoma (NSCLC) [80]). In this context, a parallel and more specific approach concerns the employment of mimetics of endogenous regulators of the JAK/STAT signaling, as SOCS proteins. Recently, we developed a mimetic of SOCS3, named KIRCONG chim, including two non-contiguous regions able to bind to JAK2 that also exhibited anti-inflammatory and antioxidant properties [51]. By employing KIRCONG chim as template compound, herein we present SAR investigations of several analogues bearing different cycles in the monomeric structure, obtained by the insertion and/or substitution of non-crucial residues of linear peptide counterpart.

In detail, four cyclic analogues of KIRCONG chim were conceived: the first two, KIRCONG amide and disulfide, contained both a head-to-tail macrocycle where the cycle was obtained through the side chains of residues located at the opposite extremities of the sequence. In them the macrocyclization does not induce a significant limitation of flexibility (as observable from CD analysis) allowing however the recognition of JAK2 catalytic domain even if with lower affinity with respect to KIRCONG chim (Table 1). As expected, these kinds of cycles, more "natural" with respect to the others two, do not produce a significant improvement of stability to proteases after 24 h of analysis.

On the other hand, in KIRCONG i/i+5 and i/i+7, the stapled linkages arise from side-chain-to side-chain non native bonds and are shorter with respect to *amide* and *disulfide* and involve different stretches of the chimeric structure: in the first analogue the cycle encompass only CONG region (its N-terminal residue is located in the β-alanines linker) while in the second both KIR and CONG fragments are involved in the staple.

From a conformational point of view, by comparing NMR structures of KIRCONG i/i+7 with KIRCONG chim [51] (both in H_2O/TFE 60/40 v/v), the linear reference peptide exhibited a more ordered helical conformation (Figure S3). This result indicates that the co-existence of flexible β-alanines in the linking portion and of the hydrocarbon staple does not generate a rigid helical arrangement, generally typical of most stapled peptides with i/i+7 pattern [81]. On the other hand, the *i/i+5* arrangement in KIRCONG fails to generate a well folded structure and the compound assumes multiple conformational states.

The structural stiffening of KIRCONG chim in both stapled cases, provided a negative effect on the ability of the SOCS3 mimetic to recognize JAK2: MST assay, in the investigated concentration range (0.015–500 μM on average), did not provide a significant dose-response variation of the signal and the very limited water solubility hampered to increase ligands' amounts. In addition, the same binding assay employing the linear version of KIRCONG i/i+5 peptide, reported in Figure S5, provided a good dose-response of the MST signal and a K_D value (8 μM) that is quite similar to that of KIRCONG chim (11 μM) (Table 1). This experiment confirmed that the absence of cyclization restores the ability to bind JAK2. It is important to point out that the presence of stapled cycles provided greater stabilities to proteases till 42 h.

In conclusion, overall data indicate, how in the cyclization of bioactive peptides, a crucial role is exerted by the balance between limitation of flexibility of the backbone and right exposure of hot spots of interaction. The clear modulating effect by the length and the position of the cycle within the chimeric structure reported by this study, could help, in the future, the design of more soluble mimetics of SOCS3 as anti-inflammatory agents.

Supplementary Materials: The following supporting information can be downloaded at: https://www.mdpi.com/article/10.3390/ph15040458/s1, Figure S1: 1D [^{1}H] and 2D [^{1}H-^{1}H] NOESY 300 spectra of KIRCONG i/i+5 in H_2O/TFE 85/15 v/v and in H_2O/TFE 60/40 v/v.; Figure S2: 2D [^{1}H-^{1}H] TOCSY and NOESY 300 spectra of KIRCONG i/i+7 in H_2O/TFE 60/40 v/v; Figure S3: NMR structures of KIRCONG i/i+7 in H_2O/TFE (60/40, v/v) and KIRCONG chim in H_2O/TFE (60/40, v/v); Figure S4: Capillary shape and capillary scan of designed analogues; Figure S5: Binding isotherms for MST signals versus linear KIRCONG i/i+5 concentrations; Table S1: Deconvolution of CD spectra; Table S2: ^{1}H chemical shifts of KIRCONG i/i+5 in H_2O/TFE (60/40, v/v); Table S3: ^{1}H chemical shifts of KIRCONG i/i+5 in H_2O/TFE (85/15, v/v); Table S4: ^{1}H chemical shifts of KIRCONG i/i+7 in H_2O/TFE (60/40, v/v); Table S5: Structure statistics of KIRCONG i/i+7 NMR conformers in H_2O/TFE (60/40figurre, v/v).

Author Contributions: S.L.M. synthesized the compounds and performed shake flask assay and CD and MST experiments. S.L.M. and D.F. performed serum stability assay. M.L. and F.A.M. performed NMR experiments and calculated NMR structures. D.M. designed the concept and supervised the experiments. S.L.M. and D.M. wrote the manuscript. All authors have read and agreed to the published version of the manuscript.

Funding: This research was partially funded by FISM, grant number 2020/PRSingle/015, to D.M. Sara La Manna was supported by AIRC fellowship for Italy.

Institutional Review Board Statement: Not applicable.

Informed Consent Statement: Not applicable.

Data Availability Statement: The data is contained within the article and supplementary files.

Acknowledgments: Sara La Manna was supported by the AIRC fellowship for Italy.

Conflicts of Interest: The authors declare no conflict of interest.

References

1. Yoshimura, A.; Ito, M.; Mise-Omata, S.; Ando, M. SOCS: Negative regulators of cytokine signaling for immune tolerance. *Int. Immunol.* **2021**, *33*, 711–716. [CrossRef] [PubMed]
2. Hu, X.; Li, J.; Fu, M.; Zhao, X.; Wang, W. The JAK/STAT signaling pathway: From bench to clinic. *Signal Transduct Target Ther.* **2021**, *6*, 402. [CrossRef] [PubMed]
3. Galic, S.; Sachithanandan, N.; Kay, T.W.; Steinberg, G.R. Suppressor of cytokine signalling (SOCS) proteins as guardians of inflammatory responses critical for regulating insulin sensitivity. *Biochem. J.* **2014**, *461*, 177–188. [CrossRef] [PubMed]
4. Yoshimura, A.; Yasukawa, H. JAK's SOCS: A mechanism of inhibition. *Immunity* **2012**, *36*, 157–159. [CrossRef]
5. Croker, B.A.; Kiu, H.; Nicholson, S.E. SOCS regulation of the JAK/STAT signalling pathway. *Semin. Cell Dev. Biol.* **2008**, *19*, 414–422. [CrossRef]
6. Liau, N.P.D.; Laktyushin, A.; Lucet, I.S.; Murphy, J.M.; Yao, S.; Whitlock, E.; Callaghan, K.; Nicola, N.A.; Kershaw, N.J.; Babon, J.J. The molecular basis of JAK/STAT inhibition by SOCS1. *Nat. Commun.* **2018**, *9*, 1558. [CrossRef]
7. Kershaw, N.J.; Murphy, J.M.; Liau, N.P.; Varghese, L.N.; Laktyushin, A.; Whitlock, E.L.; Lucet, I.S.; Nicola, N.A.; Babon, J.J. SOCS3 binds specific receptor-JAK complexes to control cytokine signaling by direct kinase inhibition. *Nat. Struct. Mol. Biol.* **2013**, *20*, 469–476. [CrossRef]
8. Kishore, R.; Verma, S.K. Roles of STATs signaling in cardiovascular diseases. *JAKSTAT* **2012**, *1*, 118–124. [CrossRef]
9. Banerjee, S.; Biehl, A.; Gadina, M.; Hasni, S.; Schwartz, D.M. JAK-STAT Signaling as a Target for Inflammatory and Autoimmune Diseases: Current and Future Prospects. *Drugs* **2017**, *77*, 521–546. [CrossRef]
10. Blumer, T.; Coto-Llerena, M.; Duong, F.H.T.; Heim, M.H. SOCS1 is an inducible negative regulator of interferon lambda (IFN-lambda)-induced gene expression in vivo. *J. Biol. Chem.* **2017**, *292*, 17928–17938. [CrossRef]
11. Gao, A.H.; Hu, Y.R.; Zhu, W.P. IFN-gamma inhibits ovarian cancer progression via SOCS1/JAK/STAT signaling pathway. *Clin. Transl. Oncol.* **2022**, *24*, 57–65. [CrossRef] [PubMed]
12. Davey, G.M.; Starr, R.; Cornish, A.L.; Burghardt, J.T.; Alexander, W.S.; Carbone, F.R.; Surh, C.D.; Heath, W.R. SOCS-1 regulates IL-15-driven homeostatic proliferation of antigen-naive CD8 T cells, limiting their autoimmune potential. *J. Exp. Med.* **2005**, *202*, 1099–1108. [CrossRef] [PubMed]
13. Eyles, J.L.; Metcalf, D.; Grusby, M.J.; Hilton, D.J.; Starr, R. Negative regulation of interleukin-12 signaling by suppressor of cytokine signaling-1. *J. Biol. Chem.* **2002**, *277*, 43735–43740. [CrossRef] [PubMed]
14. Tamiya, T.; Kashiwagi, I.; Takahashi, R.; Yasukawa, H.; Yoshimura, A. Suppressors of Cytokine Signaling (SOCS) Proteins and JAK/STAT Pathways Regulation of T-Cell Inflammation by SOCS1 and SOCS3. *Arter. Throm. Vas.* **2011**, *31*, 980–985. [CrossRef]
15. Yoshimura, A.; Suzuki, M.; Sakaguchi, R.; Hanada, T.; Yasukawa, H. SOCS, Inflammation, and Autoimmunity. *Front. Immunol.* **2012**, *3*, 20. [CrossRef] [PubMed]
16. Madonna, S.; Scarponi, C.; De Pita, O.; Albanesi, C. Suppressor of cytokine signaling 1 inhibits IFN-gamma inflammatory signaling in human keratinocytes by sustaining ERK1/2 activation. *FASEB J.* **2008**, *22*, 3287–3297. [CrossRef] [PubMed]
17. Jung, S.H.; Kim, S.M.; Lee, C.E. Mechanism of suppressors of cytokine signaling 1 inhibition of epithelial-mesenchymal transition signaling through ROS regulation in colon cancer cells: Suppression of Src leading to thioredoxin up-regulation. *Oncotarget* **2016**, *7*, 62559–62571. [CrossRef]
18. Schuett, J.; Kreutz, J.; Grote, K.; Vlacil, A.K.; Schuett, H.; Oberoi, R.; Schmid, A.; Witten, A.; Stoll, M.; Schieffer, B.; et al. Suppressor of Cytokine Signaling 1 is Involved in Gene Regulation Which Controls the Survival of Ly6C(low) Monocytes in Mice. *Cell. Physiol. Biochem.* **2019**, *52*, 336–353. [CrossRef]
19. Miklossy, G.; Hilliard, T.S.; Turkson, J. Therapeutic modulators of STAT signalling for human diseases. *Nat. Rev. Drug Discov.* **2013**, *12*, 611–629. [CrossRef]

20. Torella, D.; Curcio, A.; Gasparri, C.; Galuppo, V.; De Serio, D.; Surace, F.C.; Cavaliere, A.L.; Leone, A.; Coppola, C.; Ellison, G.M.; et al. Fludarabine prevents smooth muscle proliferation in vitro and neointimal hyperplasia in vivo through specific inhibition of STAT-1 activation. *Am. J. Physiol. Heart Circ. Physiol.* **2007**, *292*, H2935–H2943. [CrossRef]
21. Du, Y.; Yang, F.; Wang, Q.; Xu, N.; Xie, Y.; Chen, S.; Qin, T.; Peng, D. Influenza a virus antagonizes type I and type II interferon responses via SOCS1-dependent ubiquitination and degradation of JAK1. *Virol. J.* **2020**, *17*, 74. [CrossRef] [PubMed]
22. Seong, R.K.; Lee, J.K.; Shin, O.S. Zika Virus-Induction of the Suppressor of Cytokine Signaling 1/3 Contributes to the Modulation of Viral Replication. *Pathogens* **2020**, *9*, 163. [CrossRef] [PubMed]
23. Khan, M.G.M.; Ghosh, A.; Variya, B.; Santharam, M.A.; Kandhi, R.; Ramanathan, S.; Ilangumaran, S. Hepatocyte growth control by SOCS1 and SOCS3. *Cytokine* **2019**, *121*, 154733. [CrossRef] [PubMed]
24. Chu, Q.; Shen, D.; He, L.; Wang, H.; Liu, C.; Zhang, W. Prognostic significance of SOCS3 and its biological function in colorectal cancer. *Gene* **2017**, *627*, 114–122. [CrossRef]
25. Shang, A.Q.; Wu, J.; Bi, F.; Zhang, Y.J.; Xu, L.R.; Li, L.L.; Chen, F.F.; Wang, W.W.; Zhu, J.J.; Liu, Y.Y. Relationship between HER2 and JAK/STAT-SOCS3 signaling pathway and clinicopathological features and prognosis of ovarian cancer. *Cancer Biol. Ther.* **2017**, *18*, 314–322. [CrossRef]
26. Dai, L.; Li, Z.; Tao, Y.; Liang, W.; Hu, W.; Zhou, S.; Fu, X.; Wang, X. Emerging roles of suppressor of cytokine signaling 3 in human cancers. *Biomed. Pharm.* **2021**, *144*, 112262. [CrossRef]
27. Barclay, J.L.; Anderson, S.T.; Waters, M.J.; Curlewis, J.D. SOCS3 as a tumor suppressor in breast cancer cells, and its regulation by PRL. *Int. J. Cancer* **2009**, *124*, 1756–1766. [CrossRef]
28. Hill, G.R.; Kuns, R.D.; Raffelt, N.C.; Don, A.L.; Olver, S.D.; Markey, K.A.; Wilson, Y.A.; Tocker, J.; Alexander, W.S.; Clouston, A.D.; et al. SOCS3 regulates graft-versus-host disease. *Blood* **2010**, *116*, 287–296. [CrossRef]
29. Lin, Y.C.; Lin, C.K.; Tsai, Y.H.; Weng, H.H.; Li, Y.C.; You, L.; Chen, J.K.; Jablons, D.M.; Yang, C.T. Adenovirus-mediated SOCS3 gene transfer inhibits the growth and enhances the radiosensitivity of human non-small cell lung cancer cells. *Oncol. Rep.* **2010**, *24*, 1605–1612. [CrossRef]
30. Speth, J.M.; Penke, L.R.; Bazzill, J.D.; Park, K.S.; de Rubio, R.G.; Schneider, D.J.; Ouchi, H.; Moon, J.J.; Keshamouni, V.G.; Zemans, R.L.; et al. Alveolar macrophage secretion of vesicular SOCS3 represents a platform for lung cancer therapeutics. *JCI Insight* **2019**, *4*, e131340. [CrossRef]
31. Tang, H.; Long, Q.; Zhuang, K.; Yan, Y.; Han, K.; Guo, H.; Lu, X. miR-665 promotes the progression of gastric adenocarcinoma via elevating FAK activation through targeting SOCS3 and is negatively regulated by lncRNA MEG3. *J. Cell Physiol.* **2020**, *235*, 4709–4719. [CrossRef] [PubMed]
32. Ying, M.Z.; Li, D.W.; Yang, L.J.; Wang, M.; Wang, N.; Chen, Y.; He, M.X.; Wang, Y.J. Loss of SOCS3 expression is associated with an increased risk of recurrent disease in breast carcinoma. *J. Cancer Res. Clin.* **2010**, *136*, 1617–1626. [CrossRef] [PubMed]
33. Dominguez, E.; Mauborgne, A.; Mallet, J.; Desclaux, M.; Pohl, M. SOCS3-mediated blockade of JAK/STAT3 signaling pathway reveals its major contribution to spinal cord neuroinflammation and mechanical allodynia after peripheral nerve injury. *J. Neurosci.* **2010**, *30*, 5754–5766. [CrossRef] [PubMed]
34. Tao, H.; Shi, P.; Zhao, X.D.; Xuan, H.Y.; Gong, W.H.; Ding, X.S. DNMT1 deregulation of SOCS3 axis drives cardiac fibroblast activation in diabetic cardiac fibrosis. *J. Cell Physiol.* **2021**, *236*, 3481–3494. [CrossRef]
35. Wang, T.; Zhou, P.; Xie, X.; Tomita, Y.; Cho, S.; Tsirukis, D.; Lam, E.; Luo, H.R.; Sun, Y. Myeloid lineage contributes to pathological choroidal neovascularization formation via SOCS3. *EBioMedicine* **2021**, *73*, 103632. [CrossRef]
36. La Manna, S.; De Benedictis, I.; Marasco, D. Proteomimetics of Natural Regulators of JAK-STAT Pathway: Novel Therapeutic Perspectives. *Front. Mol. Biosci.* **2021**, *8*, 792546. [CrossRef]
37. La Manna, S.; Lopez-Sanz, L.; Bernal, S.; Fortuna, S.; Mercurio, F.A.; Leone, M.; Gomez-Guerrero, C.; Marasco, D. Cyclic mimetics of kinase-inhibitory region of Suppressors of Cytokine Signaling 1: Progress toward novel anti-inflammatory therapeutics. *Eur. J. Med. Chem.* **2021**, *221*, 113547. [CrossRef]
38. La Manna, S.; Lopez-Sanz, L.; Bernal, S.; Jimenez-Castilla, L.; Prieto, I.; Morelli, G.; Gomez-Guerrero, C.; Marasco, D. Antioxidant Effects of PS5, a Peptidomimetic of Suppressor of Cytokine Signaling 1, in Experimental Atherosclerosis. *Antioxidants* **2020**, *9*, 754. [CrossRef]
39. La Manna, S.; Lopez-Sanz, L.; Leone, M.; Brandi, P.; Scognamiglio, P.L.; Morelli, G.; Novellino, E.; Gomez-Guerrero, C.; Marasco, D. Structure-activity studies of peptidomimetics based on kinase-inhibitory region of suppressors of cytokine signaling 1. *Biopolymers* **2017**, *110*, e23082. [CrossRef]
40. Doti, N.; Scognamiglio, P.L.; Madonna, S.; Scarponi, C.; Ruvo, M.; Perretta, G.; Albanesi, C.; Marasco, D. New mimetic peptides of the kinase-inhibitory region (KIR) of SOCS1 through focused peptide libraries. *Biochem. J.* **2012**, *443*, 231–240. [CrossRef]
41. Ahmed, C.M.; Massengill, M.T.; Brown, E.E.; Ildefonso, C.J.; Johnson, H.M.; Lewin, A.S. A cell penetrating peptide from SOCS-1 prevents ocular damage in experimental autoimmune uveitis. *Exp. Eye Res.* **2018**, *177*, 12–22. [CrossRef] [PubMed]
42. Ahmed, C.M.; Patel, A.P.; Ildefonso, C.J.; Johnson, H.M.; Lewin, A.S. Corneal Application of R9-SOCS1-KIR Peptide Alleviates Endotoxin-Induced Uveitis. *Transl. Vis. Sci. Technol.* **2021**, *10*, 25. [CrossRef] [PubMed]
43. Recio, C.; Oguiza, A.; Lazaro, I.; Mallavia, B.; Egido, J.; Gomez-Guerrero, C. Suppressor of cytokine signaling 1-derived peptide inhibits Janus kinase/signal transducers and activators of transcription pathway and improves inflammation and atherosclerosis in diabetic mice. *Arter. Thromb. Vasc. Biol.* **2014**, *34*, 1953–1960. [CrossRef] [PubMed]

44. Bernal, S.; Lopez-Sanz, L.; Jimenez-Castilla, L.; Prieto, I.; Melgar, A.; La Manna, S.; Martin-Ventura, J.L.; Blanco-Colio, L.M.; Egido, J.; Gomez-Guerrero, C. Protective effect of suppressor of cytokine signalling 1-based therapy in experimental abdominal aortic aneurysm. *Br. J. Pharm.* **2020**, *178*, 564–581. [CrossRef] [PubMed]
45. Lopez-Sanz, L.; Bernal, S.; Recio, C.; Lazaro, I.; Oguiza, A.; Melgar, A.; Jimenez-Castilla, L.; Egido, J.; Gomez-Guerrero, C. SOCS1-targeted therapy ameliorates renal and vascular oxidative stress in diabetes via STAT1 and PI3K inhibition. *Lab. Investig.* **2018**, *98*, 1276–1290. [CrossRef]
46. Opazo-Rios, L.; Sanchez Matus, Y.; Rodrigues-Diez, R.R.; Carpio, D.; Droguett, A.; Egido, J.; Gomez-Guerrero, C.; Mezzano, S. Anti-inflammatory, antioxidant and renoprotective effects of SOCS1 mimetic peptide in the BTBR ob/ob mouse model of type 2 diabetes. *BMJ Open Diabetes Res. Care* **2020**, *8*, e001242. [CrossRef]
47. Recio, C.; Lazaro, I.; Oguiza, A.; Lopez-Sanz, L.; Bernal, S.; Blanco, J.; Egido, J.; Gomez-Guerrero, C. Suppressor of Cytokine Signaling-1 Peptidomimetic Limits Progression of Diabetic Nephropathy. *J. Am. Soc. Nephrol.* **2017**, *28*, 575–585. [CrossRef]
48. Chen, H.; Wu, Y.; Li, K.; Currie, I.; Keating, N.; Dehkhoda, F.; Grohmann, C.; Babon, J.J.; Nicholson, S.E.; Sleebs, B.E. Optimization of Phosphotyrosine Peptides that Target the SH2 Domain of SOCS1 and Block Substrate Ubiquitination. *ACS Chem. Biol.* **2022**, *17*, 449–462. [CrossRef]
49. Madonna, S.; Scarponi, C.; Morelli, M.; Sestito, R.; Scognamiglio, P.L.; Marasco, D.; Albanesi, C. SOCS3 inhibits the pathological effects of IL-22 in non-melanoma skin tumor-derived keratinocytes. *Oncotarget* **2017**, *8*, 24652–24667. [CrossRef]
50. La Manna, S.; Lee, E.; Ouzounova, M.; Di Natale, C.; Novellino, E.; Merlino, A.; Korkaya, H.; Marasco, D. Mimetics of suppressor of cytokine signaling 3: Novel potential therapeutics in triple breast cancer. *Int. J. Cancer* **2018**, *143*, 2177–2186. [CrossRef]
51. La Manna, S.; Lopez-Sanz, L.; Mercurio, F.A.; Fortuna, S.; Leone, M.; Gomez-Guerrero, C.; Marasco, D. Chimeric Peptidomimetics of SOCS 3 Able to Interact with JAK2 as Anti-inflammatory Compounds. *ACS Med. Chem. Lett.* **2020**, *11*, 615–623. [CrossRef] [PubMed]
52. La Manna, S.; Scognamiglio, P.L.; Di Natale, C.; Leone, M.; Mercurio, F.A.; Malfitano, A.M.; Cianfarani, F.; Madonna, S.; Caravella, S.; Albanesi, C.; et al. Characterization of linear mimetic peptides of Interleukin-22 from dissection of protein interfaces. *Biochimie* **2017**, *138*, 106–115. [CrossRef] [PubMed]
53. OECD. Guidelines for the Testing of Chemicals, Test No. 107: Partition Coefficient (n-Octanol/Water): Shake Flask Method. Available online: https://www.oecd.org/chemicalsafety/testing/21047299.pdf (accessed on 2 March 2022).
54. Griesinger, C.; Otting, G.; Wuthrich, K.; Ernst, R.R. Clean TOCSY for proton spin system identification in macromolecules. *J. Am. Chem. Soc.* **1988**, *110*, 7870–7872. [CrossRef]
55. Kumar, A.; Ernst, R.R.; Wuthrich, K. A two-dimensional nuclear Overhauser enhancement (2D NOE) experiment for the elucidation of complete proton-proton cross-relaxation networks in biological macromolecules. *Biochem. Biophys. Res. Commun.* **1980**, *95*, 1–6. [CrossRef]
56. Bax, A.; Davis, D.G. Practical aspects of two-dimensional transverse NOE spectroscopy. *J. of Magn. Reson. 1969* **1985**, *63*, 207–213. [CrossRef]
57. Piantini, U.; Sorensen, O.W.; Ernst, R.R. Multiple quantum filters for elucidating NMR coupling networks. *J. Am. Chem. Soc.* **1982**, *104*, 6800–6801. [CrossRef]
58. Hwang, T.L.; Shaka, A.J. Water suppression that works. Excitation sculpting using arbitrary waveforms and pulsed field gradients. *J. Magn. Reson. Ser. A* **1995**, *112*, 275–279. [CrossRef]
59. Wuthrich, K. *NMR of Proteins and Nucleic Acids*; Wiley: New York, NY, USA, 1986.
60. Bartels, C.; Xia, T.; Billeter, M.; Güntert, P.; Wüthrich, K. The program XEASY for computer-supported NMR spectral analysis of biological macromolecules. *J. Biomol. NMR* **1995**, *6*, 1–10. [CrossRef]
61. Kjaergaard, M.; Brander, S.; Poulsen, F.M. Random coil chemical shift for intrinsically disordered proteins: Effects of temperature and pH. *J. Biomol. NMR* **2011**, *49*, 139–149. [CrossRef]
62. Herrmann, T.; Guntert, P.; Wuthrich, K. Protein NMR structure determination with automated NOE assignment using the new software CANDID and the torsion angle dynamics algorithm DYANA. *J. Mol. Biol.* **2002**, *319*, 209–227. [CrossRef]
63. Yilmaz, E.M.; Guntert, P. NMR structure calculation for all small molecule ligands and non-standard residues from the PDB Chemical Component Dictionary. *J. Biomol. NMR* **2015**, *63*, 21–37. [CrossRef] [PubMed]
64. Laskowski, R.A.; Rullmannn, J.A.; MacArthur, M.W.; Kaptein, R.; Thornton, J.M. AQUA and PROCHECK-NMR: Programs for checking the quality of protein structures solved by NMR. *J. Biomol. NMR* **1996**, *8*, 477–486. [CrossRef] [PubMed]
65. Koradi, R.; Billeter, M.; Wuthrich, K. MOLMOL: A program for display and analysis of macromolecular structures. *J. Mol. Graph.* **1996**, *14*, 51–55. [CrossRef]
66. Hellinen, L.; Bahrpeyma, S.; Rimpela, A.K.; Hagstrom, M.; Reinisalo, M.; Urtti, A. Microscale Thermophoresis as a Screening Tool to Predict Melanin Binding of Drugs. *Pharmaceutics* **2020**, *12*, 554. [CrossRef]
67. Scognamiglio, P.L.; Di Natale, C.; Perretta, G.; Marasco, D. From peptides to small molecules: An intriguing but intricated way to new drugs. *Curr. Med. Chem.* **2013**, *20*, 3803–3817. [CrossRef] [PubMed]
68. La Manna, S.; Di Natale, C.; Florio, D.; Marasco, D. Peptides as Therapeutic Agents for Inflammatory-Related Diseases. *Int. J. Mol. Sci.* **2018**, *19*, 2714. [CrossRef] [PubMed]
69. Russo, A.; Aiello, C.; Grieco, P.; Marasco, D. Targeting "Undruggable" Proteins: Design of Synthetic Cyclopeptides. *Curr. Med. Chem.* **2016**, *23*, 748–762. [CrossRef] [PubMed]

70. Moiola, M.; Memeo, M.G.; Quadrelli, P. Stapled Peptides—A Useful Improvement for Peptide-Based Drugs. *Molecules* **2019**, *24*, 3654. [CrossRef]
71. Fodje, M.N.; Al-Karadaghi, S. Occurrence, conformational features and amino acid propensities for the pi-helix. *Protein Eng.* **2002**, *15*, 353–358. [CrossRef]
72. Cooley, R.B.; Arp, D.J.; Karplus, P.A. Evolutionary origin of a secondary structure: Pi-helices as cryptic but widespread insertional variations of alpha-helices that enhance protein functionality. *J. Mol. Biol.* **2010**, *404*, 232–246. [CrossRef]
73. Andersson, D.; Carlsson, U.; Freskgard, P.O. Contribution of tryptophan residues to the CD spectrum of the extracellular domain of human tissue factor: Application in folding studies and prediction of secondary structure. *Eur. J. Biochem.* **2001**, *268*, 1118–1128. [CrossRef]
74. Mercurio, F.A.; Pirone, L.; Di Natale, C.; Marasco, D.; Pedone, E.M.; Leone, M. Sam domain-based stapled peptides: Structural analysis and interaction studies with the Sam domains from the EphA2 receptor and the lipid phosphatase Ship2. *Bioorg. Chem.* **2018**, *80*, 602–610. [CrossRef] [PubMed]
75. Wishart, D.S.; Sykes, B.D.; Richards, F.M. Relationship between nuclear magnetic resonance chemical shift and protein secondary structure. *J. Mol. Biol.* **1991**, *222*, 311–333. [CrossRef]
76. Wishart, D.S.; Sykes, B.D.; Richards, F.M. The chemical shift index: A fast and simple method for the assignment of protein secondary structure through NMR spectroscopy. *Biochemistry* **1992**, *31*, 1647–1651. [CrossRef] [PubMed]
77. Asmari, M.; Ratih, R.; Alhazmi, H.A.; El Deeb, S. Thermophoresis for characterizing biomolecular interaction. *Methods* **2018**, *146*, 107–119. [CrossRef] [PubMed]
78. Jerabek-Willemsen, M.; Wienken, C.J.; Braun, D.; Baaske, P.; Duhr, S. Molecular interaction studies using microscale thermophoresis. *Assay Drug Dev. Technol.* **2011**, *9*, 342–353. [CrossRef]
79. Verstovsek, S.; Kantarjian, H.; Mesa, R.A.; Pardanani, A.D.; Cortes-Franco, J.; Thomas, D.A.; Estrov, Z.; Fridman, J.S.; Bradley, E.C.; Erickson-Viitanen, S. Safety and efficacy of INCB018424, a JAK1 and JAK2 inhibitor, in myelofibrosis. *N. Engl. J. Med.* **2010**, *363*, 1117–1127. [CrossRef]
80. Park, J.S.; Hong, M.H.; Chun, Y.J.; Kim, H.R.; Cho, B.C. A phase Ib study of the combination of afatinib and ruxolitinib in EGFR mutant NSCLC with progression on EGFR-TKIs. *Lung Cancer* **2019**, *134*, 46–51. [CrossRef]
81. Walensky, L.D.; Bird, G.H. Hydrocarbon-stapled peptides: Principles, practice, and progress. *J. Med. Chem.* **2014**, *57*, 6275–6288. [CrossRef]

Article

Cu and Zn Interactions with Peptides Revealed by High-Resolution Mass Spectrometry

Monica Iavorschi [1,2,†], Ancuța-Veronica Lupăescu [1,2,†], Laura Darie-Ion [2], Maria Indeykina [3], Gabriela Elena Hitruc [4] and Brîndușa Alina Petre [2,5,*]

1. Department of Biomedical Sciences, Faculty of Medicine and Biological Sciences, Stefan cel Mare University of Suceava, 13 University, 720229 Suceava, Romania
2. Faculty of Chemistry, Al. I. Cuza University of Iasi, 11 Carol I, 700506 Iasi, Romania
3. Emanuel Institute for Biochemical Physics, Russian Academy of Sciences, 119334 Moscow, Russia
4. Physical Chemistry of Polymers Department, Petru Poni Institute of Macromolecular Chemistry, 41A Gr. Ghica Voda Alley, 700487 Iasi, Romania
5. Center for Fundamental Research and Experimental Development in Translation Medicine—TRANSCEND, Regional Institute of Oncology, 2-4 General Henri Mathias Berthelot, 700483 Iași, Romania
* Correspondence: brindusa.petre@uaic.ro
† These authors contributed equally to this work.

Citation: Iavorschi, M.; Lupăescu, A.-V.; Darie-Ion, L.; Indeykina, M.; Hitruc, G.E.; Petre, B.A. Cu and Zn Interactions with Peptides Revealed by High-Resolution Mass Spectrometry. *Pharmaceuticals* 2022, 15, 1096. https://doi.org/10.3390/ph15091096

Academic Editor: Klaus Kopka

Received: 13 July 2022
Accepted: 29 August 2022
Published: 31 August 2022

Publisher's Note: MDPI stays neutral with regard to jurisdictional claims in published maps and institutional affiliations.

Copyright: © 2022 by the authors. Licensee MDPI, Basel, Switzerland. This article is an open access article distributed under the terms and conditions of the Creative Commons Attribution (CC BY) license (https://creativecommons.org/licenses/by/4.0/).

Abstract: Alzheimer's disease (AD) is a progressive neurodegenerative disease characterized by abnormal extracellular amyloid-beta (Aβ) peptide depositions in the brain. Among amorphous aggregates, altered metal homeostasis is considered a common risk factor for neurodegeneration known to accelerate plaque formation. Recently, peptide-based drugs capable of inhibiting amyloid aggregation have achieved unprecedented scientific and pharmaceutical interest. In response to metal ions binding to Aβ peptide, metal chelation was also proposed as a therapy in AD. The present study analyzes the interactions formed between NAP octapeptide, derived from activity-dependent neuroprotective protein (ADNP), amyloid Aβ(9–16) fragment and divalent metal ions such as Cu and Zn. The binding affinity studies for Cu and Zn ions of synthetic NAP peptide and Aβ(9–16) fragment were investigated by matrix-assisted laser desorption/ionization mass spectrometry (MALDI-MS), electrospray ion trap mass spectrometry (ESI-MS) and atomic force microscopy (AFM). Both mass spectrometric methods confirmed the formation of metal–peptide complexes while the AFM technique provided morphological and topographical information regarding the influence of metal ions upon peptide crystallization. Our findings showed that due to a rich histidine center, the Aβ(9–16) fragment is capable of binding metal ions, thus becoming stiff and promoting aggregation of the entire amyloid peptide. Apart from this, the protective effect of the NAP peptide was found to rely on the ability of this octapeptide to generate both chelating properties with metals and interactions with Aβ peptide, thus stopping its folding process.

Keywords: neurological peptides; metal ion interactions; MALDI-MS; ESI-MS; AFM

1. Introduction

Addition of metal ions to peptides or proteins normally involves chelation of the metal ion at several amino acids that expose Lewis-basic properties. Lewis bases are nucleophile species that donate an electron pair to an electrophile to form a chemical bond. Thus, in the case of peptides, the nucleophile sites are found both in the covalent amide linkages and the amino acid side chains. The chelation provided by the amide carbonyl oxygen, known as charge-solvation binding, is a characteristic of alkali metal ions and Ca^{2+} or Mg^{2+} ions. Alternatively, the deprotonation of the amide nitrogen motif is considered to be distinctive for more active metal ions such as Co^{2+}, Ni^{2+}, Cu^{2+}, Zn^{2+}, Pd^{2+} and Cd^{2+} [1].

Matrix-assisted laser desorption/ionization mass spectrometry (MALDI-MS) is a powerful analytical tool for the analysis of various biomolecules and synthetic polymers.

When studying biological systems with MALDI-MS, it is very important to know whether the non-covalently bound complexes that are stable under physiological conditions are also able to survive laser desorption and ionization processes, which occur inside the ionization source. The electrospray ionization mass spectrometry (ESI-MS) is a powerful tool that in recent years has provided considerable information on non-covalent interactions between proteins and ligands, including the stoichiometry and affinity of complexes. This "soft" ionization technique allows non-covalent complexes of proteins to be admitted to the gas phase for detection and investigation according to their "charge-state families" [2]. Furthermore, different research reports have shown that during the electrospray ionization mass spectrometry ESI-MS, the non-covalent protein–metal interactions in-solution are maintained during desolation and sample transfer to the gas phase [3]. Until now, a few studies were able to compare solution and gas phase chemistries of non-covalent and metal ion–biomolecule complexes using both, ESI and MALDI ionization sources [4–6].

Copper (Cu^+ and Cu^{2+}) and zinc (Zn^{2+}) ions play important roles in many chemical and biochemical processes, such as oxidation, dioxygen transport and electron transfer [7–9]. Both metal ions have also been implicated in human neurodegenerative diseases, such as Alzheimer's disease [10,11]. Proteins associated with neurodegenerative diseases interact with several transition metal ions, which leads to an aggregation process. Furthermore, redox-active metal ions such as iron and copper favor the occurrence of reactive oxygen species (ROS) promoted by Fenton reactions. This process leads to the formation of the highly reactive hydroxyl radical ($OH^•$) responsible for stimulating oxidative stress in the cell by lipid membrane peroxidation, DNA damage and protein oxidation or misfolding [12].

Previous MALDI investigations using Aβ model peptides showed that peptide–copper ion complexes are formed by a reductive process which yields primarily $[M + Cu(I)]^+$ ions [13], whereas the electrospray ionization method generates both $[M + Cu(I)]^+$ and $[M + Cu(II)-H]^+$ ions [14]. In principle, the formation of $[M + Cu(I)]^+$ ions is provided by the basic amino acids such as arginine (Arg), lysine (Lys) and histidine (His) [13,15]. For NAP neuroprotective peptide and amyloidal fragment, the interaction between the Cu and the basic residues is described in terms of competitive binding [16,17]. However, few studies have been performed to determine Cu and Zn binding sites preference and their neurological analogues. So far, we have not found research literature reports on the competition between metal ions and different binding sites. Addressing these issues will elucidate (i) the chemistry of metal binding to the octapeptide NAP and the amyloid fragment Aβ(9–16); and (ii) the challenges of Cu and Zn ions as specific and competitive binding to a specific site in a biological mixture of two peptides. In this study we investigated the interactions formed between NAP neuroprotective peptide, Aβ(9–16) fragment and divalent metal ions such as Cu and Zn. Matrix-assisted laser desorption/ionization mass spectrometry (MALDI-MS), electrospray ion trap mass spectrometry (ESI-MS) and atomic force microscopy (AFM) were used to investigate the stoichiometries and affinity toward Cu and Zn ions of model synthetic NAP peptide and Aβ(9–16) fragment.

2. Results

2.1. Mass Spectrometric Analysis

Characterization of zinc and copper ions interaction with NAP neuroprotective peptide and Aβ(9–16) fragments was performed by ESI and MALDI-ToF MS in combination with collision-induced dissociation (CID) fragmentation experiments. As expected, in ESI-MS as well as MALDI-MS analysis, the obtained spectra highlighting the formation of metal–peptide complex. Also, peak assignments were achieved based on the theoretical expected ion values (m/z) and validated by isotopic pattern information.

Figure 1 presents the ESI and MALDI-ToF mass spectra of the octapeptide NAP incubated with Zn ions. Both mass spectra revealed the anticipated signal of metal–peptide complex, $[M + Zn-H]^+$ at m/z 887.4. More precisely, $[M + Zn-H]^+$ can be written as $[M + H + Zn-2H]^+$ in order to account for the fact that the peptide loses two protons upon Zn^{2+} complexation, and acquires an additional proton to form a singly charged species [18,19]. The most intense signals depicted in the MALDI spectrum were assigned to the sodium adduct ion ($[M + Na]^+$, m/z 847.6) and deaminated $[M-16 + H]^+$ ion (m/z 809.6), while the molecular ion $[M + H]^+$ generated a small signal at m/z 825.6. The mechanism of deamination is photo chemically promoted by the laser light source and is favored by the C-terminus glutamine residue, where a rearrangement such as cyclization leads to pyroglutamic acid formation [20,21]. Additional signals observed in the mass spectrum at m/z 832.5, m/z 863.6 and m/z 869.6 correspond to different adducts with sodium and potassium ($[M-16 + Na]^+$, $[M + K]^+$, $[M + 2Na-H]^+$). In contrast, ESI mass spectra generated a high abundant peak at m/z 825.5, which was attributed to the molecular ion $[M + H]^+$. Furthermore, due to the use of a milder ionization technique such as ESI, only a small amount of peptide underwent a deamidation process and generated a small signal at m/z 809.6 $[M-16 + H]^+$ ions. Signals assigned to the Zn^{2+}–NAP metal complex were observed both in single- ($[M + Zn-H]^+$ ion, m/z 887.4) and double-charged peptide fragments ($[2M + Zn]^{2+}$ ion, m/z 856.4; $[M + Zn]^{2+}$ ion, m/z 444.2). Additional doubly charged peptide ions generated during the ionization process were visualized at m/z 405.2, m/z 413.2 and m/z 421.7 and were assigned to the doubly protonated peptide ($[M-16 + H]^+$, $[M + H]^+$) and ammonium $[M + NH_4-H]^{2+}$ ion.

In the case of Aβ(9–16) peptide, the MALDI ToF MS (Figure 2A) spectra contained a signal only for the single-charged peptide while the ESI MS spectra (Figure 2B) presented two peaks in the double-charged region. Thus, the zinc affinity toward the amyloidal peptide was confirmed by the presence of a small signal at m/z 1058.4 and m/z 529.7 corresponding to the zinc–peptide complex single- and double-charged species: $[M + Zn-H]^+$ and $[M + Zn]^{2+}$ ions. However, both ionization techniques showed an intense signal for the protonated 9–16 amyloid fragment that was observed at m/z 996.5 in the MALDI spectra and m/z 498.8 in the ESI experiment. The absence of single-charged species during electrospray ionization is due to the ability of histidine's imidazole nitrogen atoms to act either as electron donors or acceptors in different cases [22] and thus, seize the second proton on its structure to form the doubly protonated ion [23].

Peptide interaction with copper (II) ions was also investigated by mass spectrometry. Figure 3 presents the MALDI and ESI MS spectra recorded after NAP octapeptide incubation with Cu^{2+} ions. At first glance, it can be easily observed that the peptides form stable complexes with metal ions. However, during matrix-assisted laser desorption ionization, species assigned to the copper (I)–peptide ions were observed at m/z 887.9 for the $[M + Cu]^+$ species and m/z 949.8 for the $[M + 2Cu-H]^{2+}$ containing two copper ions. The reduction of divalent metal ions is a process favored by both gas-phase charge exchange with matrix molecules and free electron capture [24]. Contrary to the MALDI spectra, where copper is present in reduced form almost exclusively, during ESI ionization, copper (II) ions maintain their charge state. Other signals observed in the mass spectra at m/z 825.9, m/z 847.9 and m/z 863.8 were assigned to the protonated peptide ($[M + H]^+$) and adducts with sodium ($[M + Na]^+$), potassium ($[M + K]^+$). Furthermore, using ESI–MS, we found that NAP peptide was capable of binding only one copper ion. Thus, low-intensity signals attributed to the copper (II)–peptide complex were observed at m/z 886.4 corresponding to the single-charged $[M + Cu-H]^+$ ions and m/z 444.2 assigned to the double-charged $[M + Cu]^{2+}$ species. In addition to the peaks assigned to the metal ion–peptide complexes, the MALDI-MS spectra present an intense signal at m/z 809.9 characteristic of the deaminated peptide $[M-16 + H]^+$, while the ESI experiment revealed two powerful peaks at m/z 825.4 and m/z 413.2 generated by the mono- and double-protonated peptide species. Additionally, during electrospray ionization, the peptide formed beside the single-charged $[M-16 + H]^+$ and $[M + Na]^+$ ions observed at m/z 809.9 and m/z 847.4, and double-charged species that were assigned to the sodium ($[M + Na + H]^{2+}$, m/z 421.7) and ammonium adducts ($[M + NH_4 + H]^{2+}$, m/z 424.2).

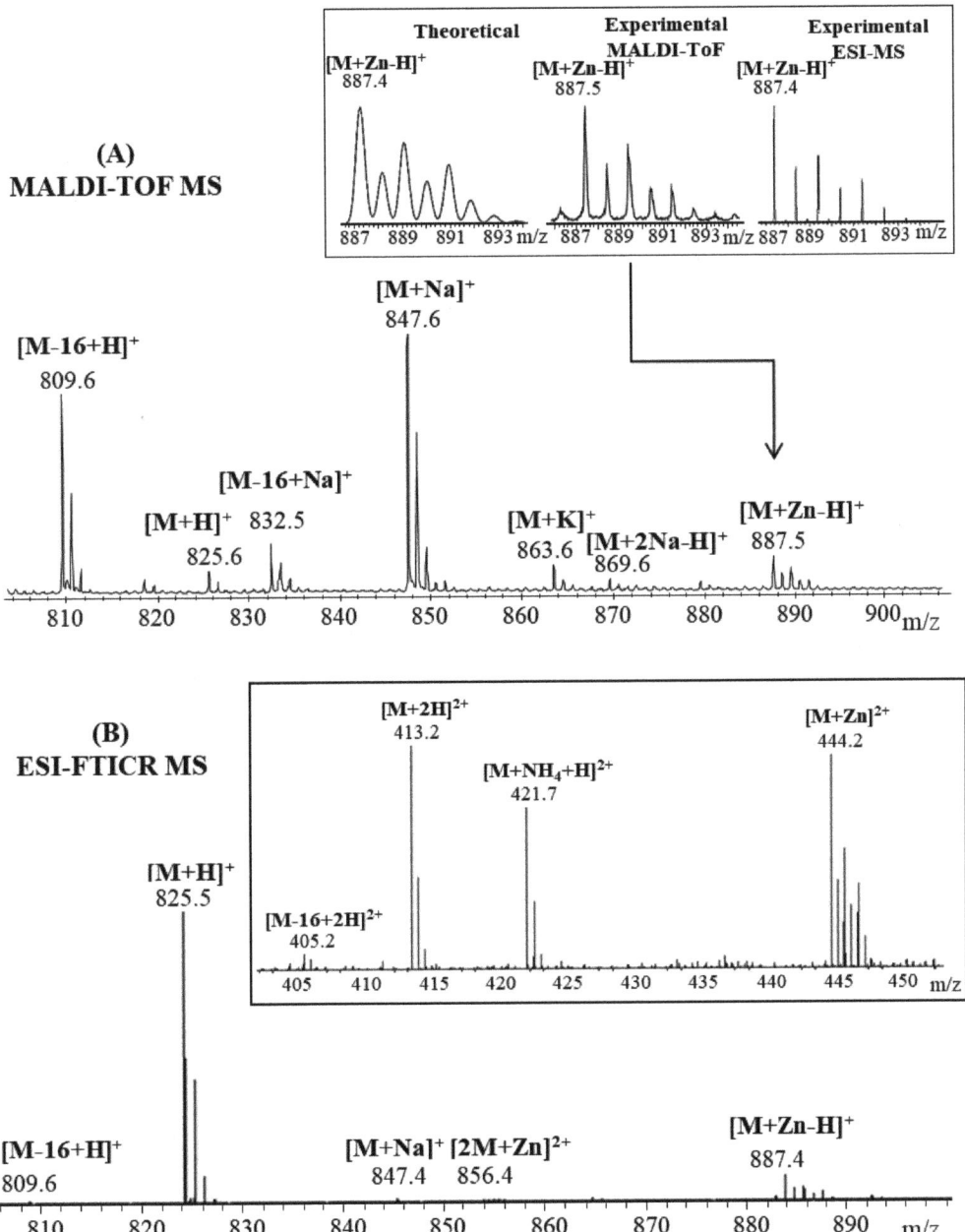

Figure 1. Enlarged region of mass spectra of NAP peptide incubated with Zn^{2+} ions acquired by (**A**) MALDI ToF (reflectron mode) with CHCA matrix and (**B**) ESI-FTICR, both in positive mode. Inserts: (**A**) details of theoretical and experimental isotopic patterns of [M + Zn-H]$^+$ ion; (**B**) detail of the double-positive-charged peptide fragments.

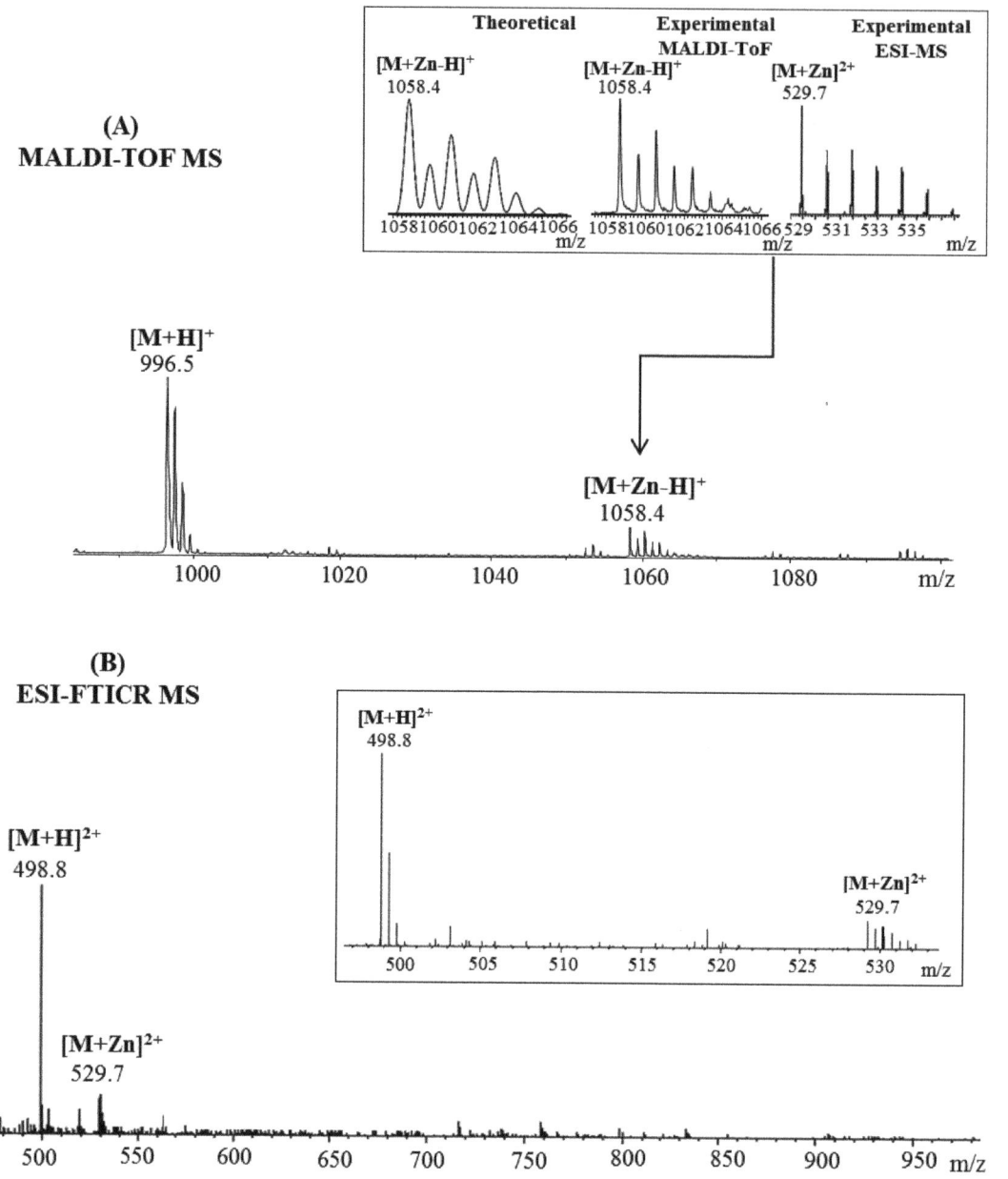

Figure 2. Enlarged region of mass spectra of Aβ(9–16) peptide incubated with Zn^{2+} ions acquired by (**A**) MALDI ToF (reflectron mode) with CHCA matrix and (**B**) ESI-FTICR, both in positive mode. Inserts: (**A**) details of theoretical and experimental isotopic patterns of $[M + Zn-H]^+$ and $[M + Zn]^{2+}$ ions; (**B**) detail of the double-positive-charged peptide fragments.

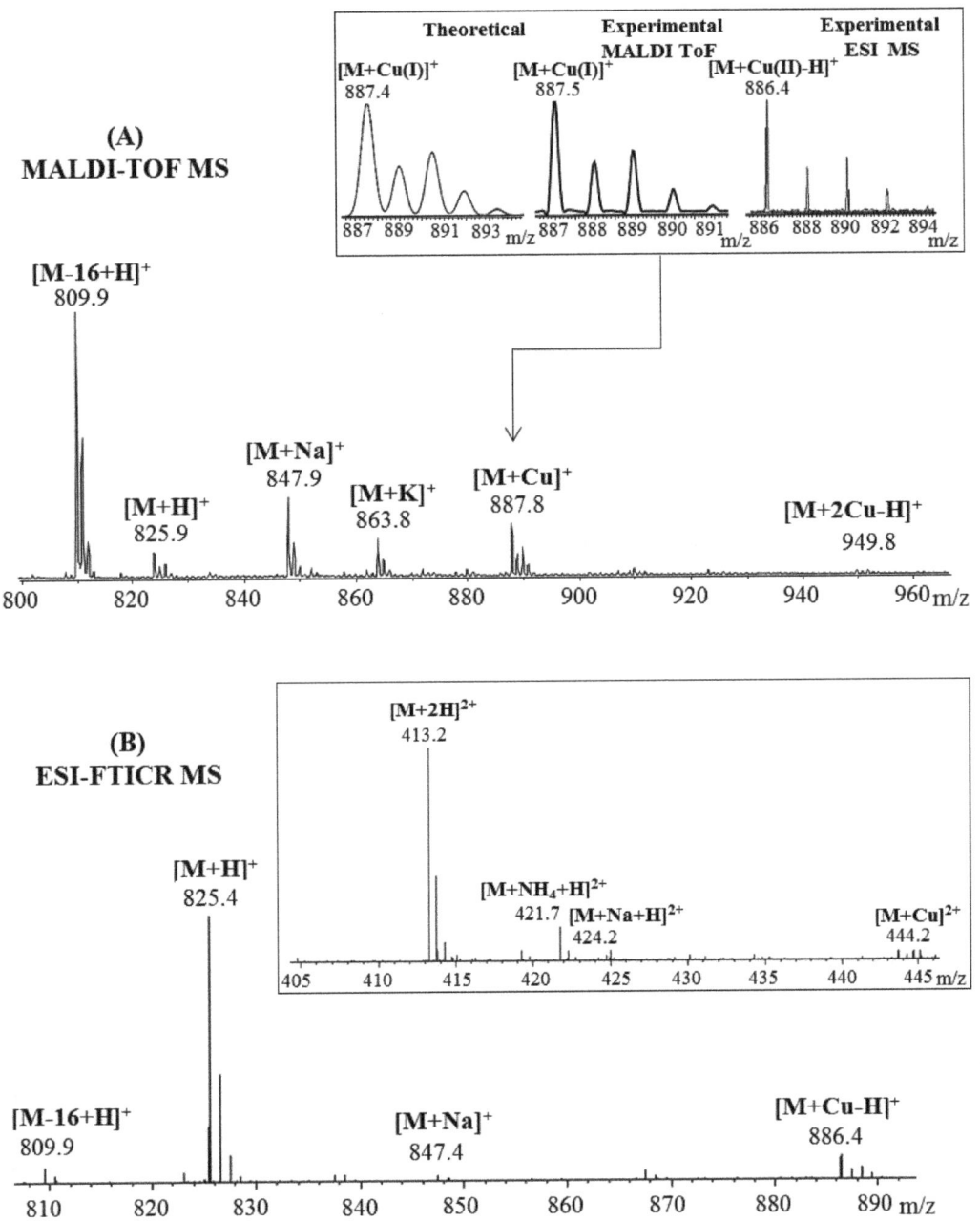

Figure 3. Enlarged region of mass spectra of NAP peptide incubated with Cu^{2+} ions acquired by (**A**) MALDI ToF (reflectron mode) with CHCA matrix and (**B**) ESI-FTICR, both in positive mode. Inserts: (**A**) details of theoretical and experimental isotopic patterns of $[M + Cu-H]^+$ ion; (**B**) detail of the double-positive-charged peptide fragments.

The expected copper–peptide complex was also observed in the mass spectra of amyloid Aβ(9–16) peptide fragment (Figure 4). Thus, a strong signal observed at m/z 1058.2 in the MALDI MS spectrum corresponded to the $[M + Cu]^+$. Similar to the previous case, the ionization technique which involves a mechanism of gas-phase charge exchange with matrix molecules favored the reduction of copper (II) ions to copper (I). However, this process did not restrain the metal ions affinity toward Aβ(9–16) peptide. Beside the Cu–Aβ(9–16) complex species, another intense signal visible at m/z 996.4, generated by the non-complexed peptide, was present in the MS spectrum. Furthermore, no single-charged ions were observed after electrospray peptide's ionization. Thereby, the only peaks observed at m/z 498.8, assigned to the protonated peptide $[M + 2H]^{2+}$ and m/z 529.2, corresponding to the Cu–peptide species $[M + Cu]^{2+}$, belonged to the double-charged domain. Furthermore, to confirm the presence of a metal–peptide complex that involves copper(II) species, the peak region was zoomed (Figure 4A insert, right spectra) and compared to the theoretical one. Thus, by comparing the isotopic pattern, it was highlighted that the copper ion is found in complexes in both oxidation states.

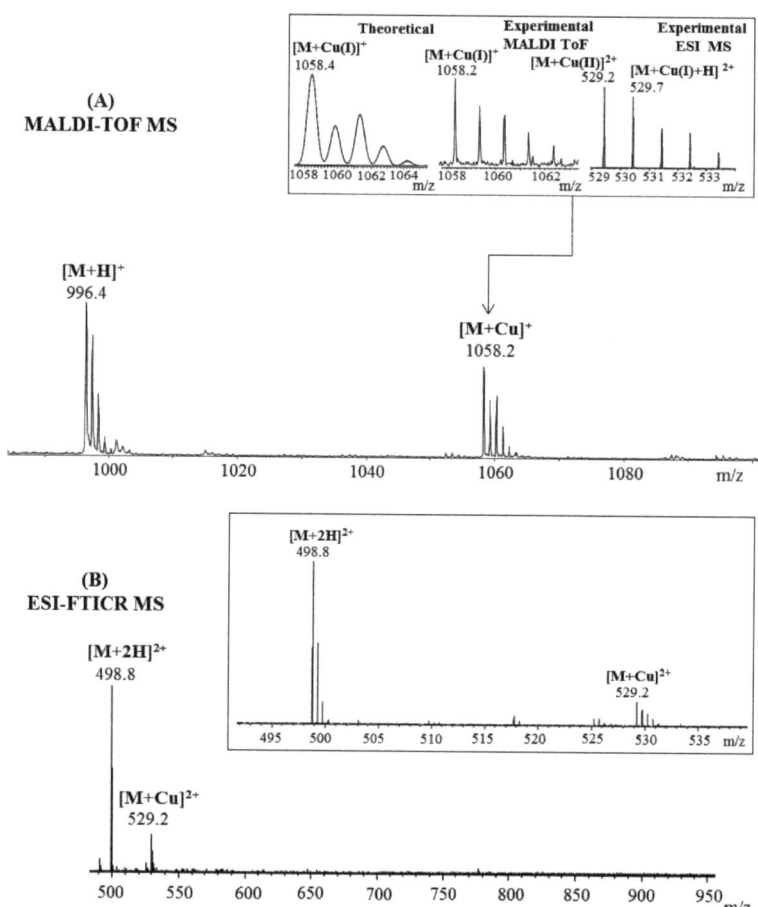

Figure 4. Enlarged region of mass spectra of Aβ(9–16) peptide incubated with Cu^{2+} ions acquired by (**A**) MALDI ToF (reflectron mode) with CHCA matrix and (**B**) ESI-FTICR, both in positive mode. Inserts: (**A**) details of theoretical and experimental isotopic patterns of $[M + Cu]^+$ ion; (**B**) detail of the double-positive-charged peptide fragments.

However, by looking at the intensity of the isotopes, the highest influence seems to come from [M + Cu(II)]$^{2+}$ species visible at m/z 529.2, while the reduced [M + Cu(I) + H]$^{2+}$ ions observed at m/z 529.7 overlap with the isotopic distribution of the precedent complex. These results are in accordance with previous findings [25] where the authors identified a mixture of Cu(I)– and Cu(II)–Aβ(9–16) complexes by electrospray ionization mass spectrometry.

2.2. Peptides Affinity toward Metal Ions: Competition Study

Figure 5 presents the MS spectra recorded after in-solution NAP and Aβ(9–16) peptides incubation with zinc ions. As observed in the (A) spectrum, both peptides formed non-covalent complexes with zinc ions. Thus, in addition to the signals assigned to the protonated [M + H]$^+$, deaminated [M-16 + H]$^+$ or sodium ion adducts [M + Na]$^+$, the peptides generated peaks characteristic of the zinc–peptide complex [M + Zn-H]$^+$. This result indicates that there is no relevant interference between the amyloid fragment and neuroprotective NAP peptide and their metal ions binding sites.

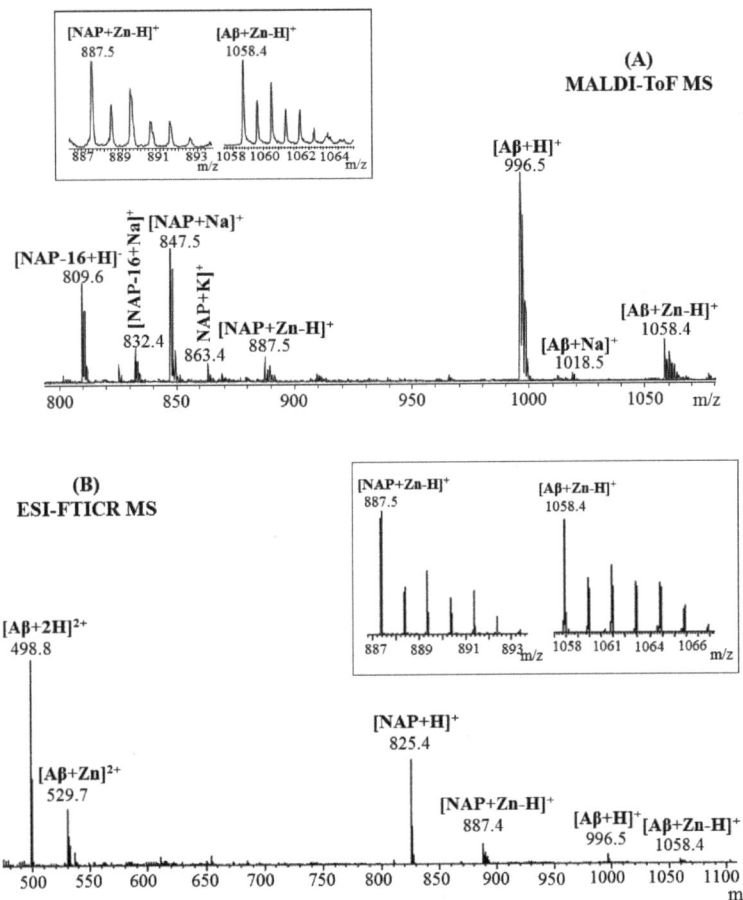

Figure 5. Enlarged region of mass spectra of both NAP and Aβ(9–16) peptide incubated with Zn^{2+} ions acquired by (**A**) MALDI ToF (reflectron mode) with CHCA matrix and (**B**) ESI-FTICR, both in positive mode. Inserts: (**A**) details of theoretical and experimental isotopic patterns of [M + Zn-H]$^+$ ion; (**B**) details of the double-positive-charged peptide fragments.

Additionally, an ESI-MS experiment was performed in order to reflect, with good accuracy, the solution conditions. The zinc–peptide interaction was easily confirmed by the presence of strong signals at m/z 1058.4, m/z 887.4 and m/z 529.7 corresponding to the single- and double-charged [M + Zn] complex. Therefore, the obtained spectrum (Figure 5B) pointed out that there is no interaction between peptides that would disrupt their affinity for metal ions. Interestingly, Aβ(9–16) peptide was observed predominantly in its doubly protonated forms while the NAP peptide preferred the singly charged ions.

In the presence of copper (II) ions, the peptides formed non-covalent complexes with the metal ion. The Cu–peptide complexes were visible in both MALDI and ESI mass spectra (Figure 6), confirming the strong affinity of octapeptides for copper ions. Interestingly, the MALDI ToF/ToF analysis provided evidence of a copper (I)–peptide interaction for both amyloidal and neuroprotective fragments. Further, the analysis of the ESI-MS soft ionization method highlighted the presence of M + Cu(II) ions for NAP peptide while the Aβ9–16 peptide generated signals for both M + Cu(II) and M + Cu(I) species.

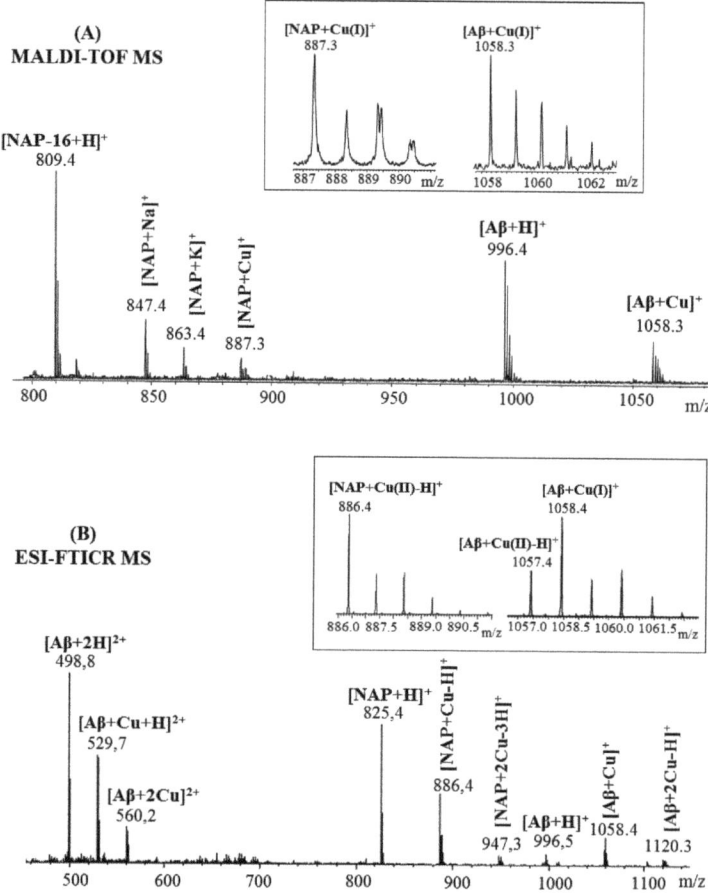

Figure 6. Enlarged region of mass spectra of both NAP and Aβ(9–16) peptide incubated with Cu^{2+} ions acquired by (**A**) MALDI ToF (reflectron mode) with CHCA matrix and (**B**) ESI-FTICR, both in positive mode. Inserts: (**A**) details of theoretical and experimental isotopic patterns of [M + Cu-H]$^+$ ion; (**B**) details of the double-positive-charged peptide fragments.

In addition, the ability of NAP peptide to bind two copper ions and display signals at m/z 886.4 ([NAP + Cu-H]$^+$) and m/z 947.3 ([NAP + 2Cu-3H]$^+$) confirms the preference of neuroprotective octapeptide for divalent copper ions. Overall, the NAP peptide may play a role in copper homeostasis and/or metabolism since it is able to assimilate the copper redox couple and despite the structural constraints imposed by the proline residues. A similar behavior was also remarked at Aβ(9–16) peptide. However, the reduced Cu+ species was found to be the metal ions predominant in the Cu–Aβ complexes. Furthermore, in the presence of excess Cu, the amyloid fragment showed strong binding of only one Cu+ ion in the MALDI spectrum. In the ESI spectrum, beside the fact that the binding of the second copper ion was substantially weaker, the two peaks assigned to the metal–peptide complex include copper–Aβ complexes with both oxidation states.

2.3. Confirmation of Metal Ion Binding by Tandem Mass Spectrometry

In order to determine the metal-binding site for the zinc and copper ions, ESI tandem (MS/MS) experiments were carried out. Tandem mass spectrometry [26] employs two stages of mass analysis in order to examine selectively the fragmentation of a particular ion in a mixture of ions generated in a mass analyzer (in our case, the LIFT cell) [27] by using a collision-neutral gas such as nitrogen. Thus, by analyzing the molecular masses of generated fragments from a selected parent ion, it is easy to identify the specific residues that interact with the metal ions due to their corresponding increase in mass. Thereby, the main metal–peptide signal ions in the mass spectra were selected as parent ions and fragmented in the CID chamber.

Figure 7 presents the tandem MS spectra of [M + Zn-H]$^+$ (m/z 887.4) and [M + Cu-H]$^+$ (m/z 886.4) ions of octapeptide NAP complex that resulted from collision-induced dissociation. The presence of a strong signal assigned to the deaminated Zn–peptide complex [M + Zn-NH$_4$]$^+$ provides evidence that the binding site is not located within the N-terminal region of the peptide. The most intense fragments observed in both spectra were those assigned to the y6-b6 pair (Table 1). Meanwhile, the smallest fragments identified to bind Zn^{2+} were a5 and x5, respectively, suggesting that the metal ion interacts with the serine residue located at position 5 in the ^{3}PVS^5 peptide fragment. A related signal was observed in the case of copper ions where the presence of an y5 Cu-binding fragment confirms the role of Ser as a metal-binding site. Based on these results, it is likely that coordination in both cases involves the side-chains of the P, S and Q residues present in this region, as previously suggested by other methods [28,29].

Similar to the NAP peptide, the metal–Aβ(9–16) doubly charged species preserved the protein–metal interactions, while the amide backbone of the peptide was cleaved. As observed in Figure 8, the most intense signal, registered in both spectra, was assigned to the b6 metal-binding fragment. The presence of a strong signal at m/z 1058.4, characteristic of the [M + Zn-H]$^+$ ions, and absence of other relevant fragments, beside the b6, indicates that zinc forms a strong binding with the histidine-rich fragment and stabilizes the peptide structure. By comparison, in a previous research study, it was observed that in the absence of the metal ion, the peptide fragmentation allows the division of the two histidine residues [30]. Likewise, the copper–peptide complex formed mostly b6 fragments carrying Cu(I) species. This interaction observed in the CID spectrum of [M + Cu]$^{2+}$ ions was most likely mediated through the histidine residues in this region. In addition, besides the fragments holding the reduced copper ions, the MS^2 spectrum generated signals, characteristic of the y6, y5 and the unfragmented molecular ion, containing the Cu(II) species (Table 2). However, the intensity of those peaks is insignificant compared to the ones attributed to Cu$^+$-associated fragments.

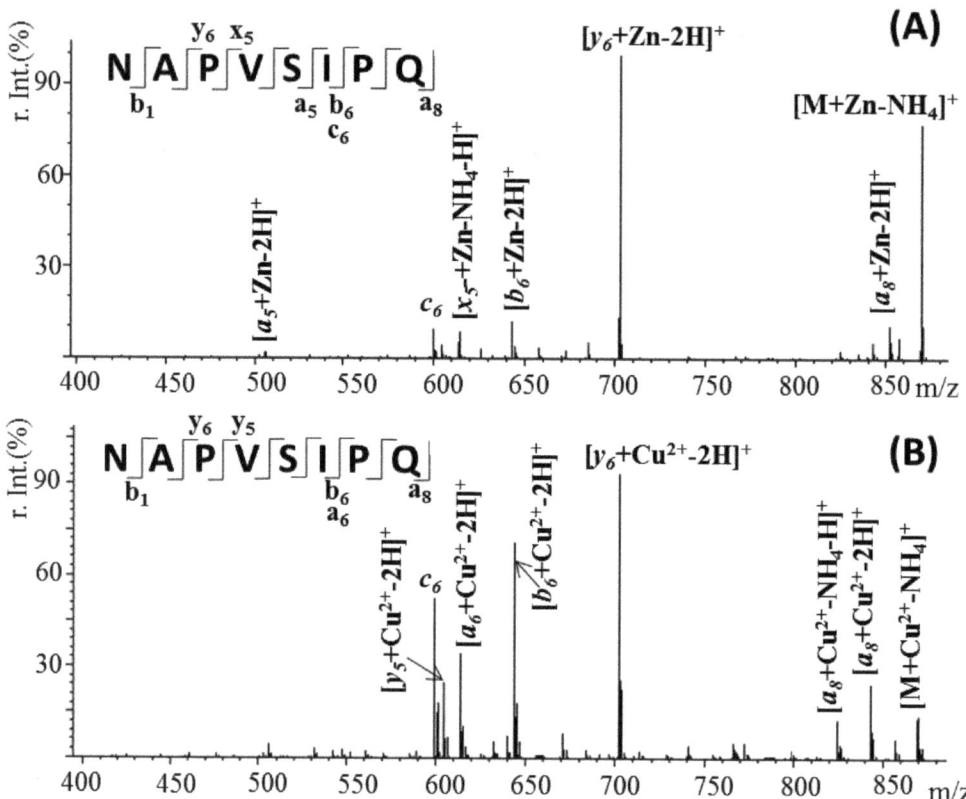

Figure 7. Tandem mass spectrometry (MS/MS) spectrum of the singly charged (**A**) [M + Zn-H]$^+$ion at m/z 887.4 and (**B**) [M + Cu-H]$^+$ ion at m/z 886.4, showing the x, y, z and a, b fragments of the peptide sequence.

Table 1. Collision-induced dissociation of [M + Zn-H]$^+$ and [M + Cu-H]$^+$ ions.

Fragmented Ion	Sequence	Predicted (m/z)	Observed (m/z)
[M+ Zn-H]$^+$	[M + Zn-NH$_4$]$^+$	871.1988	871.2012
	[a$_8$ + Zn-2H]$^+$	842.4358	842.4398
	[y$_6$ + Zn-2H]$^+$	703.3127	703.3145
	[b$_6$ + Zn-2H]$^+$	645.2395	645.2457
	[x$_5$ + Zn-NH$_4$-H]$^+$	615.4617	615.4673
	c$_6$	599.3511	599.3515
	[a$_5$ + Zn-2H]$^+$	504.3096	504.3138
[M+ Cu-H]$^+$	[M + Cu^{2+}-NH$_4$]$^+$	869.3648	869.3651
	[a$_8$ + Cu^{2+}-2H]$^+$	840.4818	840.4827
	[a$_8$ + Cu^{2+}-NH$_4$-2H]$^+$	823.4428	823.443
	[y$_6$ + Cu^{2+}-2H]$^+$	701.3787	701.3782
	[b$_6$ + Cu^{2+}-2H]$^+$	643.2024	643.1998
	[a$_6$ + Cu^{2+}-2H]$^+$	615.1836	615.1843
	[y$_5$ + Cu^{2+}-2H]$^+$	604.1872	604.1902
	c$_6$	599.3511	599.3524

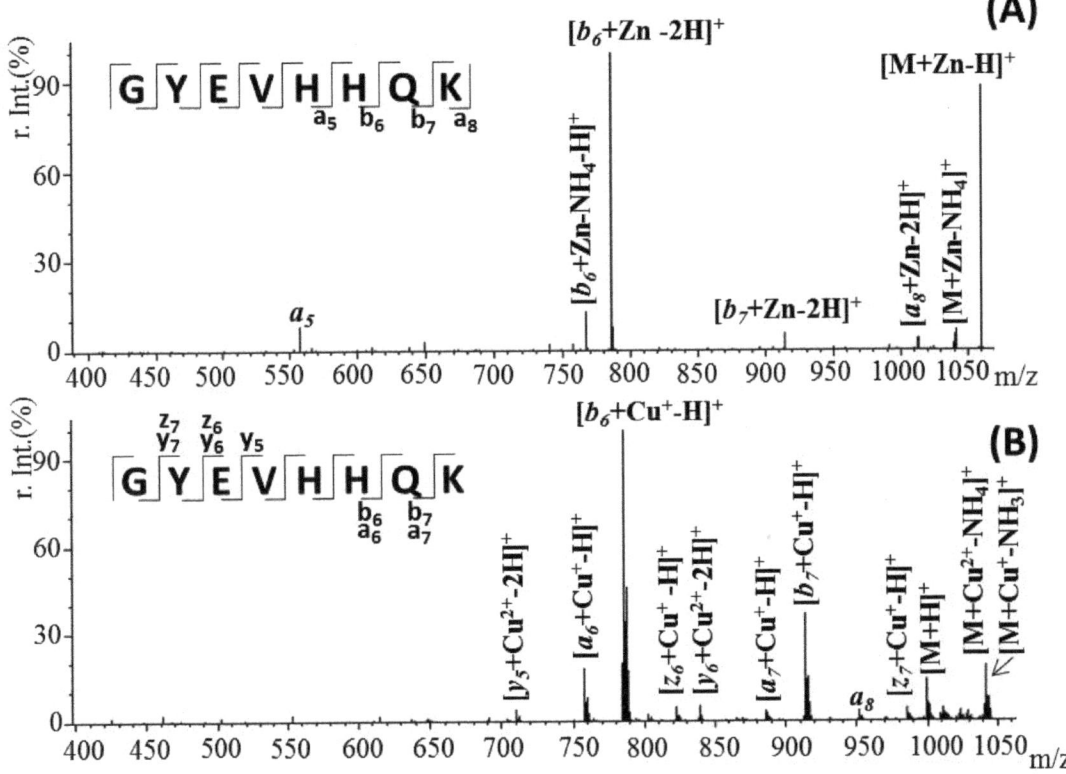

Figure 8. Tandem mass spectrometry (MS/MS) spectrum of the doubly charged (**A**) $[M + Zn]^{2+}$ ion at m/z 529.7 and (**B**) $[M + Cu]^{2+}$ ion at m/z 529.2, showing the y, z and a, b fragments of the peptide sequence.

Table 2. Collision-induced dissociation of $[M + Zn]^{2+}$ and $[M + Cu]^{2+}$ ions.

Fragmented Ion	Sequence	Predicted (m/z)	Observed (m/z)
$[M + Zn-H]^+$	$[M + Zn-H]^+$	1060.3988	1060.4022
	$[M + Zn-NH_4]^+$	1043.4514	1043.4437
	$[a_8 + Zn-2H]^+$	1014.2408	1014.2487
	$[b_7 + Zn-2H]^+$	914.3472	914.3408
	$[b_6 + Zn-2H]^+$	786.2281	786.2245
	$[b_6 + Zn-NH_4-H]^+$	769.2502	769.2429
	a_5	558.2040	558.1998
$[M + Cu-H]^+$	$[M + Cu^+-NH_3]^+$	1043.3867	1043.3816
	$[M + Cu^{2+}-NH_4]^+$	1042.3859	1042.3842
	$[z_7 + Cu^+-H]^+$	985.4288	985.4268
	a_8	951.4534	951.4452
	$[b_7 + Cu^+-H]^+$	913.3502	913.3427
	$[a_7 + Cu^+-H]^+$	885.2957	885.2904
	$[y_6 + Cu^{2+}-2H]^+$	838.3398	838.3342
	$[z_6 + Cu^+-H]^+$	822.3448	822.3403
	$[b_6 + Cu^+-H]^+$	785.2418	785.2442
	$[a_6 + Cu^+-H]^+$	757.2336	757.2369
	$[y_5 + Cu^{2+}-2H]^+$	709.3309	709.3378

2.4. Morphological and Topographical Characterization/Atomic Force Microscopy Investigation

Atomic force microscopy (AFM) is a powerful profilometry technique that gives quantitative information about the surface microstructure of thin films. AFM also provides morphological information in terms of roughness and heights of the structures [31]. In addition to characterizing different films, this method has utility in characterizing self-assembled nanostructures of peptides [32,33]. Hane F. et al. previously demonstrated by AFM that Cu^{2+} influences amyloid-(1-42) aggregation by increasing peptide–peptide binding forces [34].

AFM experiments conducted on fresh solution of peptides formed a thin film on the microscope slide. As observed in Figure 9A, NAP peptide forms, in its native state, homogeneous elongated crystals with a maximum height of 22 nm.

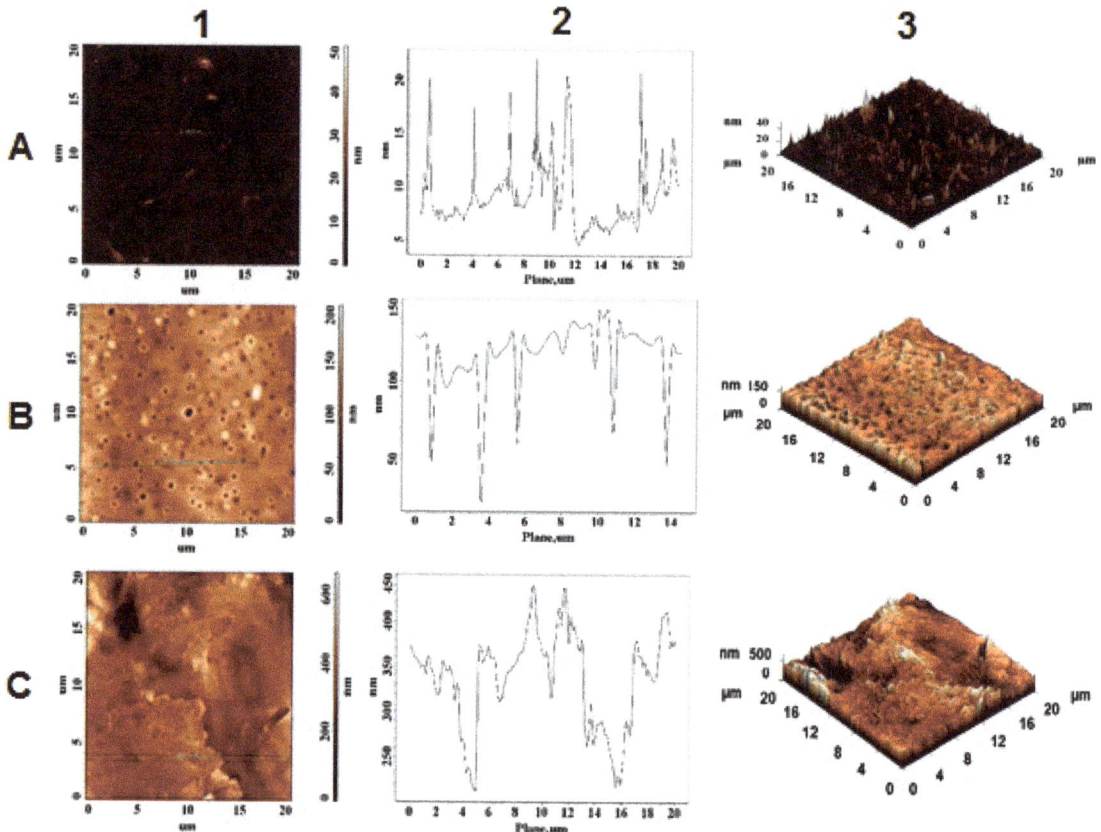

Figure 9. AFM images of NAP peptide (**A**); in the presence of copper ions (**B**) and zinc ions (**C**); (1) height of AFM images; (2) plot height distribution of particles along the profile line; and (3) 3D AFM images.

Moreover, Aβ(9–16) peptide (Figure 10A) formed mostly monodisperse spherical crystals characterized by a more convex shape than NAP peptide, having maximum height of 38.5 nm. As anticipated, the presence of metal ions significantly influenced the nanostructures of peptides in aqueous solution. Thus, a strong contrast between peptide crystallization in the absence and presence of peptide ions was recorded by the AFM equipment. For example, AFM images of NAP peptide carried out in the presence of Cu^{2+} ions (Figure 9B) showed the formation of a porous layer having a thickness of approximately 130–150 nm whereas the Aβ(9–16) fragment (Figure 10B) displayed, on the

0.5–1.2 μm thick layer, a reduction in the density of native Aβ crystals with amorphous and flattened tendency. The difference in topography may be related to the peptides' structure and their affinity toward copper ions. Meanwhile, the addition of Zn^{2+} ions induced the formation of a solid layer having a maximum height of approximately 450 nm. In addition, according to AFM images, the irregular layer formed in the presence of zinc ions preserves some of the peptide crystals, suggesting a lower interaction with the peptides.

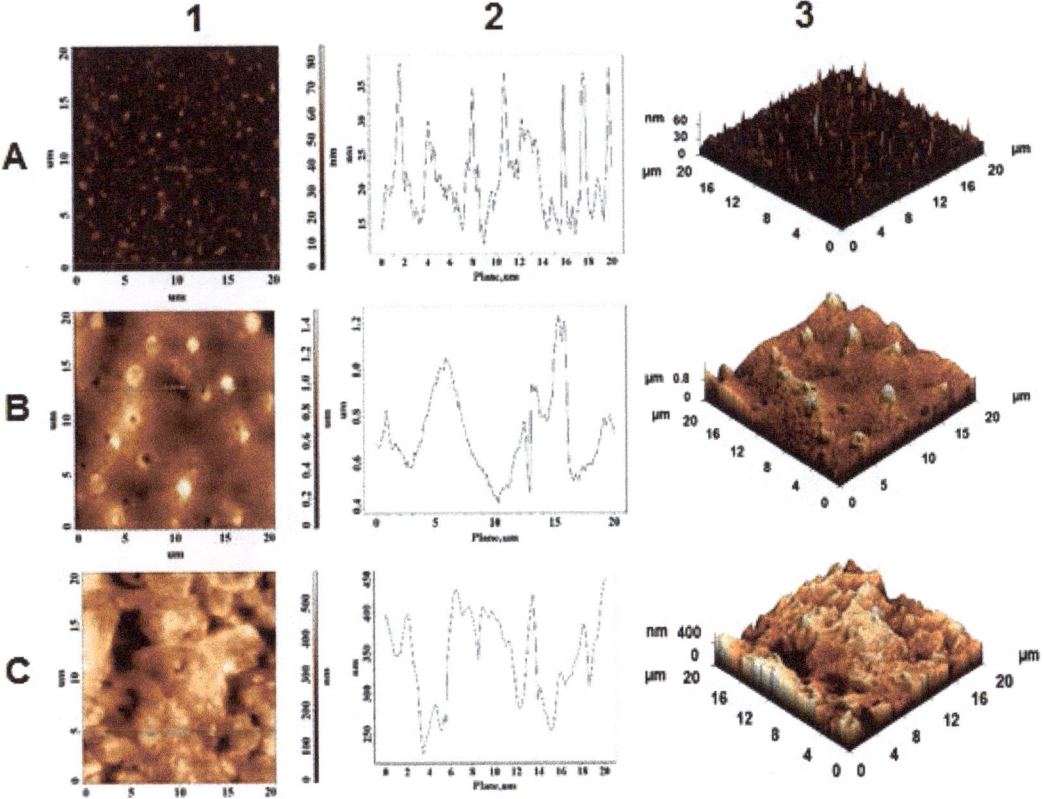

Figure 10. AFM images of Abeta9–16 peptide (**A**); in the presence of copper ions (**B**) and zinc ions (**C**); (1) height of AFM images; (2) plot height distribution of particles along the profile line; and (3) 3D AFM images.

The interaction between the two peptides was also investigated. Thus, as observed in Figure 11A, the mixture system formed a slim layer with a maximum height of 13 nm. Compared to pure peptides, shrinkage in crystal height was noticed. Furthermore, spherical crystals of Aβ(9–16) peptide were found to encapsulate NAP peptide, thus blocking amino acids that may interact with metals. As observed in Figure 11B, in the presence of copper(II) ions, the peptide system formed a porous layer characterized by a maximum height of 200 nm and a pore diameter of 0.5 μm. Meanwhile, the intensities of peptides in the presence of Zn^{2+} ions were significantly lower than those registered on individual samples, which means that both peptides could chelate zinc ions and alleviate the reduction in intensities of the smooth film.

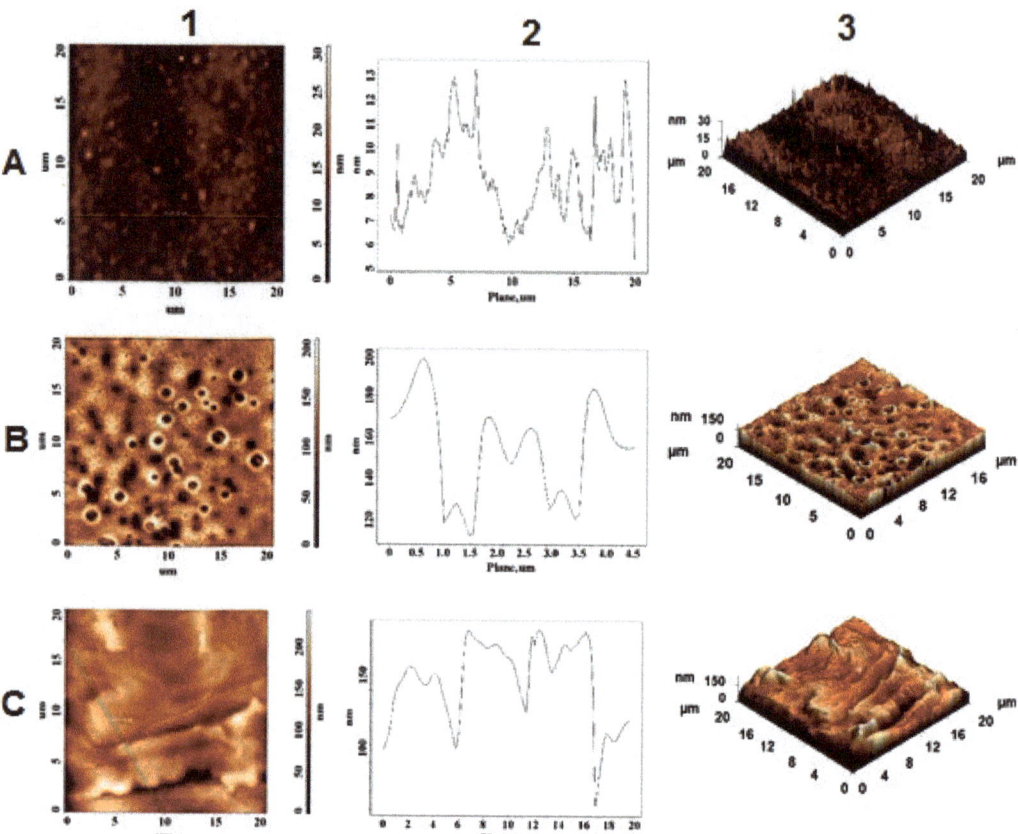

Figure 11. AFM images of NAP and Aβ(9–16) peptide mixture (**A**) in the presence of copper ions (**B**) and zinc ions (**C**); (1) height of AFM images; (2) plot height distribution of particles along the profile line; and (3) 3D AFM images.

3. Discussion

Alzheimer's disease has been a priority among research efforts into neurodegenerative diseases, and due to the fact that there is still no treatment, but only methods to alleviate or delay severe symptoms, research in the field continues through different approaches. Lately, there is growing interest in identifying bioactive compounds capable of preventing amyloid fibrils formation associated with neurodegenerative diseases. Among cell-penetrating peptides [35], peptide–drug conjugates hold great promise as an exciting area of research for targeted therapeutic approaches [36]. Various hybrid drugs were designed by combining the therapeutic characteristics of bioactive peptides with small organic molecules for treating different pathologies such as neurodegenerative diseases [37,38], cancer [39,40] and others [41].

In the present study, we examined the interaction between metal ions and two peptides that were previously synthesized [21,42] and are known to be involved in neurodegenerative pathogenesis. NAP exhibits a neuroprotective role and the Aβ(9–16) model peptide fragment represents the N-terminal part of aggregating amyloidogenic peptides, responsible for metal complexation, but not for the formation of neurotoxic fibrils. Both peptides exhibit an affinity to forming complexes with both metal ions (Zn^{2+} and Cu^{2+}) in independent experiments, showing a clear mass spectrometric signature. Both peptides, NAP and Aβ(9–16), formed stable ion complexes with Zn^{2+}, as shown by MALDI and ESI mass spectrometric measurements, incorporating a single Zn atom into the peptide sequence. In contrast, Cu–peptide interaction studies by the ESI-MS soft ionization method highlighted the presence of M + Cu(II) ions for NAP peptide while the Aβ(9–16) peptide generated signals for both M + Cu(II) and M + Cu(I) species. In addition, the ability of NAP peptide to bind two copper ions confirms the preference of neuroprotective octapeptide for divalent copper ion interactions.

Further, tandem mass spectrometry using the LIFT fragmentation approach was used to specifically identify the metal ion binding site within the two peptides. In the case of the NAP peptide, we observed that the Zn metal ion interacts with the serine residue located in the $^3PVS^5$ peptide fragment, while Cu interactions need involvement of the next proline residues. Tandem MS of Aβ(9–16)–zinc or copper complexes showed strong binding with the histidine-rich fragment that stabilizes the peptide structure.

Moreover, in-solution incubation of both peptides with zinc ions showed that there is no interaction between peptides that would disrupt their affinity for metal ions, suggesting that NAP and Aβ(9–16) peptide form independent zinc complexes. In the case of concomitant copper–peptides interaction, we observed a strong affinity and preference for divalent copper ions of NAP octapeptides and strong binding of only one Cu+ ion in the case of the amyloidogenic fragment.

Our findings show that the N-terminal Aβ(9–16) fragment can bind metal ions due to its rich histidine center, causing it to stiffen in comparison to its normal in-solution flexibility. The Aβ(9–16)–metal interactions help to promote the aggregating species of the pathophysiological Aβ(1-40/42).

The morphological and topographical characterization of both peptides in the absence and in the presence of zinc and copper ions was investigated by atomic force microscopy. As discussed in Section 2.4, the peptides showed a specific morphological and topographical signature that is reflected in their in-solution interaction with the zinc and copper ions. These in vitro studies show the structural diversity that peptides adopt in simple experimental work environments, suggesting their complexity in in vivo experiments.

We are convinced that natural or synthetic peptides offer enormous growth as future therapeutics in numerous human pathologies, such as development in cell-penetrating peptides, peptide drug conjugates and peptide metal chelators.

4. Materials and Methods

4.1. Materials

Peptide synthesis was performed using as solid support Fmoc-Gln(Trt)-Wang Resin (0.4–0.8 mmol amino acid/g of resin) purchased from NovaBiochem (Darmstadt, Germany). The amino acids required for synthesizing the desired peptides sequence were protected at N-terminal with Fmoc group (9-fluorenylmethyloxycarbonyl) and were obtained from GL Biochem (Shanghai, China). Other materials used for peptide synthesis were obtained from Merck (Germany). The metal salts were supplied by Sigma Aldrich and used for the preparation of in-solution metal–peptide complexes. The experimental solutions were prepared using deionized water (18.2 MΩ·cm) produced by a Milli-Q system (Millipore, Bedford, MA, USA). All other reagents were used without further purification.

4.2. Peptide Synthesis

The synthesis of the required peptides was carried out manually using classical solid phase Fmoc/tBu solid phase synthesis (SPPS). The NAP octapeptide (NH_2-NAPVSIPQ-COOH) was prepared starting from an Fmoc-Gln(Trt)-Wang resin, while for the Aβ(9–16) fragment, Fmoc Rink Amide resin (H_2N-GYEVHHQK-$CONH_2$) was used. Summarily, the peptides synthesis was performed in a fritted plastic reactor connected to a vacuum pump to remove washing solutions by successive cycles of Fmoc/tert-butyl deprotection and amino acid addition. All reactions were performed in dimethylformamide (DMF) medium: protecting groups such as Fmoc or tert-butyl were removed with 20% piperidine while activation of the new amino acid was accomplished in the presence of benzotriazol-1-yl-oxytripyrrolidinophosphonium hexafluorophosphate (PyBOP) and N-methylmorpholine (NMM). Final cleavage of the peptides from the resin was performed with a TFA:TIS:H_2O solution, in a ratio of 95:2.5:2.5 ($v/v/v$). The precipitation of the peptide was performed cold (-20 °C) in a solution of ethyl ether. Finally, the resulting fractions, eluted with 5% acetic acid solution in ultrapure water, were lyophilized and stored at -20 °C until further use.

4.3. Mass Spectrometry and Peptide Complex Formation

For MALDI-ToF experiments, 4 mM stock peptide solutions, prepared in deionized water at a pH of 7.4 adjusted by drop-wise addition of aqueous NaOH, were incubated in the presence of metal salts ($CuSO_4$, $ZnSO_4$) at a 1:10 peptide to metal molar ratio. The resulting mixture was gently mixed for 20 h at 24 °C and 350 rpm in a thermo-shaker (Thermomixer Compact Eppendorf AG 22331, Hamburg, Germany). MALDI-ToF MS was performed on a Bruker Ultraflex MALDI ToF/ToF mass spectrometer equipped with a pulsed nitrogen UV laser. The instrument was run in positive ionization mode and measurements were performed in the reflectron mode using α-cyano-4-hydroxycinnamic acid (HCCA) as matrix. The samples were loaded onto a 384-spot target plate of the MALDI-ToF instrument using the dried-droplet method by spotting 0.5 µL of matrix solution and 0.5 µL of sample solution, which were mixed directly on the target and allowed to dry in the ambient air. Calculations conducted with the ChemCalc online program were employed for theoretical monoisotopic mass determination. The spectra were recorded using the following parameters: 20 kV acceleration voltage, 40% grid voltage, 140 ns delay, low mass gate of 500 Da and an acquisition mass range of 600–3500 Da. In the final mass spectrum, 300 shots per acquisition were accumulated. External calibration was carried out using the monoisotopic masses of singly protonated ion signals of Bradykinin (1–7) (m/z 757.4), Angiotensin II (m/z 1046.5), Angiotensin I (m/z 1296.7), Substance P (m/z 1347.7), Bombesin (m/z 1619.8), Renin Substrate (m/z 1758.9), ACTH clip (1–17) (m/z 2093.1), ACTH clip (18–39) (m/z 2465.2) and Somatostatin (m/z 3147.5). The obtained spectra were processed using Bruker Flex Analysis 3.4 software.

ESI experiments were performed using sample solutions containing 12.5 µM of peptide and 125 µM metal salt. To preserve native peptide–metal interactions, samples were prepared in Milli-Q H_2O containing 25 mM ammonium acetate (pH 7.4). The ESI MS measurements were performed using a high-resolution mass spectrometer with Fourier transform and ion cyclotronic resonance FT-ICR MS (7T Thermo Finnigan LTQ FT, Bremen, Germany) coupled with an Agilent 1100 nano-HPLC system. The samples were initially diluted in 50% methanol solution and directly injected using a Hamilton syringe into the interface using a Harvard 11 PLUS pump, while the flow was set at 10 µL/min. The ions, obtained using the nano-ESI ionization source in positive ion mode, were accumulated externally in a hexapole collision cell before being transferred to the LTQ cell and subjected to a CID. The parent ions were isolated in a window with a width of 5–10 m/z with the center mass upshifted by 0.5–1 from the parent. CID fragmentation was conducted with normalized collision energy in the range of 15 to 25, and fragment spectra were acquired in the FTICR cell with a resolution of 25–50 k in the mass range 240–1100 with 1 micro scan and 500 ms max injection time at an AGC setting of 2×10^5. The spray voltage used for positive ionization was ~2800 V. The mass spectra were obtained within the mass range m/z 200–2000 with a resolution of 50,000 at AGC 1×10^6 and with an injection time of maximum 500 THX. The standard FT calibration mixture was used to calibrate the external mass. Fragmentation of precursor ions was accomplished by LIFT cell in MALDI–ToF/ToF mass analyzer [27]. The obtained MALDI spectra were processed using the FlexAnalysis program while for ESI spectra we used mMass software. Moreover, the correct peptide mass, theoretical fragmentation ions, hydrophobicity plots and peptide-charged ions were theoretically calculated using the GPMAW program.

4.4. Atomic Force Microscopy (AFM)

AFM experiments were carried out at a 1:10 peptide: metal molar ratio. In total, 5 μL of each sample was deposited onto a 1 × 1 cm cleaved mica surface and allowed to dry for 24 h at RT covered by a Petri dish to avoid dust. The surface images were obtained with an NTEGRA scanning probe microscope (NT-MDT Spectrum Instruments, Moscow, Russia), in AFM configuration. Rectangular silicon cantilevers NSG10 (NT-MDT, Moscow, Russia) with tips of high aspect ratio (sharpened pyramidal tip, angle of nearly 20°, tip curvature radius of 10 nm and height of 14–16 μm) were used in order to minimize convolution effects. All images were acquired in air, at RT, in tapping mode, with the velocity of 6 mm/s. For image acquisition, the Nova v.19891 for Solver software was used. All AFM images were obtained at a resolution of 256 × 256 pixels on a scale of 20 μm × 20 μm.

Author Contributions: Conceptualization, writing—original draft preparation: A.-V.L. and M.I. (Monica Iavorschi); AFM measures: G.E.H. and L.D.-I.; ESI-FTICR measures: M.I. (Maria Indeykina); writing—review and editing: L.D.-I. and B.A.P.; visualization and supervision: B.A.P. All authors have read and agreed to the published version of the manuscript.

Funding: This research was funded by Ministry of Research, Innovation and Digitalization within Program 1, Development of national research and development system, Subprogram 1.2, Institutional Performance—RDI excellence funding projects, under contract 10PFE/2021. L.D.I. also acknowledges the financial support from the Ministry of Research, Innovation and Digitization, CNCS/CCCDI–UEFISCDI, project number PN-III-P1-1.1-PD-2019-0442, within PNCDI III.

Institutional Review Board Statement: Not applicable.

Informed Consent Statement: Not applicable.

Data Availability Statement: Data is contained within the article.

Acknowledgments: A.V.L. and B.A.P. acknowledge the financial support of EU FT-ICR MS project, www.eu-fticr-ms.eu, for the short research stage at Emanuel Institute for Biochemical Physics, Russian Academy of Sciences, Moscow, Russia during 15–20 September 2019. Special thanks to Alexey Kononikhin and Eugene Nikolaev for making the collaboration possible.

Conflicts of Interest: The authors declare no conflict of interest.

References

1. Dunbar, R.C.; Polfer, N.C.; Berden, G.; Oomens, J. Metal Ion Binding to Peptides: Oxygen or Nitrogen Sites? *Int. J. Mass Spectrom.* **2012**, *330–332*, 71–77. [CrossRef]
2. Chitta, R.K.; Gross, M.L. Electrospray Ionization-Mass Spectrometry and Tandem Mass Spectrometry Reveal Self-Association and Metal-Ion Binding of Hydrophobic Peptides: A Study of the Gramicidin Dimer. *Biophys. J.* **2004**, *86*, 473–479. [CrossRef]
3. Carlton, D.D.; Schug, K.A. A Review on the Interrogation of Peptide–Metal Interactions Using Electrospray Ionization-Mass Spectrometry. *Anal. Chim. Acta* **2011**, *686*, 19–39. [CrossRef] [PubMed]
4. Giampà, M.; Sgobba, E. Insight to functional conformation and noncovalent interactions of protein-protein assembly using MALDI mass spectrometry. *Molecules* **2020**, *25*, 4979. [CrossRef]
5. Chen, G.; Fan, M.; Liu, Y.; Sun, B.; Liu, M.; Wu, J.; Li, N.; Guo, M. Advances in MS based strategies for probing ligand-target interactions: Focus on soft ionization mass spectrometric techniques. *Front. Chem.* **2019**, *7*, 703. [CrossRef]
6. Lehmann, E.; Zenobi, R.; Vetter, S. Matrix-Assisted Laser Desorption/Ionization Mass Spectra Reflect Solution-Phase Zinc Finger Peptide Complexation. *J. Am. Soc. Mass Spectrom.* **1999**, *10*, 27–34. [CrossRef]
7. Bertz, S.H.; Cope, S.; Dorton, D.; Murphy, M.; Ogle, C.A. Organocuprate Cross-Coupling: The Central Role of the Copper(III) Intermediate and the Importance of the Copper(I) Precursor. *Angew. Chem.* **2007**, *119*, 7212–7215. [CrossRef]
8. Roohani, N.; Hurrell, R.; Kelishadi, R.; Schulin, R. Zinc and Its Importance for Human Health: An Integrative Review. *J. Res. Med. Sci.* **2013**, *18*, 144–157.
9. Brown, K.H.; Wuehler, S.E.; Peerson, J.M. The Importance of Zinc in Human Nutrition and Estimation of the Global Prevalence of Zinc Deficiency. *Food Nutr. Bull.* **2001**, *22*, 113–125. [CrossRef]
10. Hasselbalch, S.G.; Madsen, K.; Svarer, C.; Pinborg, L.H.; Holm, S.; Paulson, O.B.; Waldemar, G.; Knudsen, G.M. Reduced 5-HT2A Receptor Binding in Patients with Mild Cognitive Impairment. *Neurobiol. Aging* **2008**, *29*, 1830–1838. [CrossRef]
11. Kozlowski, H.; Luczkowski, M.; Remelli, M.; Valensin, D. Copper, Zinc and Iron in Neurodegenerative Diseases (Alzheimer's, Parkinson's and Prion Diseases). *Coord. Chem. Rev.* **2012**, *256*, 2129–2141. [CrossRef]

12. Zatta, P.; Drago, D.; Bolognin, S.; Sensi, S.L. Alzheimer's Disease, Metal Ions and Metal Homeostatic Therapy. *Trends Pharmacol. Sci.* **2009**, *30*, 346–355. [CrossRef]
13. Wu, Z.; Fernandez-Lima, F.A.; Perez, L.M.; Russell, D.H. A New Copper Containing MALDI Matrix That Yields High Abundances of [Peptide + Cu] + Ions. *J. Am. Soc. Mass Spectrom.* **2009**, *20*, 1263–1271. [CrossRef]
14. Prudent, M.; Girault, H.H. On-Line Electrogeneration of Copper-Peptide Complexes in Microspray Mass Spectrometry. *J. Am. Soc. Mass Spectrom.* **2008**, *19*, 560–568. [CrossRef]
15. Bluhm, B.K.; Shields, S.J.; Bayse, C.A.; Hall, M.B.; Russell, D.H. Determination of Copper Binding Sites in Peptides Containing Basic Residues: A Combined Experimental and Theoretical Study. *Int. J. Mass Spectrom.* **2001**, *204*, 31–46. [CrossRef]
16. Lupaescu, A.V.; Jureschi, M.; Ciobanu, C.I.; Ion, L.; Zbancioc, G.; Petre, B.A.; Drochioiu, G. FTIR and MS Evidence for Heavy Metal Binding to Anti-Amyloidal NAP-Like Peptides. *Int. J. Pept. Res. Ther.* **2019**, *25*, 303–309. [CrossRef]
17. Murariu, M.; Habasescu, L.; Ciobanu, C.-I.; Gradinaru, R.V.; Pui, A.; Drochioiu, G.; Mangalagiu, I. Interaction of Amyloid Aβ(9–16) Peptide Fragment with Metal Ions: CD, FT-IR, and Fluorescence Spectroscopic Studies. *Int. J. Pept. Res. Ther.* **2019**, *25*, 897–909. [CrossRef]
18. Wang, C.; Wang, C.; Li, B.; Li, H. Zn(II) Chelating with Peptides Found in Sesame Protein Hydrolysates: Identification of the Binding Sites of Complexes. *Food Chem.* **2014**, *165*, 594–602. [CrossRef]
19. Zoroddu, M.A.; Medici, S.; Peana, M.; Anedda, R. NMR Studies of Zinc Binding in a Multi-Histidinic Peptide Fragment. *Dalton Trans.* **2010**, *39*, 1282–1294. [CrossRef]
20. Chen, X.; Drogaris, P.; Bern, M. Identification of Tandem Mass Spectra of Mixtures of Isomeric Peptides. *J. Proteome Res.* **2010**, *9*, 3270–3279. [CrossRef]
21. Lupaescu, A.-V.; Ciobanu, C.-I.; Humelnicu, I.; Petre, B.A.; Murariu, M.; Drochioiu, G. Design and Synthesis of New Anti-Amyloid NAP-Based/like Peptides. *Rev. Roum. Chim.* **2019**, *64*, 535–546. [CrossRef]
22. Kessler, A.T.; Raja, A. Biochemistry, Histidine. In *StatPearls*; StatPearls Publishing: Treasure Island, FL, USA, 2021.
23. Willard, B.B.; Kinter, M. Effects of the Position of Internal Histidine Residues on the Collision-Induced Fragmentation of Triply Protonated Tryptic Peptides. *J. Am. Soc. Mass Spectrom.* **2001**, *12*, 1262–1271. [CrossRef]
24. Zhang, J.; Frankevich, V.; Knochenmuss, R.; Friess, S.D.; Zenobi, R. Reduction of Cu(II) in Matrix-Assisted Laser Desorption/Ionization Mass Spectrometry. *J. Am. Soc. Mass Spectrom.* **2003**, *14*, 42–50. [CrossRef]
25. Lu, Y.; Prudent, M.; Qiao, L.; Mendez, M.A.; Girault, H.H. Copper(i) and Copper(Ii) Binding to β-Amyloid 16 (Aβ16) Studied by Electrospray Ionization Mass Spectrometry. *Metallomics* **2010**, *2*, 474. [CrossRef] [PubMed]
26. Roepstorff, P.; Fohlman, J. Proposal for a common nomenclature for sequence ions in mass spectra of peptides. *Biomed. Mass Spectrom.* **1984**, *11*, 601–605. [CrossRef] [PubMed]
27. Suckau, D.; Resemann, A.; Schuerenberg, M.; Hufnagel, P.; Franzen, J.; Holle, A. A novel MALDI LIFT-TOF/TOF mass spectrometer for proteomics. *Anal. Bioanal. Chem.* **2003**, *376*, 952–965. [CrossRef]
28. Trzaskowski, B.; Adamowicz, L.; Deymier, P.A. A Theoretical Study of Zinc(II) Interactions with Amino Acid Models and Peptide Fragments. *JBIC J. Biol. Inorg. Chem.* **2007**, *13*, 133–137. [CrossRef]
29. Deschamps, P.; Zerrouk, N.; Martens, T.; Charlot, M.-F.; Girerd, J.J.; Chaumeil, J.C.; Tomas, A. Copper Complexation by Amino Acid: L-Glutamine–Copper(II)– L-Histidine Ternary System. *J. Trace Microprobe Tech.* **2003**, *21*, 729–741. [CrossRef]
30. Mocanu, C.S.; Jureschi, M.; Drochioiu, G. Aluminium binding to modified amyloid-β peptides: Implications for alzheimer's disease. *Molecules* **2020**, *25*, 4536. [CrossRef]
31. Joshi, J.; Homburg, S.V.; Ehrmann, A. Atomic Force Microscopy (AFM) on Biopolymers and Hydrogels for Biotechnological Applications—Possibilities and Limits. *Polymers* **2022**, *14*, 1267. [CrossRef]
32. Habasescu, L.; Jureschi, M.; Petre, B.-A.; Mihai, M.; Gradinaru, R.-V.; Murariu, M.; Drochioiu, G. Histidine-Lacked Aβ(1–16) Peptides: PH-Dependent Conformational Changes in Metal Ion Binding. *Int. J. Pept. Res. Ther.* **2020**, *26*, 2529–2546. [CrossRef]
33. Drochioiu, G.; Manea, M.; Dragusanu, M.; Murariu, M.; Dragan, E.S.; Petre, B.A.; Mezo, G.; Przybylski, M. Interaction of β-Amyloid(1–40) Peptide with Pairs of Metal Ions: An Electrospray Ion Trap Mass Spectrometric Model Study. *Biophys. Chem.* **2009**, *144*, 9–20. [CrossRef]
34. Hane, F.; Tran, G.; Attwood, S.J.; Leonenko, Z. Cu2+ affects amyloid-β (1–42) aggregation by increasing peptide-peptide binding forces. *PLoS ONE* **2013**, *8*, e59005. [CrossRef]
35. Österlund, N.; Wärmländer, S.K.T.S.; Gräslund, A. Cell-Penetrating Peptides with Unexpected Anti-Amyloid Properties. *Pharmaceutics* **2022**, *14*, 823. [CrossRef]
36. Lupaescu, A.-V.; Iavorschi, M.; Covasa, M. The Use of Bioactive Compounds in Hyperglycemia- and Amyloid Fibrils-Induced Toxicity in Type 2 Diabetes and Alzheimer's Disease. *Pharmaceutics* **2022**, *14*, 235. [CrossRef]
37. Ilina, A.; Khavinson, V.; Linkova, N.; Petukhov, M. Neuroepigenetic Mechanisms of Action of Ultrashort Peptides in Alzheimer's Disease. *Int. J. Mol. Sci.* **2022**, *23*, 4259. [CrossRef]
38. Chernick, D.; Zhong, R.; Li, L. The Role of HDL and HDL Mimetic Peptides as Potential Therapeutics for Alzheimer's Disease. *Biomolecules* **2020**, *10*, 1276. [CrossRef]

39. Ciobanasu, C. Peptides-Based Therapy and Diagnosis. Strategies for Non-Invasive Therapies in Cancer. *J. Drug Target.* **2021**, *29*, 1063–1079. [CrossRef]
40. Liu, R.; Li, X.; Xiao, W.; Lam, K.S. Tumor-Targeting Peptides from Combinatorial Libraries. *Adv. Drug Deliv. Rev.* **2017**, *110–111*, 13–37. [CrossRef]
41. Wang, L.; Wang, N.; Zhang, W.; Cheng, X.; Yan, Z.; Shao, G.; Wang, X.; Wang, R.; Fu, C. Therapeutic Peptides: Current Applications and Future Directions. *Signal Transduct. Target. Ther.* **2022**, *7*, 48. [CrossRef]
42. Jureschi, M.; Humelnicu, I.; Petre, B.A.; Ciobanu, C.I.; Murariu, M.; Drochioiu, G. Solid Phase Synthesis of Four Analogs of Amyloid-β(9–16) Peptide: MS and FT-IR Characterization. *Rev. Roum. Chim.* **2019**, *64*, 433–443. [CrossRef]

Review

Fmoc-Diphenylalanine Hydrogels: Optimization of Preparation Methods and Structural Insights

Carlo Diaferia, Elisabetta Rosa, Giancarlo Morelli and Antonella Accardo *

Department of Pharmacy and Interuniversity Research Centre on Bioactive Peptides (CIRPeB), University of Naples "Federico II", Via Montesano 49, 80131 Naples, Italy
* Correspondence: antonella.accardo@unina.it; Tel.: +39-081-2532045

Abstract: Hydrogels (HGs) are tri-dimensional materials with a non-Newtonian flow behaviour formed by networks able to encapsulate high amounts of water or other biological fluids. They can be prepared using both synthetic or natural polymers and their mechanical and functional properties may change according to the preparation method, the solvent, the pH, and to others experimental parameters. Recently, many short and ultra-short peptides have been investigated as building blocks for the formulation of biocompatible hydrogels suitable for different biomedical applications. Due to its simplicity and capability to gel in physiological conditions, Fmoc-FF dipeptide is one of the most studied peptide hydrogelators. Although its identification dates to 15 ago, its behaviour is currently studied because of the observation that the final material obtained is deeply dependent on the preparation method. To collect information about their formulation, here are reported some different strategies adopted until now for the Fmoc-FF HG preparation, noting the changes in the structural arrangement and behaviour in terms of stiffness, matrix porosity, and stability induced by the different formulation strategy on the final material.

Keywords: Fmoc-FF; peptide hydrogels; peptide materials; hydrogel preparation; diphenylalanine

Citation: Diaferia, C.; Rosa, E.; Morelli, G.; Accardo, A. Fmoc-Diphenylalanine Hydrogels: Optimization of Preparation Methods and Structural Insights. *Pharmaceuticals* 2022, 15, 1048. https://doi.org/10.3390/ph15091048

Academic Editors: Giovanni N. Roviello and Rosanna Palumbo

Received: 18 July 2022
Accepted: 18 August 2022
Published: 25 August 2022

Publisher's Note: MDPI stays neutral with regard to jurisdictional claims in published maps and institutional affiliations.

Copyright: © 2022 by the authors. Licensee MDPI, Basel, Switzerland. This article is an open access article distributed under the terms and conditions of the Creative Commons Attribution (CC BY) license (https://creativecommons.org/licenses/by/4.0/).

1. Introduction

Researchers have directed their gaze towards nature since the analysis of biologically relevant structures, elements, and processes can embody a motivating well for inspiration. On the evidence that numerous structures are the consequence of natural self-organization of proteinaceous materials, peptide-based building blocks and their analogues have begun to be studied [1–4]. In addition, supramolecular aggregates and misfolding protein materials were found as pathological hallmarks of major human illnesses too, including bovine spongiform encephalopathy (mad cow disease), Creutzfeldt–Jakob disease, Alzheimer's disease, Huntington's, and Parkinson's diseases [5]. Although first identified as pathological entities, a new research line has shown that amyloid proteinaceous materials contribute to support and conduct complex biological functions [6]. For all these reasons, amyloids, once exclusively affiliated as pathological and toxic structures, are now raising interest as biomimicry self-assembling class of biological elements for artificial materials production. Their potential engineering is a consequence of the study and comprehension of the supramolecular structuration [7].

Identified in the middle of the primary sequences of $A\beta_{1-40}$ and $A\beta_{1-42}$, diphenylalanine (FF) was recognized as the crucial aggregative motif in the Alzheimer's β-amyloid polypeptides [8]. This simple homopeptide revealed the capability to self-organize efficiently into well-ordered tubular architectures with a long persistence length (~100 μm) because of π-π staking and H-bonding interaction networks [9]. Due to its chemical simplicity and versatility, FF rapidly became the paradigm for the study for peptide self-assembly. Moreover, modifying its very simple chemical structure, a plethora of FF analogues was proposed during in recent years [10–18]. Cationic FF differs from FF (or zwitterionic

FF) for the amidation of the C-terminus. In the amide form, the C-terminus is unable to generate head-to-tail hydrogen-bound interactions and dipeptide preferentially self-assembles into nanowires [19]. With the rationale of generating a covalent attachment to fabricated gold electrodes for application in nano-devices, Gazit et al. designed the tripeptide Cys-FF in which the thiol group contributes to cross-linking phenomena and to the self-aggregation into nanotubes [20]. Other diphenylalanine analogues able to aggregate in micro-structures, such as flat plates, flattened micro-planks, and micro-rods, were obtained for progressive elongation of the Phe side chain with methylene groups (dihomophenylalanine (DiHpa); di-2-amino-5-phenylpentanoic acid (DiApp); di-2-amino-6-phenylhexanoic acid (DiAph)) [21]. With the aim to elucidate the supposed role of electrostatic interactions in the peptide backbone aggregation, in 2005 an FF derivative in which the N-terminal amine and the C-terminal carboxyl of the peptide are respectively acetylated and amidated (Ac-Phe-Phe-NH$_2$) [20] were designed and synthetized. This uncharged peptide showed the capability to efficiently self-assemble into tubular structures. Consequently, the authors analysed a small library of analogues, Ac-Phe-Phe-OH, Boc-Phe-Phe-OH, Cbz-Phe-Phe-OH, and Fmoc-Phe-Phe-OH, in which only the amino group was blocked by using conventional solid phase peptide synthesis (SPPS) protecting groups [22]. The tert-Butyloxycarbonyl (Boc)-FF monomer analogue allows to obtain peptide spheres or small peptide particles when its HFIP stock solution is diluted in ethanol [23,24]. Instead, due to the additional stacking between the Benzyloxycarbonyl (Cbz) or 9-fluorenylmethoxycarbonyl (Fmoc) aromatic moieties, the two correspondent FF protected compounds self-assembled into fibrillary structures. Fibrillary structures of Fmoc-FF-OH showed an ultrastructure and dimensions extremely similar to the amyloid fibrils. Fmoc-FF dipeptide (see Figure 1 for the chemical structure) is one of the most studied ultra-short peptides for hydrogel (HG) preparation [25,26]. The main reason for this large interest is related to the possibility to obtain stable self-supporting hydrogels at pH values compatible with physiological applications, including tissue engineering and drug delivery. Additional applicative areas cover chemical catalysis, nanoreactors development, optical engineering, wound treatments, ophthalmic preparations, energy harvesting, antifouling, and biocompatible coating applications, optoelectronics, potential immuno-responsive agents, and absorbents systems for oil/water separation [27].

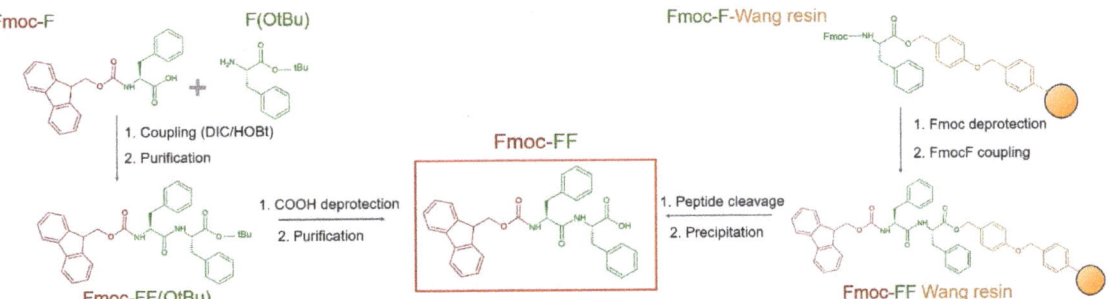

Figure 1. Scheme for the synthesis of Fmoc-FF (red rectangle) using liquid phase strategy (on the left) and solid phase peptide synthesis (SPPS) on the right. DIC: N,N'-Diisopropylcarbodiimide; HOBt: 1-Hydroxybenzotriazole.

The chemical accessibility, biodegradability, biofunctionality and the possibility to adopt specific secondary, tertiary, or quaternary architectures represent additional advantages for this simple peptide building block, alone, in combination with different chemical entities or in different morphological shapes [28,29]. Moreover, it was observed that matrix structural and functional properties can be tuned opportunely by simply modifying the gelation kinetic and other experimental parameters (such as pH, temperature, and used solvent) [30,31]. This observation pushed the research towards the development of novel and alternative methods to achieve HG generation. On the other hand, novel peptide analogues, containing codified or non-codified amino acids, have been proposed as building blocks for the preparation of hydrogels with enhanced properties [32,33]. It is worth noting that although the identification of Fmoc-FF as a hydrogelator dates to 15 years ago, its behaviour is continuously studied. The aim of this review is to provide an overview of the different preparation methods reported until now for the fabrication of Fmoc-FF HGs. This study also notes how the used method can affect both the structural and mechanical properties of the obtained material.

2. Applicative and Biomedical Relevance of Fmoc-FF Hydrogel

In the plethora of peptide-based hydrogels, Fmoc-FF becomes a relevant system due to some specific advantages. With respect to other peptide-hydrogelators (e.g., RADA peptides, [34] MAX-1, [35], amphiphilic and no-natural containing sequences [36,37]), Fmoc-FF represents a simple and accessible chemical entity, commercially available from different companies. The structural simplicity of the molecules additionally allows its production in high purity and with contained costs, both related to a solid phase peptide synthesis (SPPS) [38] and to the synthesis in solution (Figure 1). In the SPPS, the product is obtained into a three-step synthetic route. Specifically, a preloaded Fmoc-F-Wang resin undergoes a Fmoc-F-OH coupling after a Fmoc-deprotection. The cleavage of peptide and its precipitation make available the final powder with an efficient scalability. Alternatively, the same molecule can be obtained in solution, taking advantage of the non-requested protection of the Phe side chain. In this case, a C-protected phenylalanine is coupled with a Fmoc-F-OH residue. Then, after the deprotection of the carboxylic acid, Fmoc-FF is precipitated and purified.

Another advantage of Fmoc-FF as a building block for hydrogel preparation is its fast kinetics of formation, which occurs in a few minutes. Moreover, the gel can be prepared by using both in vitro and in vivo friendly solvents (water, buffers, and cell media) [39]. Additionally, the optical transparency of the Fmoc-FF matrix is a benefit for a preliminary macroscopical evaluation in terms of homogeneity and correct formulation. Moreover, the mechanical properties and the tunability exhibited by the Fmoc-FF matrix make it compatible with extrusion, electrospinning, and filming deposition procedures [40,41]. Finally, the resulting hydrogel exhibits a good shelf stability. All these advantages make Fmoc-FF highly accessible for application in many research fields.

Early investigations carried out by Gazit and co-workers noted only the capability of the dipeptide to self-assemble into a rigid material with macroscopic characteristics of a self-supporting gel [26]. The hydrogel formation was achieved with a "solvent switch" methodology (*vide intra*), which consists into the dilution of an organic peptide stock solution (generally at 100 mg/mL) in water. In this specific case, HFIP (1,1,1,3,3,3-hexafluoro-2-propanol) was used as an organic solvent, diluted in water at a final concentration of 5 mg/mL (0.5 wt%) [26]. The pre-dissolution in HFIP is currently used as aggregative step for other aromatic peptide-based materials [42–44]. The so prepared materials were found syringable, shaped, and stable across a broad range of temperatures, over a wide pH range and in presence of aggressive chemical agents, such as guanidinium and urea. Due to the presence of hollow cavities in the supramolecular architecture, the possible use of Fmoc-FF as a reservoir was scrutinized by the encapsulation and release study of model drugs such as fluorescein (FITC) and insulin-FITC, indicating a retain of molecules of 5 kDa, and a slow release for small chemical entities [26]. The remarkable mechanical rigidity of this hydrogel compared with others physically cross-linked ones, the direct correlation between rheological properties and peptide concentration, and its capability to support Chinese Hamster Ovarian (CHO) cell adhesion led authors to suggest Fmoc-FF as a promising, tunable, and versatile scaffold for tissue engineering too. Simultaneously, the evidence reported that some Fmoc-protected dipeptides and amino acids are able to form scaffold materials [45–47], Ulijn and co-workers synthetized am Fmoc-based dipeptide library built as combination of glycine (Gly), leucine (Leu), phenylalanine (Phe), and alanine (Ala).

Using a progressive decrease in the pH (from 8 to < 4, procedure that are reported as "pH switch" method), the authors noted that all the library members were able to form fibrous-based hydrogels. Only Fmoc-FF produced low-concentration HGs after lowering the pH to a physiologically relevant value [25]. Fmoc-FF gel was also tested for its ability to support proliferation and retention of phenotype bovine chondrocytes, both in 2D and 3D experiments.

Beyond tissue engineering and drug delivery applications, Fmoc-FF nanomaterials have been investigated as innovative materials in industrial and biotechnological fields. For example, in 2010 Rosenman et al. proposed self-assembled Fmoc-FF materials as biosensors for the detection of amyloid fibrils. From the analysis of the optical properties, it was proposed a shift from a 0D-quantum dot to a 2D-quantum well with a thickness around 1 nm. This observation demonstrates the capability of this self-assembled peptide to exhibit the same quantum physical phenomenon previously observed only for semiconductor crystals [48]. Moreover, to improve the biofunctionality and the biocompatibility of bio-based materials such as silica wafers, Fmoc-FF peptide was covalently anchored on its the surface. In this case, the AFM characterization pointed out the tendency of the aromatic dipeptide to self-assemble in nanorods with a mean radius ranging from 10 to 30 nm. The structural arrangement of the peptide on the silica surface allows a modification of the functional properties of the material in terms of angle of contact. As expected, the density of nanorods on the silica surface is strictly related to the concentration of the immobilized peptide [49]. Analogously to FF-based nanomaterials, dried Fmoc-FF peptide hydrogels revealed piezoelectricity under piezoresponse force microscopy. Due to their biomimicry nature, coupled to these electric properties, Fmoc-FF gels were also proposed as advantageous tools in application in which electrical stimuli are required (e.g., axonal regeneration). The non-centrosymmetric topology of the β-sheet rich fibres were quoted as the ascribable structural reason of the electric features of the material. Indeed, an overall polarization along the fibre axis is related to dipoles, running perpendicularly to the direction of the β-strands [50]. Recently, Fmoc-FF hydrogels and some of its cationic variants (Fmoc-FFKK, Fmoc-FFFKK, and Fmoc-FFOO, in which O is the one code symbol for ornithine) have been studied for the first time as antibacterial materials [51]. Gel preparation was achieved at a final peptide concentration of 2.0 wt%. Unsurprisingly, self-assembled cationic variants exhibited high ordered antiparallel β-sheet structures and low rigidity and viscosity. Fmoc-FF hydrogels showed significant antibiofilm activity preventing the growth of bacteria and biofilm of Gram-positive (*Staphylococcus aureus* and *epidermis*) and Gram-negative (*Pseudomonas aeruginosa* and *Escherichia Coli*) bacteria that can be localized on the medical devise surface.

Recently, Adler-Abramovich and co-workers reported the prominent capability of Fmoc-FF hydrogel to specifically encage oxygen molecules and restrict its movement in absence of metal ions [52]. Molecular dynamics computations highlighted that the binding of O_2 is originated by specific interactions that take place between O_2 and the interior surface of Fmoc-FF fibrils. The ability of the gel to encage oxygen suggests its potential application as a passive mean to maintain hydrogen production by the O_2-hypersensitive enzyme [FeFe]-hydrogenase. These results leave envisage potential utilization of Fmoc-FF hydrogels in a wide range of O_2-sensitive applications.

3. Structural Organization and Proposed Model

The growing interest around Fmoc-FF as innovative material also encouraged structural studies aimed to identify the model of aggregation into fibrillary networks. In 2008, Uljin and co-workers proposed for the first time the aggregation model for Fmoc-FF [53]. The model was designed on the basis of their experimental data collected by Circular Dicrohism (CD) and Fourier Transformed Infrared spectroscopies (FT-IR). Structural characterization highlighted an anti-parallel β-sheets arrangement of the peptide building blocks and anti-parallel π-stacking of the fluorenyl groups (Figure 2A,B). In more detail, in disassembled systems such as Fmoc-Phe, Fmoc group shows two dichroic prints at 307 nm and below 214 nm, without contribution in the far-UV region. On the contrary, in Fmoc-FF gels a negative pick centred at 218 nm is consistent with a β-sheet structure, implying that an ordered supramolecular structure is associated with the gel matrix. Gels caused a signal in the range of 304–308 nm, attributed to the $\pi \to \pi^*$ transition in the fluorenyl-moiety too. Two other CD prints (a local maximum at 192 nm and a minimum at 202 nm) are indicative for α-helix transition associate to $\pi \to \pi^*$ transition. The predominant β-arrangements were supported by FT-IR analysis too. Indeed, Fmoc-FF gels are characterized by two peaks in the amide I region (1630 and 1685 cm^{-1}), compatible with an antiparallel organization of the peptides. All these structural requirements were well-satisfied by the model proposed by the authors: this model was based on a nanocylindrical structure (with an external diameter of ~3.0 nm, Figure 2C) formed by the interlocking through lateral π-π interactions of four twisted anti-parallel β-sheets. These fibrils, forming J-aggregates, were then shown to further self-assemble laterally, forming large flat ribbons under specific pH conditions, visible in TEM microscopy (Figure 2D). The scattering pattern of the Fmoc-FF-dried gel showed a series of diffractions compatible with the structural organization of the proposed model [53]. Based on the evidence that pK_a can significantly change as a consequence of protein and peptide self-organization phenomenon, Saiani et al. studied effect of pH on the Fmoc-FF self-assembly process [54]. Their results suggested that the self-assembly of Fmoc-FF building blocks is prompted by a lowering of the pH as consequence of two apparent pK_a shifts (of ≈6.4 and 2.2 pH units, respectively) above the theoretical pK_a value (3.5) (see Figure 2E). At high pH, where the Fmoc-FF building blocks are in their ionized form, the self-assembly is mainly forbidden. Then in correspondence of the first pK_{a1}, at pH 10.2–9.5, both protonated and non-protonated molecules begin to self-assemble into paired fibrils (Figure 2F). The further pH lowering from 9.5 to 6.2 causes a decrease in the fibre surface charge that in turn brings to the formation of large rigid ribbons due to the lateral interactions of the fibres. Below pH 6.2 (between 6.2 and 5.2) the second apparent pK_a shift is observed, where a further aggregation of the ribbons occurs [54]. Recently, Yan et al. speculated on the possibility to achieve a conformational transition from a β-sheet structure to a helical one by a charge-induced strategy. To demonstrate their hypothesis, they studied the structural behaviour of Fmoc-FF hydrogel before and after the addition of $Na_2B_4O_7$-EDTA buffer (pH = 8.5) [55]. Under these basic conditions, a structural transformation was detected from antiparallel β-sheet structures to a parallel helical one, imputable to the electrostatic repulsion between the negatively charged peptide molecules. This transition was further confirmed by computational simulations studies. The interaction energy analysis also highlighted the crucial role performed from the water molecule in stabilizing the Fmoc-FF$^-$ helix nanofibril. Successively, the same authors also demonstrated that the addition of metal ions can induce structural transformation in Fmoc-FF from an amyloid-like β-sheet into a superhelix or random coil [56]. The peptide structural organization was found dependent from the types and ratios of metal ion/dipeptide through metal.

The inner size of the hydrogel cavity was estimated by Huppert et al. by a simple and non-destructive method based on the reversible photoproteolytic cycle of the photoacid 8-hydroxypyrene-1,3,6-trisulfonate (HPTS, pyranine) [57]. HPTS, a photoacid molecule with a triple negative nature, is able to transfer H$^+$ ions with a time constant around 100 ps. The study was conducted on the experimental evidence that the parameter affecting the photocycles are related to water sphere radius in very small, confined volumes of media, as in hydrogel matrices. In the Fmoc-FF hydrogel the water cavity dimension between the fibril walls was found to be around 100 Å. This narrow size, inaccessible for living cells, suggests that in tissue engineering applications the scaffold peptide material is directly formed around the cells. Recently, the Point Accumulation for Imaging in Nanoscale Topography (PAINT) technique was used for the first time to 3D-image Fmoc-FF hydrogels in native conditions (no dry gel) and in absence of a direct labelling of the gel. PAINT images revealed the presence in the gel of fibres with a diameter of 50 nm and a mesh size ranged between 20 and 40 nm^2 [58].

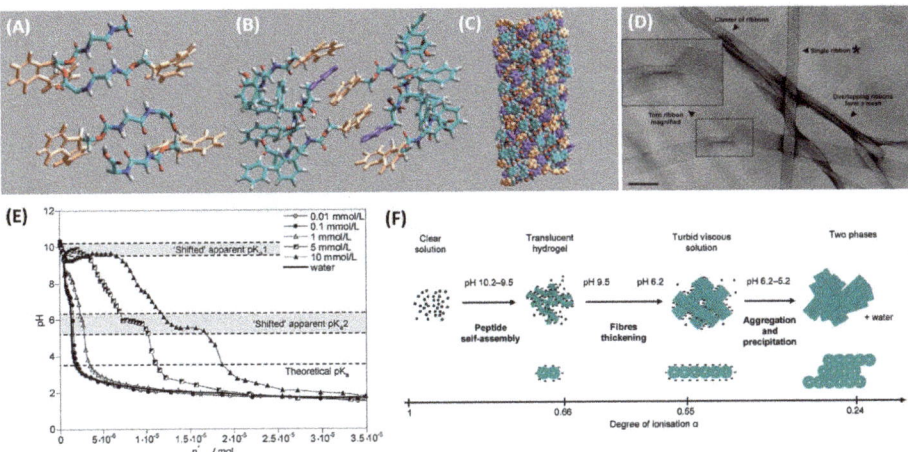

Figure 2. Structural model of Fmoc-FF peptides. (**A**) Dipeptide copies are arranged into β-sheet with an antiparallel orientation of β-strands. (**B**) π-stacked pairs due to the interlocking of fluorenyl groups from alternate β-sheets. (**C**) The final model obtained by energy minimization. In the model Fmoc and the phenyl groups are coloured in orange and in purple, respectively. (**D**) Transmission electron microscopy of Fmoc-FF xerogel (scale bar = 500 Å); the ribbon asterixed by authors was selected for other morphological analysis. (**E**) Titration curves of water and Fmoc-FF samples at different peptide concentrations (0.01, 0.1, 1, 5, and 10 mmol/L). (**F**) Mechanism proposed to explain the formation of Fmoc-FF aggregates as consequence of the pH decrease (figure adapted with permission for Refs. [53,54], Copyright 2009 American Chemical Society).

4. Preparation Methods

Fifteen years of scientific research on Fmoc-FF have been pointed out that the local organization of this peptide fragment and its structural and macroscopic architecture is deeply affected by the preparation method and by the experimental conditions used to generate the supramolecular material. After all, it is well-known and intuitively predictable that the self-assembling environmental conditions and/or the self-assembling strategy can affect both the macroscopic and the microscopic structures of the resulting hydrogel and, in turn, its functional features [59–61]. Moreover, fabrication of uniform hydrogels is often hampered by the diffusion of peptide molecules and aggregates into water medium. As a consequence of this different hierarchical organization, the physicochemical characteristic (stiffness, matrix porosity, stability, opacity, and so on) and consequently the potential applications (tissue engineering, drug delivery, biocatalysis, and biosensors) of the material change. For example, mechanical properties of the hydrogel can change up to four orders of magnitude by modification of some parameter in the formulative process, such as temperature, peptide concentration, pH of the solution, and the addition of salt/additives [62]. In this contest it is relevant to understand the key factors that can allow obtaining the material with the desired properties. To describe and distinguish between the preparation methods available until now, in this paper they were classified into three classes (Figure 3), which are named as: (i) pH-switch method, (ii) solvent-switch method and (iii) catalytic method. All of them allow gel formation as a consequence of a change/modification of the initial conditions, introducing a trigger into the peptide solution. In the first two methods, nominally pH- and solvent-switch methods, the gelation often occurs rapidly (less than a second) in the neighbourhood where the mixing of the two different solutions takes place, and in certain cases they lead to the formation of an inhomogeneous hydrogel. However, originally proposed strategies have been opportunely modified to improve the homogeneity of the gel. Analogously, catalytic methods, in which the self-aggregation process can be kinetically and thermodynamically controlled, have been proposed. It is worth noting that the kinetic control on the self-assembly process can allow driving the aggregation towards a designed pathway and in turn, allow the formation of highly homogeneous hydrogels.

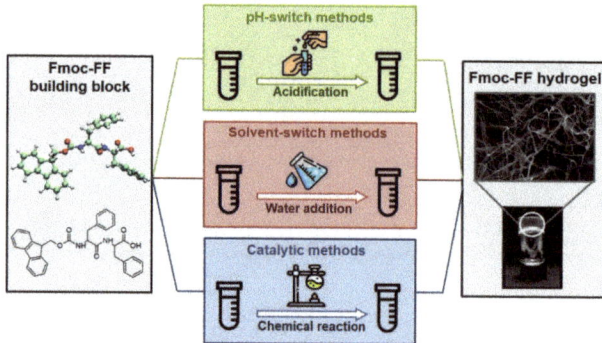

Figure 3. On the left, the molecular Fmoc-FF building block, reported both as 3D balls and sticks and as chemical formula. On the right, an inverted vial containing Fmoc-FF self-supporting gel and its TEM image. In the middle a schematic representation of three different methods (pH-switch, solvent switch, and catalytic one) commonly used for trigger gelation process.

4.1. pH-Switch Method

The pH-switch method involves the dissolution of the peptide in an aqueous solution at elevated pH (around pH 10.5) followed by a solution pH lowering via HCl addition [53]. In this preparation method, the peptide solubilisation at high pH ensures the deprotonation of the C-terminal carboxylic acid. Then the progressive acidification allows its protonation until gelation.

Due to the potential cleavage of the Fmoc protecting group under basic conditions, a careful control of the pH value and of the permanence time of the sample in NaOH is required during the peptide dissolution. It is worth noting that a negligible percentage of the fluorenyl moiety (<1%) is lost after 10 min in this basic condition. It was observed that the acidification of Fmoc-peptide solution by inorganic acids such as HCl brings to the formation of an inhomogeneous hydrogel. This result was attributed to the kinetics of mixing being slower than the initial kinetics of gelation. Indeed, at a low pH value, fibril formation and gelation are often very fast (taking place in a time lower than a second), with the consequence that it is very difficult to achieve a uniform pH in solution before the gelation process starts. Reproducible and more homogeneous hydrogels were obtained using an optimized procedure, involving a progressive addition of acid followed by a heat/cool cycle [54]. The high temperature permits to dissolve the kinetically trapped aggregates that are formed during the acidification. As described, the assembly of Fmoc-FF building blocks occurs with the appearance of two apparent pK_a: the first at pH 10.2–9.5, and the second at pH 6.2–5.2. It should be emphasised that hydrogels free from undissolved peptides exhibit a G' modulus of 1–10 Pa, which is significantly lower than G' (10^4 Pa) of hydrogels undergoing the heating step.

It is also important to pay the attention to each step during the gel preparation. Indeed, it was demonstrated that the method utilized to agitate the sample during and after the gelling transition can dramatically affect both the nano-scale morphology and the mechanical properties of the gel. The effect of the agitation (low shear and high shear) was evaluated on different pure and mixed hydrogels (100% Fmoc-FF, 70/30 Fmoc-FF/Fmoc-GG, and 50/50 Fmoc-FF/Fmoc-GG), all of them prepared according to the pH-dependent procedure [30]. In this study, low shear gels exhibited a higher storage modulus (≈4000 Pa) with respect to the high shear ones (≈1000 Pa). The different mechanical properties have been related to the morphology of gels obtained in the two shear modalities: high shear samples are more prone to lateral aggregation, whereas low shear ones to a regular form of entanglement. Successively, Adams and co-workers introduced a novel strategy in which the decrease in the pH is prompted by the slow hydrolysis of the highly soluble glucono-δ-lactone (GdL) to gluconic acid (see Figure 4A) [63]. The long timescale for the hydrolysis (≈18 h) allows a slow and homogeneous pH change, which minimizes the solvent mixing effects with the consequential obtainment of homogeneous and reproducible gel. According to this procedure, only a single apparent pK_a of 8.9 was detected. The final pH of the solution can be modulated according to the amount of GdL added to the peptide. The study demonstrated that the addition of two equivalents of GdL with respect to the peptide allows reaching a final pH of 3.6–3.9, which is very similar to the pH reached by adding HCl. However, the pH begins to equilibrate after ≈350 min and is fully equilibrated only after 24 h. The combination of the GdL-mediated pH trigger method with the cryogelation at a sub-zero temperature (−12 °C) allows the formation of macroporous Fmoc-FF hydrogels [64]. In the conditions here described, the water crystallization occurs and the Fmoc-FF molecules, concentrated in the remaining non-frozen liquid phase, gelate with a structure having a pore size in the range of 10–100 μm. Even if the pore walls of cryogels are characterized by a close packing of the fibres, they exhibit low mechanical stability with respect to classical hydrogels. This low stability was attributed to the heterogeneous structure in the cryogel. To further improve the homogeneity and the reproducibility of the hydrogel, in 2013 Ding et al. described two novel pH-triggered procedures for achieving Fmoc-FF hydrogel fabrication [65]. In the first approach, termed "the colloid method", hydrogels were obtained by a colloid-to-hydrogel transition process (Figure 4B). Initially, the amphiphilic Fmoc-FF dipeptide spontaneously self-assembled in water solution in stable, rod-like micelles with a radius and a length of ≈15 and ≈180 nm, respectively. Successively, the slow formation of hydrogel (≈20 min) was triggered by adding one equivalent of a weak base such as Na_2CO_3 to the colloid solution. The addition of the base causes the progressive deprotonation of the carboxylic functions of the dipeptides present at the surface layer of the micelles. In this context, micelles serve as a trap to regulate the amount of Fmoc-FF released in the solution and hydrogels can be prepared also at pH > 9. Instead, the second approach was based on the decomposition of the $K_2S_2O_8$ in oxygen gas and protons triggered pH decrease (reaction is reported in Figure 4C). This method exhibits several advantages compared with other methods used to lower pH such as the hydrolysis of GdL. It permits to control the amount of protons released and hence the final pH of the solution. Moreover, due to the dependence of the potassium persulfate degradation from the temperature, both the gelation kinetic (from 5 min at 90 °C to 4–5 days at 25 °C) and the structural properties of the hydrogel can be programmed, as well.

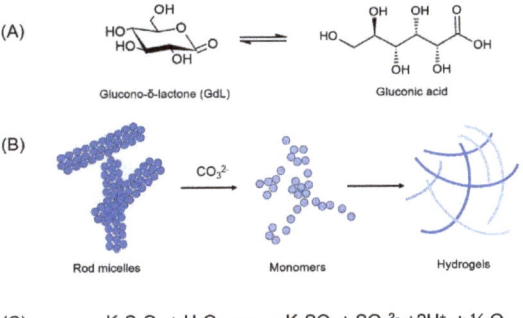

(C) $K_2S_2O_8 + H_2O \longrightarrow K_2SO_4 + SO_4^{2-} + 2H^+ + \frac{1}{2} O_2$

Figure 4. Preparation of Fmoc-FF hydrogel using three variants of the classic pH-switch method employing HCl: (**A**) decrease in pH induced by the slow hydrolysis (≈18 h) of the glucono-δ-lactone (GdL) to gluconic acid; (**B**) the colloid method in which the hydrogel is obtained by a progressive colloid-to-hydrogel transition triggered by adding of Na_2CO_3 to the colloid solution; (**C**) decrease in pH allowed by the decomposition of the $K_2S_2O_8$ in oxygen gas and protons.

4.2. Solvent-Switch Method

Beyond the pH-switch method, Fmoc-FF hydrogels can be also prepared with the solvent-switch method. This is based on the dissolution of the peptide in an organic solvent at high concentration. Then, gelation is trigged by adding water to the solution, so creating a three-component (peptide/solvent/water) system. Initially, Gazit and co-workers used HFIP as a solvent to dissolve Fmoc-FF [26]; later, other organic solvents able to solubilize Fmoc-FF peptides were used as an alternative to HFIP. Specifically, the first solvent used in place of the HFIP was dimethyl sulfoxide (DMSO). TEM images collected on Fmoc-FF gels prepared in DMSO/water confirm the presence of small-entangled fibres with a diameter of $\approx$10 nm, smaller than the wavelength of visible light. Moreover, the storage modulus value (10^4 Pa) found for these gels was similar to that one for HFIP/water preparation [66]. In 2014, Dudukovic and Zukosky established a range of concentrations under which a Fmoc-FF solution in DMSO can form hydrogels upon mixing with the water and studied the mechanical properties of the gels prepared in the different conditions [67,68]. Their results allowed to delineate a well-defined line of gel transition in a plot of water concentration as a function of ϕ (namely the particle volume fraction) and, pointed out that rigid gels can be obtained at a low Fmoc-FF volume fraction ($\phi < 1\%$) and under addition of a small amount of water to DMSO (Figure 5A). The capability of the peptide to gel also under addition of reduced quantity of water was recently demonstrated to be due to the existence of disordered oligomers and profibrils into the DMSO solution [69]. Moreover, a deep investigation on the Fmoc-FF gelation kinetics revealed that upon addition of water to DMSO solution of Fmoc-FF, initially the dipeptide self-assembles into a metastable non equilibrium state composed of spherical clusters of diameters of 2 μm, followed by a rapid rearrangement (below 5 min) into a fibrous network. The aging of the sample (up to 4 h) allows a further evolution of the gel towards a steady state in which there is the formation of a highly uniform network composed of thin fibres with a mean diameter between 5 and 10 nm. These dimensions are smaller than the wavelength of the visible light and are the reason why the gel appears transparent. The high uniformity of the fibrillary network causes an increase in the gel rigidity and provides long-term stability (years) to the gel. These studies also demonstrated the mechanical and thermal reversibility of the gels over time. Successively, the same authors speculated that the fibres in the gel can be treated as an equilibrium crystalline state, in which the gelation process is a first order phase transition resulting in the nucleation and growth of elongated anisotropic crystals. It was observed that an increase in water with respect to DMSO brought an increase in the strength of attraction between the peptide molecules that in turn can be translated into a fast nucleation rate and quasi-one-dimensional crystal growth [70]. This transition from spherulitic structures to a fibrous network was also observed by Adams and co-workers for Fmoc-FF hydrogels prepared using other polar protic or aprotic solvents (Figure 5B) such as ethanol, acetone, and HFIP in place of DMSO [71]. However, it seems that the choice of the solvent can affect the morphology of the final network and in turn rheological properties and mechanosensitivity of the final hydrogel. These differences are originated by the control exercised by the solvent on the morphology of the fibre network. For example, at a $\phi_{solvent}$ of 0.3, gels prepared in ethanol exhibit a more uniform network with respect to gels prepared in DMSO or HFIP. This high uniformity is associated with a high rigidity and a poor capability of the hydrogel to recover their mechanical strength from shear. It is worth noting that the transition from opaque to limpid state into the peptide solution is never observed for samples prepared by the pH-switch. This evidence clearly indicates that there are significant differences in the self-assembling process occurring for the two procedures.

Later, Dudukovic et al. also explored the phase behaviour of Fmoc-FF in other solvent systems (DMSO/H_2O, MeOH/H_2O, and toluene). According to the Adam's results, the authors demonstrated that gel formation can be induced also in apolar solvents and that the Fmoc-FF phase behaviour is directly correlated to the balance between the inter- and intra-molecular interactions. Indeed, it was observed that different solvents allow the formation of fibres with a different molecular order and that Fmoc-FF exhibits polymorphism in some solvents where metastable anisotropic crystals evolve towards crystal aggregation with no preferential axis of growth. Different fibre populations can co-exist within one system and the switch between these states depends on the stability of each conformational state and on the height of barrier between the two free energy minima (Figure 5) [72]. As an alternative, hydrogels can be prepared by using buffered solutions at physiological pH. However, also the ratio of DMSO to H_2O (ϕ_{DMSO}) and the choice of the buffers used in selected systems can affect the rheological properties of the hydrogel [65].

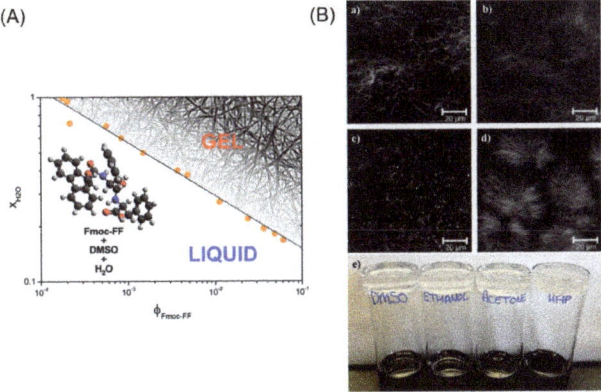

Figure 5. (**A**) Phase diagram molecular gels produced by Fmoc-FF using DMSO/H_2O solvent-switch method (figure adapted with permission from Ref. [67], Copyright 2014 American Chemical Society). (**B**) Fmoc-FF matrices stained with Nile Blue in confocal microscopy analysis (ϕsolvent = 0.3) using the solvents (**a**) DMSO (**b**) ethanol (**c**) acetone (**d**) HFIP, (**e**) macroscopic gel picture after 24 h from their preparation (at same ϕsolvent without staining) (figure reproduced from Ref. [71] with permission from the Royal Society of Chemistry).

4.3. Catalytic Methods

The catalytic process is an interesting methodology that allows directing the self-aggregation process towards structurally diverse self-assembled materials, inaccessible via classical self-assembly. This approach consists of converting precursors unable to self-assemble into building blocks able to do. Usually, this conversion can be achieved by enzymatic removal or hydrolysis of a charged or steric groups that avoids the aggregation. In this case, the nucleation site and the early stage grown mechanism are spatially confined at the site of catalytic centre. Two examples of self-assembly for Fmoc-FF and for several its analogues (FY, YL, VL and FL) were reported by the Ulijn's group [73,74]. Initially, they demonstrated the use of thermolysin, a non-specific endoprotease to catalyse in a reversible way the peptide bond formation between Fmoc-F and several dipeptides (G_2, F_2, L_2) or amino acid esters (L-OMe, F-OMe) [73]. After the bond formation, the peptide self-assembles in a spatiotemporally controlled manner, with the enzyme favouring the spatial confinement of structure grown during the early stage of the self-assembly process. On the same set of peptides, they also reported the self-assembly catalysed by subtilin, a hydrolytic enzyme from *Bacillus licheniformis*. It was observed that the enzyme concentration (ranged between 1.5 and 36 units) strongly affects the self-assembly kinetics and in turn, the supramolecular order degree, and the functional properties of the final material [74]. Chemical catalysis, achieved by starting from phenylalanine and its Fmoc-derivative and EDC/NHSS as catalysts, was described as an alternative approach with respect to the biocatalytic ones [75]. In this procedure, Fmoc-F was activated in DMSO using EDC/NHSS, then Fmoc-FF was quickly obtained by simply adding the intermediates into an aqueous solution containing phenylalanine. The continuous generation of Fmoc-FF due to the chemical reaction and its spontaneous aggregation into entangled nanofibers allows the formation of a self-supporting and a homogeneous hydrogel. At the end of the gelling process, the catalysts were removed by washing the hydrogel with deionized water. Later, several Fmoc-FFF tripeptide, containing phenylalanine residues in *L* or *D* configuration, were synthetized by lipase-catalysed reversed hydrolysis reaction between a Fmoc-amino acid and a dipeptide [76].

5. Conclusions

Most of the peptide hydrogelators described in literature are able to gelate in only one well-defined condition, which is related to their primary sequence in terms of sterical hindrance, hydrophobicity, and polarity of amino acids. On the contrary, Fmoc-FF-based hydrogels can be prepared using different conditions of solvent, pH, temperature, ionic strength, and shear. From a predictive point of view, the versatility of Fmoc-FF was not expected on the basis of its very simple chemical nature. In this contest, many attempts have been made to understand the molecular mechanisms enabling the Fmoc-FF self-aggregation. The possibility to change preparation conditions and experimental variables allows to modulate in a controlled manner the structural properties and morphology of the resulting material. This versatility envisages a relevant number of applications in a variety of fields for this ultra-short peptide.

Author Contributions: Conceptualization of the paper, A.A.; Paper topic, organization and structure, C.D. and A.A.; Collection of literature and analysis of papers, C.D., E.R., G.M. and A.A.; Writing, A.A., C.D. and G.M.; Original draft preparation, A.A. and G.M.; Image curation and editing, C.D., E.R. and A.A.; General supervision, A.A. and G.M. All authors have read and agreed to the published version of the manuscript.

Funding: This research received no external funding.

Institutional Review Board Statement: Not applicable.

Informed Consent Statement: Not applicable.

Data Availability Statement: Not applicable.

Conflicts of Interest: Authors declare no conflict of interest.

References

1. Acar, H.; Srivastava, S.; Chung, E.J.; Schnorenberg, M.R.; Barrett, J.C.; LaBelle, J.L.; Tirrell, M. Self-assembling peptide-based building blocks in medical applications. *Adv. Drug Deliv. Rev.* **2016**, *110–111*, 65–79. [CrossRef] [PubMed]
2. Nainytė, M.; Müller, F.; Ganazzoli, G.; Chan, C.Y.; Crisp, A.; Globisch, D.; Carell, T. Amino Acid Modified RNA Bases as Building Blocks of an Early Earth RNA-Peptide World. *Chem. A Eur. J.* **2020**, *26*, 14856–14860. [CrossRef] [PubMed]
3. Diaferia, C.; Gianolio, E.; Accardo, A. Peptide-based building blocks as structural elements for supramolecular Gd-containing MRI contrast agents. *J. Pept. Sci.* **2019**, *25*, e3157. [CrossRef] [PubMed]
4. Lewandowska, U.; Corra, S.; Zajaczkowski, W.; Ochs, N.A.K.; Shoshan, M.S.; Tanabe, J.; Stappert, S.; Li, C.; Yashima, E.; Pisula, W.; et al. Positional isomers of chromophore–peptide conjugates self-assemble into different morphologies. *Chem. Eur. J.* **2018**, *24*, 12623–12629. [CrossRef]
5. Eisenberg, D.; Jucker, M. The Amyloid State of Proteins in Human Diseases. *Cell* **2012**, *148*, 1188–1203. [CrossRef]
6. Otzen, D.; Rie, R. Functional amyloids. *Cold Spring Harb. Perspect. Biol.* **2019**, *11*, a033860. [CrossRef]
7. Balasco, N.; Diaferia, C.; Morelli, G.; Vitagliano, L.; Accardo, A. Amyloid-like aggregation in diseases and biomaterials: Osmosis of structural information. *Front. Bioeng. Biotechnol.* **2021**, *9*, 641372. [CrossRef]
8. Reches, M.; Gazit, E. Casting Metal Nanowires Within Discrete Self-Assembled Peptide Nanotubes. *Science* **2003**, *300*, 625–627. [CrossRef]
9. Görbitz, C.H. Nanotube Formation by Hydrophobic Dipeptides. *Chem. A Eur. J.* **2001**, *7*, 5153–5159. [CrossRef]
10. Fuentes-Caparrós, A.M.; McAulay, K.; Rogers, S.E.; Dalgliesh, R.M.; Adams, D.J. On the Mechanical Properties of N-Functionalised Dipeptide Gels. *Molecules* **2019**, *24*, 3855. [CrossRef]
11. Amdursky, N.; Molotskii, M.; Gazit, E.; Rosenman, G. Elementary Building Blocks of Self-Assembled Peptide Nanotubes. *J. Am. Chem. Soc.* **2010**, *132*, 15632–15636. [CrossRef] [PubMed]
12. Diaferia, C.; Avitabile, C.; Leone, M.; Gallo, E.; Saviano, M.; Accardo, A.; Romanelli, A. Diphenylalanine Motif Drives Self-Assembling in Hybrid PNA-Peptide Conjugates. *Chem. A Eur. J.* **2021**, *27*, 14307–14316. [CrossRef] [PubMed]
13. Reches, M.; Gazit, E. Designed aromatic homo-dipeptides: Formation of ordered nanostructures and potential nanotechnological applications. *Phys. Biol.* **2006**, *3*, S10–S19. [CrossRef] [PubMed]

14. Yang, X.; Fei, J.; Li, Q.; Li, J. Covalently assembled dipeptide nanospheres as intrinsic photosensitizers for efficient photodynamic therapy in vitro. *Chem. Eur. J.* **2016**, *22*, 1–6. [CrossRef]
15. Kumara, V.; Vijay, K.; Shruti, K.; Khashti, K.; Joshi, B. Aggregation propensity of amyloidogenic and elastomeric dipeptides constituents. *Tetrahedron* **2016**, *72*, 5369–5376. [CrossRef]
16. Mayans, E.; Casanovas, J.; Gil, A.M.; Jimenez, A.I.; Cativiela, C.; Puiggalí, J.; Aleman, C. Diversity and hierarchy in supramolecular assemblies of triphenylalanine: From laminated helical ribbons to toroids. *Langmuir* **2017**, *33*, 4036–4048. [CrossRef]
17. Diaferia, C.; Roviello, V.; Morelli, G.; Accardo, A. Self-assembly of PEGylated diphenylalanines into photoluminescent fibrillary aggregates. *ChemPhysChem* **2019**, *20*, 2774–2782. [CrossRef]
18. Creasey, R.C.G.; Louzao, I.; Arnon, Z.A.; Marco, P.; Adler-Abramovich, L.; Roberts, C.J.; Gazit, E.; Tendler, S.J.B. Disruption of diphenylalanine assembly by a Boc-modified variant. *Soft Matter* **2016**, *12*, 9451–9457. [CrossRef]
19. Yan, X.; He, Q.; Wang, K.; Duan, L.; Cui, Y.; Li, J. Transition of Cationic Dipeptide Nanotubes into Vesicles and Oligonucleotide Delivery. *Angew. Chem. Int. Ed.* **2007**, *46*, 2431–2434. [CrossRef]
20. Reches, M.; Gazit, E. Formation of Closed-Cage Nanostructures by Self-Assembly of Aromatic Dipeptides. *Nano Lett.* **2004**, *4*, 581–585. [CrossRef]
21. Pellach, M.; Mondal, S.; Shimon, L.J.W.; Adler-Abramovich, L.; Buzhansky, L.; Gazit, E. Molecular Engineering of Self-Assembling Diphenylalanine Analogues Results in the Formation of Distinctive Microstructures. *Chem. Mater.* **2016**, *28*, 4341–4348. [CrossRef]
22. Reches, M.; Gazit, E. Self-Assembly of peptide nanotubes and amyloid-like structures by charged-termini-capped diphenylalanine peptide analogues. *Isr. J. Chem.* **2005**, *45*, 363–371. [CrossRef]
23. Adler-Abramovich, L.; Gazit, E. Controlled patterning of peptide nanotubes and nanospheres using inkjet printing technology. *J. Pep. Sci.* **2007**, *14*, 217–223. [CrossRef] [PubMed]
24. Amdursky, N.; Molotskii, M.; Gazit, E.; Rosenman, G. Self-assembled bioinspired quantum dots: Optical properties. *Appl. Phys. Lett.* **2009**, *94*, 261907. [CrossRef]
25. Jayawarna, V.; Ali, M.; Jowitt, T.; Miller, A.F.; Saiani, A.; Gough, J.E.; Ulijn, R.V. Nanostructured Hydrogels for Three-Dimensional Cell Culture Through Self-Assembly of Fluorenylmethoxycarbonyl–Dipeptides. *Adv. Mater.* **2006**, *18*, 611–614. [CrossRef]
26. Mahler, A.; Reches, M.; Rechter, M.; Cohen, S.; Gazit, E. Rigid, self-assembled hydrogel composed of a modified aromatic dipeptide. *Adv. Mater.* **2006**, *18*, 1365–1370. [CrossRef]
27. Diaferia, C.; Morelli, G.; Accardo, A. Fmoc-diphenylalanine as a suitable building block for the preparation of hybrid materials and their potential applications. *J. Mater. Chem. B* **2019**, *7*, 5142–5155. [CrossRef]
28. Ghosh, M.; Bera, S.; Schiffmann, S.; Shimon, L.J.W.; Adler-Abramovich, L. Collagen-Inspired Helical Peptide Coassembly Forms a Rigid Hydrogel with Twisted Polyproline II Architecture. *ACS Nano* **2020**, *14*, 9990–10000. [CrossRef]
29. Rosa, E.; Diaferia, C.; Gallo, E.; Morelli, G.; Accardo, A. Stable Formulations of Peptide-Based Nanogels. *Molecules* **2020**, *25*, 3455. [CrossRef]
30. Helen, W.; de Leonardis, P.; Ulijn, R.V.; Gough, J.; Tirelli, N. Mechanosensitive peptide gelation: Mode of agitation controls mechanical properties and nano-scale morphology. *Soft Matter* **2011**, *7*, 1732–1740. [CrossRef]
31. Yang, X.; Xie, Y.; Wang, Y.; Qi, W.; Huang, R.; Su, R.; He, Z. Self-Assembled Microporous Peptide-Polysaccharide Aerogels for Oil–Water Separation. *Langmuir* **2018**, *34*, 10732–10738. [CrossRef]
32. Arakawa, H.; Takeda, K.; Higashi, S.L.; Shibata, A.; Kitamura, Y.; Ikeda, M. Self-assembly and hydrogel formation ability of Fmoc-dipeptides comprising α-methyl-L-phenylalanine. *Polym. J.* **2020**, *52*, 923–930. [CrossRef]
33. Wang, Z.; Li, T.; Ding, B.; Ma, X. Archieving room temperature phosphorescence from organic small molecules on amino acid skeleton. *Chi. Chem. Lett.* **2020**, *31*, 2929–2932. [CrossRef]
34. Sankar, S.; O'Neill, K.; Bagot D'Arc, M.; Rebeca, F.; Buffier, M.; Aleksi, E.; Fan, M.; Matsuda, N.; Gil, E.S.; Spirio, L. Clinical use of the self-assembling peptide RADA16: A review of current and future trends in biomedicine. *Front. Bioeng. Biotechnol.* **2021**, *9*, 679525. [CrossRef] [PubMed]
35. Nagy-Smith, K.; Moore, E.; Schneider, J.; Tycko, R. Molecular structure of monomorphic peptide fibrils within a kinetically trapped hydrogel network. *Proc. Natl. Acad. Sci. USA* **2015**, *112*, 9816–9821. [CrossRef]
36. Yaguchi, A.; Hiramatsu, H.; Ishida, A.; Oshikawa, M.; Ajioka, I.; Muraoka, T. Hydrogel-Stiffening and Non-Cell Adhesive Properties of Amphiphilic Peptides with Central Alkylene Chains. *Chem. A Eur. J.* **2021**, *27*, 9295–9301. [CrossRef]
37. Jadhav, S.P.; Amabili, P.; Stammler, H.-G.; Sewald, N. Remarkable modulation of self-assembly in short γ-Peptides by neighboring ions and orthogonal H-bonding. *Chem. Eur. J.* **2017**, *23*, 10352–10357. [CrossRef]
38. Santagada, V.; Caliendo, G. *Peptide and Peptidomimetics*; Piccin: Padova, Italy, 2012.
39. Diaferia, C.; Ghosh, M.; Sibillano, T.; Gallo, E.; Stornaiuolo, M.; Giannini, C.; Morelli, G.; Adler-Abramovich, L.; Accardo, A. Fmoc-FF and hexapeptide-based multicomponent hydrogels as scaffold materials. *Soft Matter* **2019**, *15*, 487–496. [CrossRef] [PubMed]
40. Wang, Y.; Qi, W.; Huang, R.; Su, R.; He, Z. Jet flow directed supramolecular self-assembly at aqueous liquid–liquid interface. *RSC Adv.* **2014**, *4*, 15340–15347. [CrossRef]
41. Choe, R.; Yun, S.I. Fmoc-diphenylalanine-based hydrogels as a potential carrier for drug delivery. *e-Polymers* **2020**, *20*, 458–468. [CrossRef]

42. Diaferia, C.; Balasco, N.; Sibillano, T.; Giannini, C.; Vitagliano, L.; Morelli, G.; Accardo, A. Structural Characterization of Self-Assembled Tetra-Tryptophan Based Nanostructures: Variations on a Common Theme. *ChemPhysChem* **2018**, *19*, 1635–1642. [CrossRef] [PubMed]
43. Pachahara, S.K.; Adicherla, H.; Nagaraj, R. Self-assembly of Aβ40, Aβ42 and Aβ43 peptides in aqueous mixtures of fluorinated alcohols. *PLoS ONE* **2015**, *10*, e0136567.
44. Schiattarella, C.; Diaferia, C.; Gallo, E.; Della Ventura, B.; Morelli, G.; Vitagliano, L.; Velotta, R.; Accardo, A. Solid-state optical properties of self-assembling amyloid-like peptides with different charged states at the terminal ends. *Sci. Rep.* **2022**, *12*, 759. [CrossRef] [PubMed]
45. Zhang, Y.; Gu, H.; Yang, A.Z.; Xu, B. Supramolecular Hydrogels Respond to Ligand−Receptor Interaction. *J. Am. Chem. Soc.* **2003**, *125*, 13680–13681. [CrossRef] [PubMed]
46. Yang, Z.; Gu, H.; Zhang, Y.; Wang, L.; Xu, B. Small molecule hydrogels based on a class of anti-inflammatory agents. *Chem. Commun.* **2004**, 208–209. [CrossRef]
47. Yang, Z.; Gu, H.; Fu, D.; Gao, P.; Lam, J.K.; Xu, B. Enzymatic formation of supramolecular hydrogels. *Adv. Mater.* **2004**, *16*, 1440–1444. [CrossRef]
48. Amdursky, N.; Gazit, E.; Rosenman, G. Quantum Confinement in Self-Assembled Bioinspired Peptide Hydrogels. *Adv. Mater.* **2010**, *22*, 2311–2315. [CrossRef]
49. Liu, Y.; Ding, X.; Jing, X.; Chen, X.; Cheng, H.; Zheng, X.; Ren, Z.; Zhuo, Z. Surface self-assembly of N-fluorenyl-9-methoxycarbonyl diphenylalanine on silica wafer. *Colloids Surf. B Biointerfaces* **2011**, *87*, 192–197. [CrossRef] [PubMed]
50. Ryan, K.; Beirne, J.; Redmond, G.; Kilpatrick, J.I.; Guyonnet, J.; Buchete, N.-V.; Kholkin, A.L.; Rodriguez, B.J. Nanoscale piezoelectric properties of self-assembled Fmoc-FF peptide fibrous networks. *ACS Appl. Mat. Interfaces* **2015**, *7*, 12702–12707. [CrossRef]
51. McCloskey, A.P.; Draper, E.R.; Gilmore, B.F.; Laverty, G. Ultrashort self-assembling Fmoc-peptide gelators for anti-infective biomaterial applications. *J. Pept. Sci.* **2017**, *23*, 131–140. [CrossRef]
52. Ben-Zvi, O.; Grinberg, I.; Orr, A.A.; Noy, D.; Tamamis, P.; Yacoby, I.; Adler-Abramovich, L. Protection of Oxygen-Sensitive Enzymes by Peptide Hydrogel. *ACS Nano* **2021**, *15*, 6530–6539. [CrossRef]
53. Smith, A.M.; Williams, R.J.; Tang, C.; Coppo, P.; Collins, R.F.; Turner, M.L.; Saiani, A.; Ulijn, R.V. Fmoc-diphenylalanine self-assembles to a hydrogel via a novel architecture based on π-π interlocked β-sheets. *Adv. Mater.* **2008**, *20*, 37–41. [CrossRef]
54. Tang, C.; Smith, A.M.; Collins, R.F.; Ulijn, R.V.; Saiani, A. Fmoc-diphenylalanine self-assembly mechanism induces apparent pKa shifts. *Langmuir* **2009**, *25*, 9447–9453. [CrossRef] [PubMed]
55. Xing, R.; Yuan, C.; Li, S.; Song, J.; Li, J.; Yan, X. Charge-induced secondary structure transformation of amyloid-derived dipeptide assemblies from β-sheet to α-Helix. *Angew. Chem. Int. Ed.* **2018**, *57*, 1537–1542. [CrossRef] [PubMed]
56. Ji, W.; Yuan, C.; Zilberzwige-Tal, S.; Xing, R.; Chakraborty, P.; Tao, K.; Gilead, S.; Yan, X.; Gazit, E. Metal-Ion Modulated Structural Transformation of Amyloid-Like Dipeptide Supramolecular Self-Assembly. *ACS Nano* **2019**, *13*, 7300–7309. [CrossRef] [PubMed]
57. Amdursky, N.; Orbach, R.; Gazit, E.; Huppert, D. Probing the Inner Cavities of Hydrogels by Proton Diffusion. *J. Phys. Chem. C* **2009**, *113*, 19500–19505. [CrossRef]
58. Fuentes, E.; Boháčová, K.; Fuentes-Caparrós, A.M.; Schweins, R.; Draper, E.R.; Adams, D.J.; Silvia Pujals, S.; Albertazzi, L. PAINT-ing Fluorenylmethoxycarbonyl (Fmoc)-diphenylalanine hydrogels. *Chem. Eur. J.* **2020**, *26*, 9869–9873. [CrossRef]
59. Gulrez, S.K.H.; Al-Assaf, S.; Phillips, G.O. Hydrogels: Methods of preparation, characterisation and applications. In *Progress in Molecular and Environmental Bioengineering—From Analysis and Modeling to Technology Applications*; IntechOpen: London, UK, 2011.
60. Taylor, J.; Rahimeh, B.; Alpesh, P.; Kibret, M. Fabrication of highly porous tissue-engineering scaffolds using selective spherical porogens. *BioMed. Mater. Eng.* **2010**, *20*, 107–118.
61. Annabi, N.; Nichol, J.W.; Zhong, X.; Ji, C.; Koshy, S.; Khademhosseini, A.; Dehghani, F. Controlling the porosity and microarchitecture of hydrogels for tissue engineering. *Tiss. Eng. B* **2010**, *16*, 371–383. [CrossRef]
62. Raeburn, J.; Pont, G.; Chen, L.; Cesbron, Y.; Levy, R.; Adams, D.J. Fmoc-diphenylalanine hydrogels: Understanding the variability in reported mechanical properties. *Soft Matter* **2012**, *8*, 1168–1174. [CrossRef]
63. Adams, D.J.; Butler, M.F.; Frith, W.J.; Kirkland, M.; Mullen, L.; Sanderson, P. A new method for maintaining homogeneity during liquid–hydrogel transitions using low molecular weight hydrogelators. *Soft Matter* **2009**, *5*, 1856–1862. [CrossRef]
64. Berillo, D.; Mattiasson, B.; Yu, I.; Kirsebom, G.H. Formation of macroporous self-assembled hydrogels through cryogelation of Fmoc-Phe-Phe. *J. Coll. Interface Sci.* **2012**, *368*, 226–230. [CrossRef] [PubMed]
65. Ding, B.; Li, Y.; Qin, M.; Ding, Y.; Cao, Y.; Wang, W. Two approaches for the engineering of homogeneous small-molecule hydrogels. *Soft Matter* **2013**, *9*, 4672–4680. [CrossRef]
66. Orbach, R.; Adler-Abramovich, L.; Zigerson, S.; Mironi-Harpaz, I.; Seliktar, D.; Gazit, E. Self-Assembled Fmoc-Peptides as a Platform for the Formation of Nanostructures and Hydrogels. *Biomacromolecules* **2009**, *10*, 2646–2651. [CrossRef] [PubMed]
67. Dudukovic, N.A.; Zukoski, C.F. Mechanical Properties of Self-Assembled Fmoc-Diphenylalanine Molecular Gels. *Langmuir* **2014**, *30*, 4493–4500. [CrossRef]
68. Dudukovic, N.A.; Zukoski, C.F. Evidence for equilibrium gels of valence-limited particles. *Soft Matter* **2014**, *10*, 7849–7856. [CrossRef]

69. Levine, M.S.; Ghosh, M.; Hesser, M.; Hennessy, N.; DiGuiseppi, D.M.; Adler-Abramovich, L.; Schweitzer-Stenner, R. Formation of peptide-based oligomers in dimethylsulfoxide: Identifying the precursor of fibril formation. *Soft Matter* **2020**, *16*, 7860–7868. [CrossRef]
70. Dudukovic, N.A.; Zukoski, C.F. Gelation of Fmoc-diphenylalanine is a first order phase transition. *Soft Matter* **2015**, *11*, 7663–7673. [CrossRef]
71. Raeburn, J.; Mendoza-Cuenca, C.; Cattoz, B.N.; Little, M.A.; Terry, A.E.; Cardoso, A.Z.; Griffiths, P.C.; Adams, D.J. The effect of solvent choice on the gelation and final hydrogel properties of Fmoc–diphenylalanine. *Soft Matter* **2014**, *11*, 927–935. [CrossRef]
72. Dudukovic, N.A.; Hudson, B.C.; Paravastu, A.K.; Zukoski, C.F. Self-assembly pathways and polymorphism in peptide-based nanostructures. *Nanoscale* **2017**, *10*, 1508–1516. [CrossRef]
73. Williams, R.J.; Smith, A.M.; Collins, R.; Hodson, N.; Das, A.K.; Ulijn, R.V. Enzyme-assisted self-assembly under thermodynamic control. *Nat. Nanotech.* **2009**, *4*, 19–24. [CrossRef] [PubMed]
74. Hirst, A.R.; Roy, S.; Arora, M.; Das, A.K.; Hodson, N.; Murray, P.; Marshall, S.; Javid, N.; Sefcik, J.; Boekhoven, J.; et al. Biocatalytic induction of supramolecular order. *Nat. Chem.* **2010**, *2*, 1089–1094. [CrossRef] [PubMed]
75. Huang, R.; Wang, Y.; Qi, W.; Su, R.; He, Z. Chemical catalysis triggered self-assembly for the bottom-up fabrication of peptide nanofibers and hydrogels. *Mater. Lett.* **2014**, *128*, 216–219. [CrossRef]
76. Chronopoulou, L.; Sennato, S.; Bordi, F.; Giannella, D.; Di Nitto, A.; Barbetta, A.; Dentini, M.; Togna, A.R.; Togna, G.I.; Moschini, S.; et al. Designing unconventional Fmoc-peptide-based biomaterials: Structure and related properties. *Soft Matter* **2013**, *10*, 1944–1952. [CrossRef]

 *pharmaceuticals*

Review

Usage of Synthetic Peptides in Cosmetics for Sensitive Skin

Diana I. S. P. Resende [1,2,†], Marta Salvador Ferreira [3,4,†], José Manuel Sousa-Lobo [3,4], Emília Sousa [1,2,*] and Isabel Filipa Almeida [3,4,*]

1. CIIMAR–Centro Interdisciplinar de Investigação Marinha e Ambiental, Avenida General Norton de Matos, S/N, 4450-208 Matosinhos, Portugal; dresende@ff.up.pt
2. Laboratory of Organic and Pharmaceutical Chemistry, Department of Chemical Sciences, Faculty of Pharmacy, University of Porto, 4050-313 Porto, Portugal
3. Associate Laboratory i4HB-Institute for Health and Bioeconomy, Faculty of Pharmacy, University of Porto, 4050-313 Porto, Portugal; msbferreira@ff.up.pt (M.S.F.); slobo@ff.up.pt (J.M.S.-L.)
4. UCIBIO–Applied Molecular Biosciences Unit, MedTech, Laboratory of Pharmaceutical Technology, Department of Drug Sciences, Faculty of Pharmacy, University of Porto, 4050-313 Porto, Portugal
* Correspondence: esousa@ff.up.pt (E.S.); ifalmeida@ff.up.pt (I.F.A.); Tel.: +351-220-428-621 (I.F.A.)
† These authors equally contributed to this work.

 check for updates

Citation: Resende, D.I.S.P.; Ferreira, M.S.; Sousa-Lobo, J.M.; Sousa, E.; Almeida, I.F. Usage of Synthetic Peptides in Cosmetics for Sensitive Skin. *Pharmaceuticals* 2021, 14, 702. https://doi.org/10.3390/ph14080702

Academic Editors: Giovanni N. Roviello and Rosanna Palumbo

Received: 18 June 2021
Accepted: 16 July 2021
Published: 21 July 2021

Publisher's Note: MDPI stays neutral with regard to jurisdictional claims in published maps and institutional affiliations.

Copyright: © 2021 by the authors. Licensee MDPI, Basel, Switzerland. This article is an open access article distributed under the terms and conditions of the Creative Commons Attribution (CC BY) license (https://creativecommons.org/licenses/by/4.0/).

Abstract: Sensitive skin is characterized by symptoms of discomfort when exposed to environmental factors. Peptides are used in cosmetics for sensitive skin and stand out as active ingredients for their ability to interact with skin cells by multiple mechanisms, high potency at low dosage and the ability to penetrate the stratum corneum. This study aimed to analyze the composition of 88 facial cosmetics for sensitive skin from multinational brands regarding usage of peptides, reviewing their synthetic pathways and the scientific evidence that supports their efficacy. Peptides were found in 17% of the products analyzed, namely: acetyl dipeptide-1 cetyl ester, palmitoyl tripeptide-8, acetyl tetrapeptide-15, palmitoyl tripeptide-5, acetyl hexapeptide-49, palmitoyl tetrapeptide-7 and palmitoyl oligopeptide. Three out of seven peptides have a neurotransmitter-inhibiting mechanism of action, while another three are signal peptides. Only five peptides present evidence supporting their use in sensitive skin, with only one clinical study including volunteers having this condition. Noteworthy, the available data is mostly found in patents and supplier brochures, and not in randomized placebo-controlled studies. Peptides are useful active ingredients in cosmetics for sensitive skin. Knowing their efficacy and synthetic pathways provides meaningful insight for the development of new and more effective ingredients.

Keywords: peptides; cosmetics; sensitive skin; chemical synthesis

1. Introduction

Sensitive skin is a condition characterized by the occurrence of symptoms such as tightness, stinging, burning or pruritus, which are triggered by stimuli that do not normally produce unpleasant sensations, such as cold, heat, sun, pollution, cosmetics or moisture [1]. The skin may also present erythema, dryness and desquamation, but these signs are typically absent [2]. Sensitive skin is thought to affect 71% of the general adult population, and the epidemiological studies have also shown that the symptoms are more frequent on the face [3,4]. The causes for this condition are unknown, but genetics, poor mental health, and microbiome imbalance have been proposed as contributing factors [5–7]. The pathophysiological mechanisms involved in sensitive skin remain unknown, but three hypotheses have been pointed in scientific literature: increased stratum corneum permeability, an exacerbated immune response and a hyperactivity from the somatosensory and vascular systems [8]. While the two former hypotheses have been questioned and remain poorly understood, there is growing evidence linking sensitive skin to abnormal responses from the somatosensory system. The lower sensitivity threshold in individuals with sensitive skin may be due to a dysfunction in the communication with central nervous system, leading to pain sensations and neurosensory defects, namely the hyperactivation of endothelin receptors and transient receptor potential channels (TRP), which are present in cutaneous nerve fibers such as unmyelinated C fibers and keratinocytes [4,9]. The activation of cutaneous nerve fibers by physical and chemical stimuli, such as heat, low pH solutions, or known irritants such as capsaicin, results in the release of neuropeptides, such as substance P or calcitonin gene related peptide (CGRP), which activate keratinocytes, mast cells, and antigen-presenting cells and T cells nearby, causing a burning pain sensation [10]. A lower density of unmyelinated C-fibers was detected in individuals with sensitive skin, which may be due to degeneration following the contact with the environmental factors, which are thought to be responsible for the occurrence of skin sensitivity. Paradoxically, the lower density of unmyelinated C-fibers may generate hyperreactivity of the existing ones [11]. On the other hand, the inflammatory responses associated to itching sensations are initiated by the activation of transient receptor potential vanilloid type 1 (TRPV1), which is stimulated by heat, capsaicin, and cations, therefore promoting the release of IL-23 by dendritic cells [10]. Individuals with sensitive skin are thought to present an overexpression TRPV1, thus increasing neuronal excitability [12,13]. Overall, these mechanisms may be exacerbated by an impairment in the skin barrier, which fails to protect nerve endings adequately [10].

The synthesis of glutathione in the 1930's and the isolation of oxytocin in the 1950's promoted an increase in the research on peptide synthesis, isolation, as well as their chemical, biochemical, and biological characterization [14,15]. After the surge of conformational/topographic-biological activity relationships, which allowed to determine the affinity and specificity for target receptors, peptide leads emerged, offering several advantages over small molecules (increased specificity) and antibodies (small size) [15,16]. Peptide ligands may act as agonists or antagonists at cell receptors and acceptors modulating cell function and animal behavior. This area encompasses approximately 50% of current drugs, and it is likely to keep evolving in the future. In the cosmetic industry, peptides have been used since the late 1980s, with growing notoriety during the first decade of the XXI century [17–19]. Peptides used in cosmetic products present a molecular weight lower than 500 Da and hydrophilic properties, thus achieving a moderate penetration through the stratum corneum [20]. Focusing on this challenge, chemical modifications such as esterification with alkyl chains, are usually required. Peptide leads typically derived from three sources: isolated from nature (also known as bioactive peptides); from chemical libraries, or by genetic/recombinant libraries [16]. According to Gorouhi and Maibach, peptides used in cosmetics may be classified as enzyme inhibitory, carrier, neurotransmitter-inhibitory, and signal peptides [17]. Neurotransmitter-inhibitory peptides are able to mimic amino acid sequences involved in neuron excitability, thus modulating the nervous response, while signal peptides stimulate cells' activity and growth [21]. Accordingly, these peptides may be useful for modulating the neurogenic symptoms associated with sensitive skin, as well as the synthesis of pro-inflammatory cytokines.

We have previously characterized the trends in the use of peptides in anti-aging cosmetics [22]. As the usage of these ingredients in the sensitive skin care segment remains unknown, the present study aims to fill this gap.

2. Materials and Methods
2.1. Data Collection

The composition of a pool of skin care facial cosmetic products from multinational manufacturers, marketed in Portuguese parapharmacies and pharmacies was collected in 2019, in order to access the most used active ingredients for sensitive skin. Skin care products were included in the study if they exhibited in the label one of the following expressions: "sensitive skin" OR "reactive skin" OR "intolerant skin". All the information available in the product's label was collected, along with the information available on the manufacturers' websites.

2.2. Data Analysis

The products ingredient lists were analyzed by visual inspection in order to find peptides, and they were listed according to the International Nomenclature of Cosmetic Ingredients (INCI). Data were analyzed with respect to the following parameters:

2.2.1. Peptides Usage Frequency

The relative amount of cosmetic products for sensitive skin containing peptides were evaluated and expressed in percentage.

2.2.2. Top Peptides for Sensitive Skin

The peptides were identified from INCI lists and ranked in descending order of occurrence to disclose the top.

2.2.3. Scientific Evidence Supporting the Efficacy in Sensitive Skin Care

The efficacy data of each peptide were searched on the on-line databases PubMed, Scopus, Cochrane, KOSMET, and SciFinder. Due to the lack of studies regarding the applicability of active ingredients in cosmetics for sensitive skin, a broader search was performed, using the keywords ("INCI name" OR "synonyms", when applicable).

3. Results and Discussion

Following these criteria, 88 skin care facial products were selected from 19 multinational brands. Fifteen cosmetic products contained one or more peptides in their composition, making up about 17% of products analyzed. Noteworthy, only two products contained more than one peptide in their composition.

3.1. Top Ingredients for Sensitive Skin

The peptides were identified (Figure 1) and ranked in descending order according to their relative usage (Table 1).

Table 1. Peptides found at INCI lists of cosmetic products for sensitive skin and their relative usage (%).

INCI	Classification	Relative Usage (%)
Acetyl Dipeptide-1 Cetyl Ester	Neurotransmitter-inhibiting	5.7
Palmitoyl Tripeptide-8	Neurotransmitter-inhibiting	4.5
Acetyl Tetrapeptide-15	Neurotransmitter-inhibiting	2.3
Palmitoyl Tripeptide-5	Signal	2.3
Acetyl Hexapeptide-49	Unknown	1.1
Palmitoyl Tetrapeptide-7	Signal	1.1
Palmitoyl Oligopeptide	Signal	1.1

Overall, acetyl dipeptide-1 cetyl ester was the most used ingredient in cosmetic products for sensitive skin, being present in more than 5% of all products. Palmitoyl tripeptide-8 achieved the second place, followed by acetyl tetrapeptide-15 and palmitoyl tripeptide-5. Acetyl hexapeptide-49, palmitoyl tetrapeptide-7, and palmitoyl oligopeptide were only found in the composition of one cosmetic product.

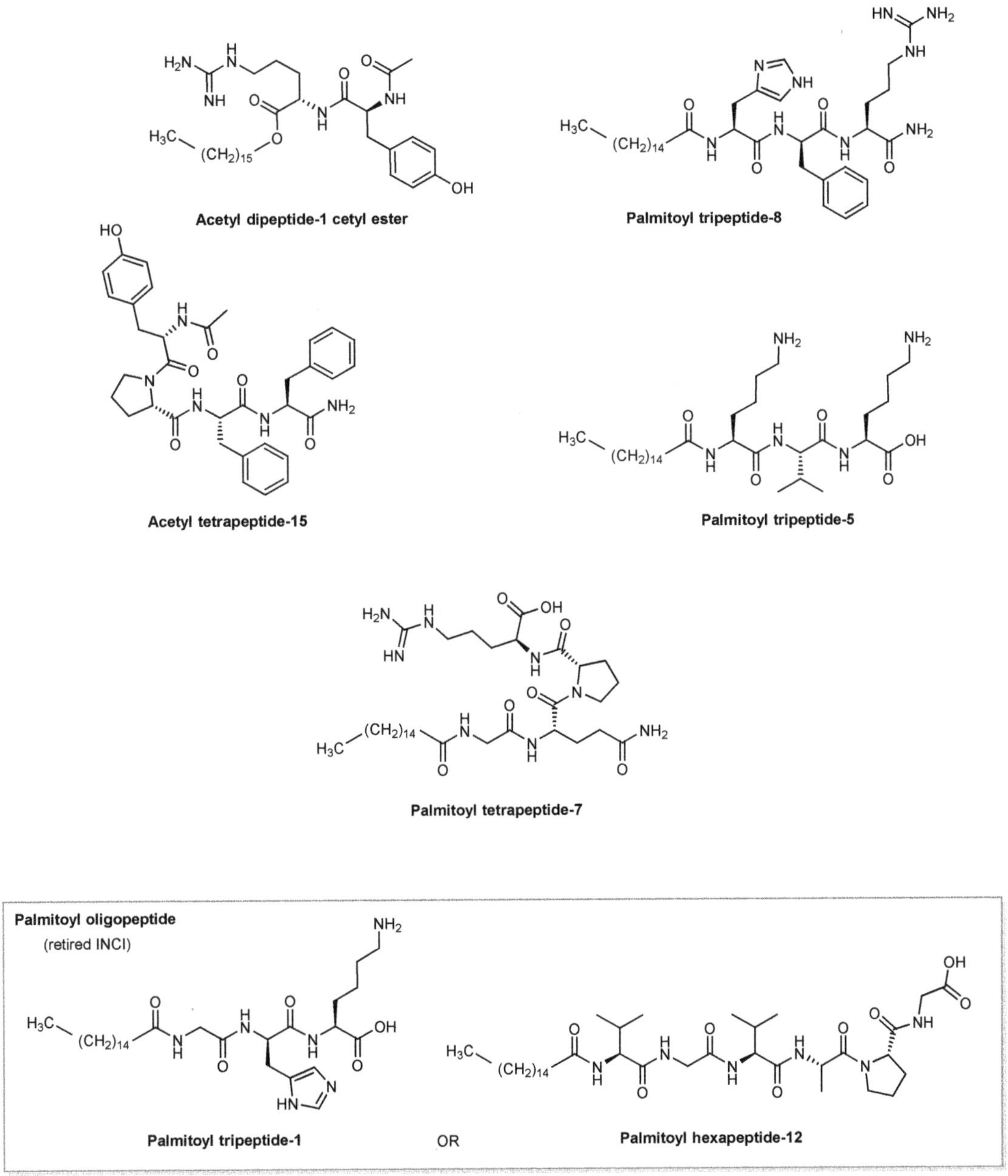

Figure 1. Structures of peptides found at INCI lists from cosmetic products for sensitive skin.

3.2. Scientific Evidence Supporting the Efficacy in Sensitive Skin Care

The search results are summarized below (Figure 2):

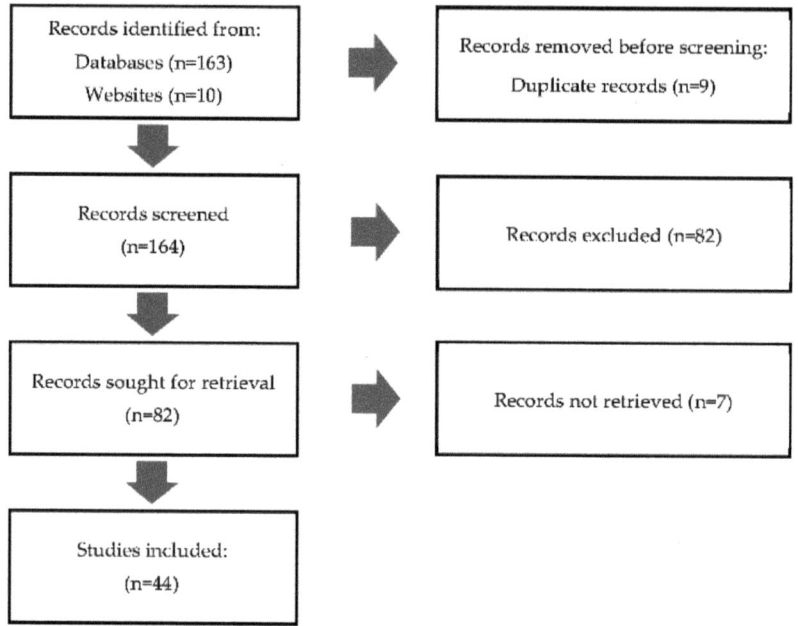

Figure 2. Flow chart of selected articles according to four different parts of the search process: identification, screening, eligibility, and inclusion.

3.2.1. Acetyl Dipeptide-1 Cetyl Ester

Acetyl dipeptide-1 cetyl ester is the INCI name for the peptide N-acetyl-L-tyrosyl-L-arginine hexadecyl ester (Figure 1). This compound was based on the bioactive dipeptide Tyr-Arg, for its alleviating and decontracting properties of muscle fibers, and is obtained by chemical synthesis, through initial esterification of L-arginine × HCl (**1**) with palmitol (**2**) [23] to give hexadecyl ether of L-arginine (**3**) (Scheme 1). Activation of N-acetyl-L-tyrosine (**4**) with N-hydroxysuccinimide (NHS) and coupling with L-arginine (**3**) allows to obtain the acetyl dipeptide-1 cetyl ester [23].

Acetyl dipeptide-1 cetyl ester promotes the pro-opiomelanocortin (POMC) gene expression. POMC incurs post-translational processing and originates the biologically active peptides melanocyte-stimulating hormones (MSHs) and adrenocorticotropin (ACTH), involved in melanin synthesis, as well as β-endorphin, which contains met-enkephalin's peptide sequence, providing an opiate activity [24,25]. α-MSH is able to bind to melanocytes melanocortin receptors, thus inducing melanin synthesis, but it also intervenes in the reduction of the inflammatory response by modulating the nuclear factor κ-β activity (NF-κβ) [26]. Noteworthy, there are α-MSH peptide fragments which do not elicit significant melanogenic activity, such those that are used in palmitoyl tripeptide-8. Furthermore the opioid β-endorphin reduces CGRP release. CGRP is able to activate TRPV1 in multiple cells thus initiating an inflammatory response [27]. Consequently, acetyl dipeptide-1 cetyl ester reduces the stinging sensation and inflammation resulting from the skin exposure to heat, contact with specific substances, such as capsaicin, and mechanical stress.

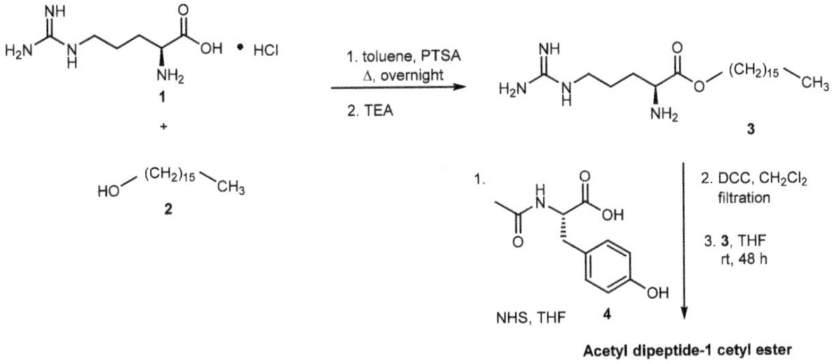

Scheme 1. Synthesis of acetyl dipeptide-1 cetyl ester. PTSA: *p*-toluenesulfonic acid; TEA: triethylamine; NHS: *N*-hydroxysuccinimide; THF: tetrahydrofuran; DCC: *N,N'*-dicyclohexylcarbodiimide; rt: room temperature; h: hours. Adapted from [23].

Furthermore, Khmaladze et al. demonstrated that acetyl dipeptide-1 cetyl ester significantly upregulates the expression of Aquaporin 3 (AQP3), Filaggrin (FLG), caspase 14, and keratin 10 genes, thus contributing to the improvement of the epidermal barrier [28]. Another study showed that acetyl dipeptide-1 cetyl ester is able to significantly reduce PGE_2 secretion and decrease NFκB signaling in vitro [29]. PGE_2 has been proposed to be associated with neurogenic inflammation in sensitive skin [2].

The ingredient supplier reports a reduced ability to perceive heating sensations [30]. The efficacy of a cream containing 3% acetyl dipeptide-1 cetyl ester was assessed regarding the interference in the ability of 21 volunteers to discriminate between four distinct levels of heat: warm, hot, very hot, and painful. The heat perception was reduced very significantly for temperatures which provide hot, very hot, and painful sensations. Additionally, the efficacy of the same cream for reducing the unpleasant sensations provoked by sandpaper aggression on one hand, using the other hand as control, was evaluated in a double-blind study including 18 volunteers. The subliminal response to discomfort was measured by a lie detector. The hand in which the cream was applied revealed greater comfort after the sandpaper aggression. Due to the above-mentioned effects, the supplier concludes acetyl dipeptide-1 cetyl ester is expected to reduce some of the unpleasant sensations of sensitive skin associated with hyperactivity of the somatosensory system.

Schoelermann et al. compared the ability of a cosmetic product containing acetyl dipeptide-1 cetyl ester with another product with 4-*t*-butylcyclohexanol for inhibiting capsaicin-induced stinging in a clinical study including 31 volunteers with sensitive to very sensitive skin. Volunteers' self-perception stinging/burning sensations and photographs of signs of skin inflammation were used for performing the evaluations. The authors concluded that the product containing 4-*t*-butylcyclohexanol presented a greater efficacy by significantly reducing neuronal activation, compared to the one with acetyl dipeptide-1 cetyl ester, which had no significant effect [31]. However, this study alone does not allow to conclude that 4-*t*-butylcyclohexanol is more efficacious than acetyl dipeptide-1 cetyl ester due to differences in the cosmetic bases containing each active ingredient, which could also interfere with study results.

Moreover, there are several manufacturers who invested in the registration of patents of cosmetic products for sensitive skin that include acetyl dipeptide-1 cetyl ester, demonstrating that researchers and cosmetic manufacturers recognize the value and usefulness of these ingredients for future applications [32–36].

3.2.2. Palmitoyl Tripeptide-8

Palmitoyl tripeptide-8 is a synthetic peptide ester based on a α-melanocyte stimulating hormone (α-MSH), originating from POMC, and it is composed by the sequence *N*-(1-

oxohexadecyl)-L-histidyl-D-phenylalanyl-L-argininamide (Figure 1). This peptide can be obtained via a solid-phase peptide synthesis using the fluorenylmethyloxycarbonyl (Fmoc) strategy on an ACT496S2 automated synthesizer with PS-Rink amide (RAM) resin (Scheme 2) [37]. The deprotection and coupling steps are carried out until the desired sequences are synthesized. Final side-chain deprotection and cleavage from the resin with a cleavage cocktail (trifluoroacetic acid/water/triisopropylsilane), affords palmitoyl tripeptide-8 [37].

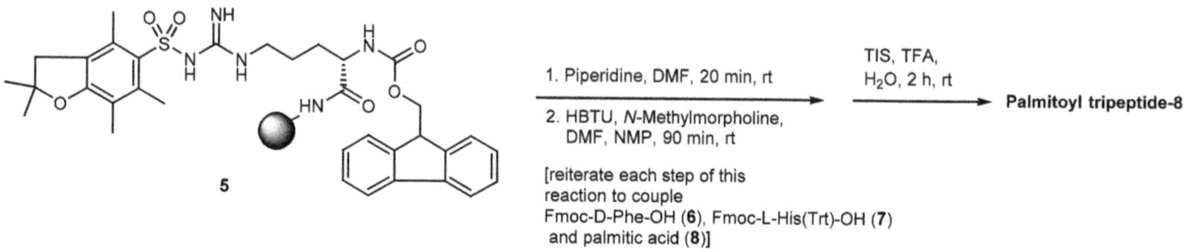

Scheme 2. Solid-phase synthesis of palmitoyl tripeptide-8. HBTU: N,N,N',N'-tetramethyl-O-(1H-benzotriazol-1-yl)uronium hexafluorophosphate; DMF: N,N-dimethylformamide; NMP: N-methyl-2-pyrrolidone; TIS: triisopropylsilane; TFA: trifluoroacetic acid. Adapted from [37].

The ingredient supplier performed several efficacy tests. In in vitro models, palmitoyl tripeptide-8 showed the ability to significantly inhibit IL-8 production up to 32% in UVB-irradiated keratinocytes, which was comparable to α-MSH, and also in IL-1 stimulated fibroblasts, reaching 64% inhibition which is greater that the achieved by α-MSH [38]. In skin explants exposed to substance P, palmitoyl tripeptide-8 significantly reduced the number of dilated capillaries and the size of dilated vessels up to 30% and 51%, respectively. Edema was also reduced by 60% due to palmitoyl tripeptide-8 [26]. The supplier also performed two clinical studies. In one study, eight individuals with no reported skin conditions applied both the control and the test formula containing palmitoyl tripeptide-8, three times a day, on separate areas of the volar side of the forearm, for eight days. After this period, single patches containing 250 μL of an aqueous 0.5% sodium dodecyl sulfate (SDS) solution were applied for 24 h, following their removal and a 24 h resting period. Then, the test areas were photographed using a video microscope and their temperature was measured using Thermovision, a temperature measurement device through infrared camera. The supplier reported that photographs demonstrated a redness reduction when palmitoyl tripeptide-8 was applied to the skin, although no quantitative measure was presented. In areas where palmitoyl tripeptide-8 was applied, a significant reduction in the skin temperature after an increase caused by SDS was found. Control results were not statistically significantly different. In another study, which included 13 individuals with no reported skin condition, the same patches containing 250 μL 0.5% SDS solution were applied for 24 h on separate areas of the volar side of the forearm. After this period, the patches were removed and both the test and control formulas were applied three times daily for two days. Again, the test areas were photographed and their temperature was measured using the same equipment. Skin redness decreased in the areas, which the test formula was applied, with no quantitative measurement, and skin temperature increase induced by SDS presented an average 78% reduction, with statistical significance, contrary to control results. Together, these results indicate that palmitoyl tripeptide-8 is able to prevent and soothe an irritative response [26].

There is only one study in the scientific literature, which addresses in vivo the efficacy of a formulation containing palmitoyl tripeptide-8 for the treatment of persistent redness in patients with rosacea who had been successfully treated with topical or oral therapy [39]. Twenty-five patients (23 women and 2 men) were asked to continue using their prior medication, while applying a lotion containing caffeine, zinc gluconate, bisabolol, *Eperua falcata* bark extract, and palmitoyl tripeptide-8 for 8 weeks. Clinical and patients' assessments for efficacy and tolerability were performed at weeks 4 and 8 using Visia CR device photographs. The evaluation of the product's efficacy showed a statistically significant improvement in redness, flushing, skin tone, and overall rosacea severity. Skin radiance, texture, and overall appearance also improved. Regarding patient's tolerance, there was a significant improvement in skin erythema, dryness, edema, and stinging. This finding may be particularly relevant for patients with sensitive skin, who also present this symptom. However, the presence of other active ingredients in the product composition and the lack of control do not allow to draw conclusions regarding palmitoyl tripeptide-8's efficacy. In addition to the cosmetic products found in our investigation, here are also several patented cosmetic formulations for sensitive skin containing palmitoyl tripeptide-8 [40–43].

3.2.3. Acetyl Tetrapeptide-15

Deriving from endomorphin-2 (Tyr-Pro-Phe-Phe-NH$_2$), a human µ-opioid agonist with selective anti-nociceptive effect, acetyl tetrapeptide-15 is a synthetic peptide constituted by the sequence N-acetyl-L-tyrosyl-L-prolyl-L-phenylalanyl-L-phenylalaninamide (Figure 1) [44,45]. Although this peptide is widely used in skincare formulations for sensitive skin, its synthesis is not fully described. However, the synthesis of a novel biologically active compound, the conjugate of jasmonic acid and of acetyl tetrapeptide-15, discloses that the synthesis of this tetrapeptide proceeds via a solid-phase method using AM RAM resin and the Fmoc/But procedure (Scheme 3) [46]. After initial treatment of the resin, the synthesis of the tetrapeptide proceeds with the addition of the resin to a mixture of hydroxybenzotriazole (HOBt) and Fmoc-L-Phe-OH (**9**). The reaction proceeds with anchoring of the Fmoc-L-Phe-OH (**9**) to the Rink amide resin (RAM) followed by protection of unreacted hydroxyl groups of the resin by capping, and deprotection of the Fmoc group. Further addition of protected amino acids Fmoc-L-Phe-OH (**9**), Fmoc-L-Pro-OH (**10**), Fmoc-L-Tyr(tBu)-OH (**11**) to the obtained amide, capping, N-Fmoc deprotection, and cleavage from the resin allows to obtain acetyl tetrapeptide-15 [46].

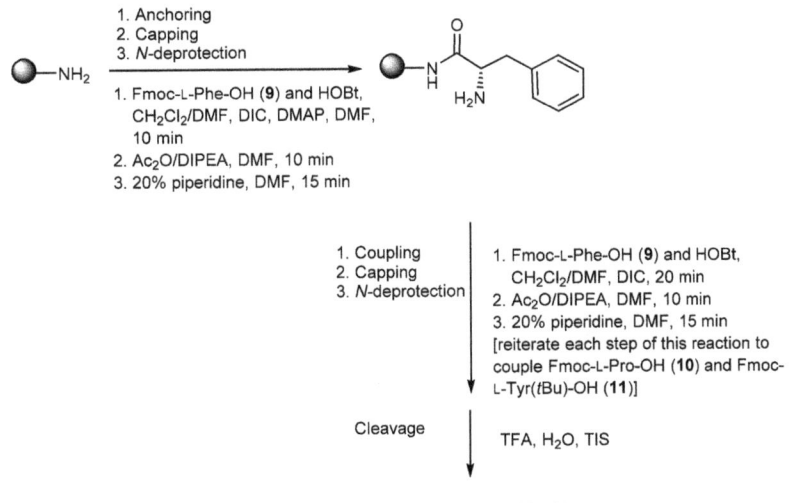

Scheme 3. Solid-phase synthesis of acetyl tetrapeptide-15. HOBt: hydroxybenzotriazole; DIC: *N,N'*-diisopropylcarbodiimide; DMAP: 4-(dimethylamino)pyridine; DIPEA: *N,N*-diisopropylethylamine. Reproduced from reference with permission from the Centre National de la Recherche Scientifique (CNRS) and the Royal Society of Chemistry [46].

Acetyl tetrapeptide-15 was developed with the aim to reduce skin hyperreactivity producing inflammatory, chronic and neuropathic pain, by increasing the threshold of neuronal excitability in μ-opioid receptor via an endorphin-like pathway [47,48]. The efficacy of this peptide was demonstrated both in vitro and in vivo by the supplier [44]. Firstly, acetyl tetrapeptide-15 was tested regarding its ability to modulate the release of CGRP. CGRP is released after the activation of TRVP1 by capsaicin, heat, or depolarizing agents such cations [45]. The test was performed by incubating sensory neurons with acetyl tetrapeptide-15 (0.0003% and 0.001%), capsazepine (10 μM, a TRPV1 antagonist), or verapamil (100 μM, a calcium channel blocker) for 6h, which were then exposed to KCl and capsaicin. Acetyl tetrapeptide-15 0.001% reduced CGRP release very significantly, both when neurons were exposed to capsaicin and KCl, performing better than capsazepine 10 μM and similarly to verapamil 100 μM. The ability of acetyl tetrapeptide-15 to activate μ-opioid receptors from cultured sensory neurons in a capsaicin media was evaluated in competition with naloxone, a receptor antagonist. Capsaicin binds to TRVP1 receptors, thus eliciting a calcium influx through the cell membrane that produces CGRP, as well as a nervous influx signaling pain and discomfort. The presence of acetyl tetrapeptide-15 significantly reduced the CGRP release by capsaicin-stimulated neurons, but this effect was compromised in the presence of naloxone, reinforcing that acetyl tetrapeptide-15 binds to μ-opioid receptors. The activation from μ-opioid receptors inhibits the TRVP1 response by reducing the phosphorylation of adenylate cyclase (ADC) to protein kinase A (PKA). Lastly, a split-faced single-blind clinical study elucidated the ability of acetyl tetrapeptide-15 to reduce skin sensitivity after the exposure to capsaicin in 20 individuals. The protocol started with the application of increasing concentrations of a capsaicin solution in the nasolabial folds, to determine the concentration, which induced discomfort. A vehicle solution was applied to the other side of the face. Then, a 0.0015% solution with acetyl tetrapeptide-15 was applied twice daily, for four days, and the application of increasing capsaicin concentrations was repeated. Overall, there was a significantly increase in the capsaicin threshold which provoked discomfort in volunteers.

There is a patent referring to the use of acetyl tetrapeptide-15 in a cosmetic product for sensitive skin [45], but no studies were found in scientific literature for this compound.

3.2.4. Palmitoyl Tripeptide-5

Palmitoyl tripeptide-5 is a fragment of Thrombospondin I (TSP-1) presenting the sequence, N-(1-oxohexadecyl)-L-lysyl-L-valyl-L-lysine (Figure 1) [21,49]. Two different liquid-phase methodologies were described for the synthesis of this tripeptide [50,51], designed to surpass some of the disadvantages associated with the solid-phase synthetic methodologies (high costs and pollution to the environment) [52]. One methodology (Scheme 4) involves a convergent synthesis with the initial formation of a N-carboxyanhydride **14** by reaction of L-valine (**12**) with phosgene (**13**) [50]. Boc-L-lysine (**15**) is then coupled to the N-carboxyanhydride **14**, forming Boc-protected dipeptide **16**. In a convergent route, N-acylated aminoacid **18** is prepared from Boc-L-lysine (**15**) and palmitoyl chloride (**17**). EDC/NHS Activation of the carboxyl group of Pal-Lys(Boc)-OH (**18**) produces Pal-Lys(Boc)-OSu (**19**). Coupling of intermediates **16** and **19** and further Boc deprotection furnishes palmitoyl tripeptide-5 [50].

The other synthetic methodology reported for the preparation of palmitoyl tripeptide-5 [51], although via a linear strategy (Scheme 5), is quite similar to the depicted in Scheme 4. Initial formation of palmitoyl chloride (**17**) from palmitic acid **8**, followed by coupling with benzyloxycarbonyl (Cbz)-L-lysine **21** forms Pal-Lys(Cbz)-OH (**22**) [51]. Further activation of the carboxyl group of Pal-Lys(Cbz)-OH (**22**) with NHS (**23**) and coupling with L-valine (**12**), followed by a second activation of the carboxyl group of **25** with NHS (**23**) and coupling with Cbz-L-lysine (**27**) furnishes Pal-Lys (Cbz)-Val-Lys(Cbz)-OH (**28**). Final deprotection of the Cbz groups allows to obtain palmitoyl tripeptide-5 [51].

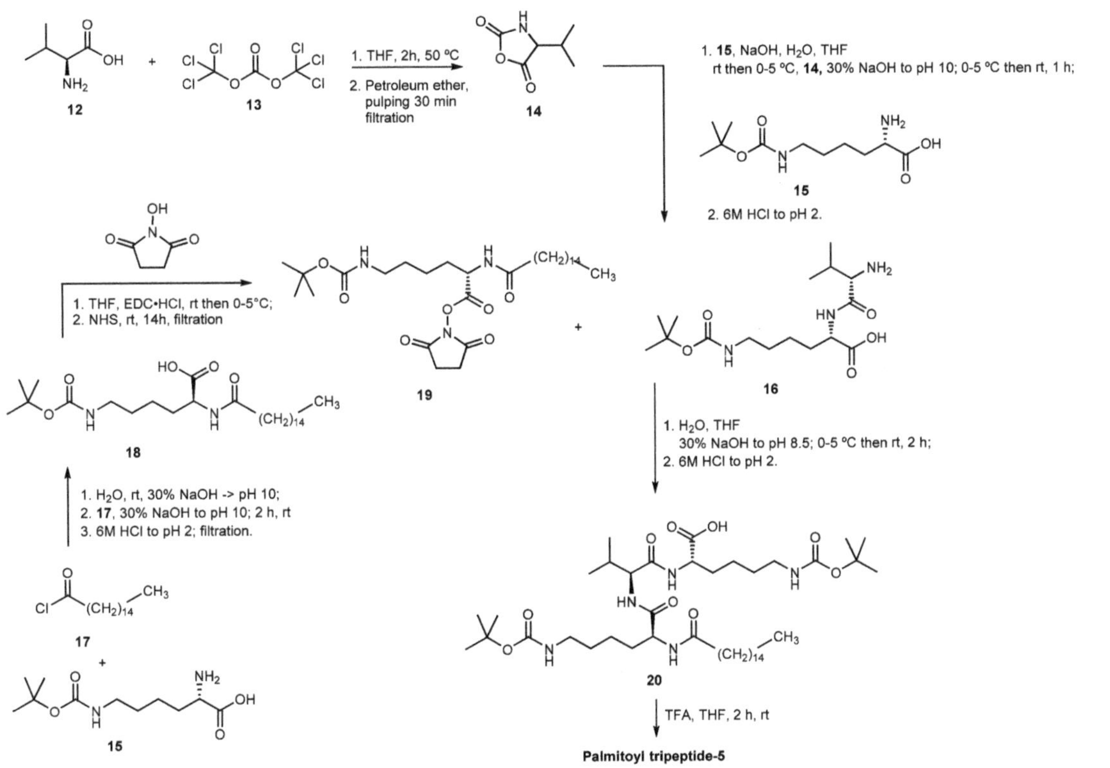

Scheme 4. Liquid-phase synthesis of palmitoyl tripeptide-5. EDC: 1-ethyl-3-(3-dimethylaminopropyl)carbodiimide. Adapted from [50].

Palmitoyl tripeptide-5 was used in a cosmetic ingredient mix from a raw material supplier, also containing spent grain wax and conjugated linoleic acid (CLA), for reducing skin redness in type I rosacea [53]. This ingredient mix is currently unavailable. Palmitoyl tripeptide-5 is proposed to reduce metalloproteases (MMP's) expression and pro-inflammatory cytokine syntheses, causing vasodilation and capillary permeability [54]. However, neither efficacy studies for the use of palmitoyl tripeptide-5 alone or in this mix in rosacea of sensitive skin are available. This peptide has also been used in patented cosmetic formulations for sensitive skin [55–57]. Palmitoyl tripeptide-5 is also used in anti-aging cosmetic products, due to its ability to reduce MMP'S and promote the synthesis of type I and type II collagen from extracellular matrix, as well as for inhibiting melanin production by reducing tyrosinase activity [51,58].

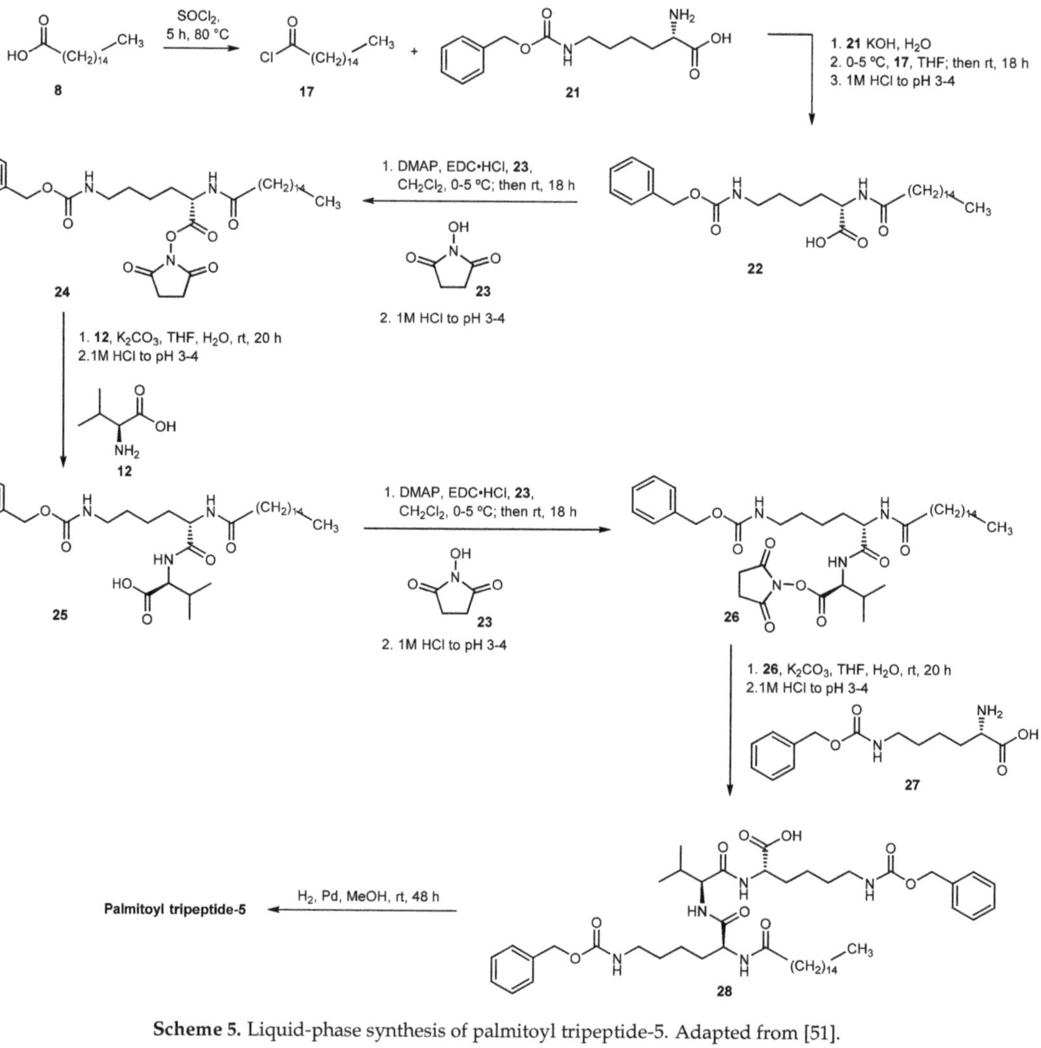

Scheme 5. Liquid-phase synthesis of palmitoyl tripeptide-5. Adapted from [51].

3.2.5. Acetyl Hexapeptide-49

Although acetyl hexapeptide-49 is widely used in the cosmetic industry, neither the structure nor the synthesis was reported to date in the literature.

This compound aims to regulate proteinase activated receptor 2 (PAR-2) from mast cells by trypsin-like serine proteases, thus reducing the inflammatory response which leads to IL-6 and IL-8 production, as well as TRVP-1 activation and subsequent CGRP release [59,60]. The supplier presents three studies for elucidating acetyl hexapeptide-49 efficacy. Primary human epidermal keratinocytes were incubated with vehicle or increasing concentrations of an acetyl hexapeptide-49 solution, and then exposed to 50 μM PAR-2 agonist. Cytokine production was determined by an ELISA test. At 0.5 mg/mL acetyl hexapeptide-49, there was a 69.6% and 71.5% decrease in IL-6 and IL-8 production, respectively. Moreover, cicatrization and barrier function recovery assays were performed using the same in vitro model. In this regard, keratinocytes treated with acetyl hexapeptide-49 solution were subject to an injury (cell-free area) induced by scraping a monolayer with a pipette tip, and then a cell proliferation assay was performed through the enzymatic conversion of the non-fluorescent calcein. The barrier function was recovered in both essays (concentrations are not disclosed). Another study in a reconstructed epidermis model was performed for evaluating the ability of a 4% acetyl hexapeptide-49 solution for reducing response to cosmetic allergens. The skin model was exposed both to hexyl cinnamal and farnesol, allergens, for 24 h, and the IL-8 expression was determined by an ELISA test. The 4% acetyl hexapeptide-49 allowed to reduce IL-8 expression in 58.2% comparing to positive control. Noteworthy, the skin model used in this study is not revealed, which would be important to evaluate its susceptibility to these allergens. Additionally, a clinical study was performed using 25 volunteers (24 to 67 years) who were selected based on their lactic acid stinging susceptibility. At the beginning of the study, volunteers applied a 10% lactic acid solution on the nasolabial fold, followed by a cream containing 2% acetyl hexapeptide-49. The soothing effect was evaluated after one hour, and volunteers reported an improvement in the stinging sensation. Then, the cream was applied twice a day for 7 days, and once again, the stinging sensation was assessed. After this period, a 32% reduction in volunteers experiencing stinging was found. Lastly, the supplier reported another clinical study including 20 volunteers (18 to 55 years) who applied a cream containing 2% acetyl hexapeptide-49 on the left leg, and a vehicle formulation on right leg twice a day for four weeks. Skin moisturization was evaluated by corneometry, and a clinical assessment of skin dryness, scaling, smoothness, softness, and suppleness was performed by a dermatologist. After four weeks, the supplier reported a significant increase in skin hydration comparing to vehicle, and the skin appeared less dry and scaly, smoother, softer, and more supple. Two patents for cosmetics with acetyl hexapeptide-49 have also been found [60,61].

3.2.6. Palmitoyl Tetrapeptide-7

Palmitoyl tetrapeptide-7 is a fragment of immunoglobulin G presenting the sequence N-(1-oxohexadecyl)glycyl-L-glutaminyl-L-prolyl-L-arginine (Figure 1) [62]. Two different solid-phase methodologies were described for the synthesis of this tetrapeptide [63,64]. In the first methodology (Scheme 6), a preloaded H-Arg(Boc)-HMPB-ChemMatrix resin (**29**) (functionalized support acylated with Riniker's super-acid-sensitive (4-hydroxymethyl-3-methoxyphenoxy)butanoic acid handle) is used and the first amino acid is attached by coupling with 2,7-disulfo-9-fluorenylmethoxycarbonyl (Smoc)-proline sodium salt (**30**). After resin wash and Smoc deprotection, coupling of the next amino acid is performed (Smoc-glutamine (**31**), and a solution of Smoc-glycine (**32**)), which was used without side-chain protecting group) until the desired peptide is completed. Oxyma [ethyl 2-cyano-2-(hydroxyimino)acetate] is an additive in the coupling medium safer than benzotriazole-based additives such as HOBt. Palmitoylation and cleavage of the peptide from the resin followed by precipitation and lyophilization gives the desired palmitoyl tetrapeptide-7 [63].

The second methodology involves the presence of a soluble fragment to improve the water solubility of the palmitoyl tetrapeptide-7, so that it is easier to purify [64]. Hence, five hydrophilic lysines are continuously coupled on the amino resin, then a connecting arm of *p*-hydroxybenzoic acid is introduced, and finally the remaining amino acids are coupled according to the peptide sequence of the palmitoyl tetrapeptide-7 (Pal-Gly-Gln-Pro-Arg-OH) [64]. HOBt/DIC methodology is adopted as a coupling approach when other amino acid residues (except the first lysine) are coupled [64]. Removal of the Fmoc protecting groups, resin cleavage, and further purification gives the refined peptide which is hydrolyzed to the target peptide palmitoyl tetrapeptide-7 [64].

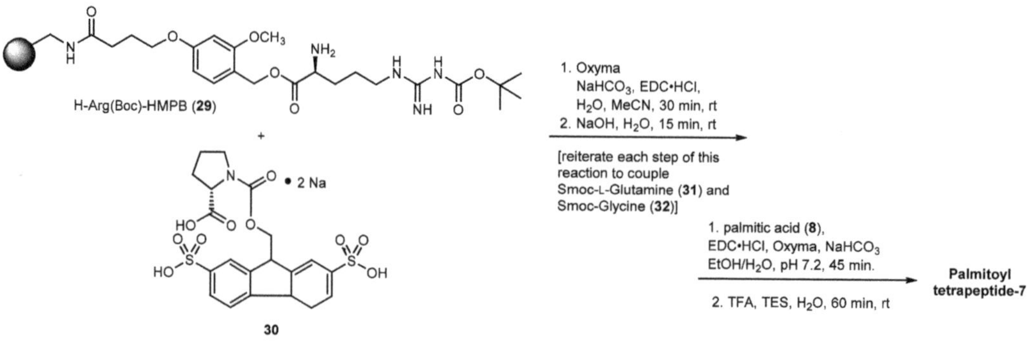

Scheme 6. Solid-phase synthesis of palmitoyl tetrapeptide-7. Adapted from [63].

Palmitoyl tetrapeptide-7 decreases IL-6 secretion, reduces inflammation after UVB exposure and stimulates laminin IV and V as well as collagen VII production [62]. In this regard, palmitoyl tetrapeptide-7 has been used in anti-aging cosmetics [65,66]. Although this mechanism of action is promising in the regulation of skin inflammation, namely for wound healing, no studies were found revealing palmitoyl tetrapeptide-7′s efficacy in this regard, nor for controlling the symptoms of sensitive skin [67]. Two patents however describe its use in cosmetic products for sensitive skin [68,69].

3.2.7. Palmitoyl Oligopeptide

The name "palmitoyl oligopeptide" was "removed" in 2013, since was used to designate two distinct molecules from the time of its development in 1994. The two compounds were renamed as palmitoyl tripeptide-1 (Pal-GHK) and palmitoyl hexapeptide-12 (Pal-KTTKS) in order to clarify the composition of cosmetic products [70].

Palmitoyl tripeptide-1 is a collagen fragment presenting the sequence N-(1-oxohexadecyl)glycyl-L-histidyl-L-lysine (Figure 1) [71]. To date, three different methodologies were reported for the synthesis of this tripeptide [63,72,73]. The first methodology (Scheme 7) [73] consists in an initial EDC-mediated coupling of H-Lys(Z)-OBzl×HCl (**33**) and Boc-His trityl(Trt)-OH (**34**) followed by removal of the trityl and Boc protecting groups affords H-His-Lys (Z)-OBzl (**36**). Coupling of this dipeptide **36** with Pal-Gly-ONb (**39**), previously synthesized from Pal-Gly-OSu (**37**) gives Pal-Gly-His-Lys (Z)-OBzl (**40**) which, after removal of Cbz and Bzl protecting groups affords palmitoyl tripeptide-1 [73].

Another reported methodology for the synthesis of this tripeptide is performed by using the same methodology as used for the synthesis of palmitoyl tetrapeptide-7 (Scheme 6). A preloaded H-Lys(Boc)-HMPB-ChemMatrix resin to which were coupled Smoc-L-Hys and Smoc-Gly is used in a solid-phase approach. Deprotection of the Smoc protecting group, palmitoylation, cleavage of the peptide from the solid support, and further lyophilization affords palmitoyl tripeptide-1 [63].

The third methodology is initiated through the protection of the carboxylic moiety of Boc-Lys-(Z)-OH (**41**) to obtain Boc-Lys(Z)-OBzl (**43**) (Scheme 8) [72]. Boc removal in acidic conditions, followed by coupling with Boc-His-OH (**44**), under usual coupling conditions (HOBt, NMM, DIPEA) gives Boc-His-Lys(Z)-OBzl (**45**) which, after deprotection/coupling with Boc-Gly-OH (**47**) and deprotection/palmitoylation with palmitic acid (**8**), and final deprotection of the carboxylic moiety with Pd/C furnishes the desired palmitoyl tripeptide-1.

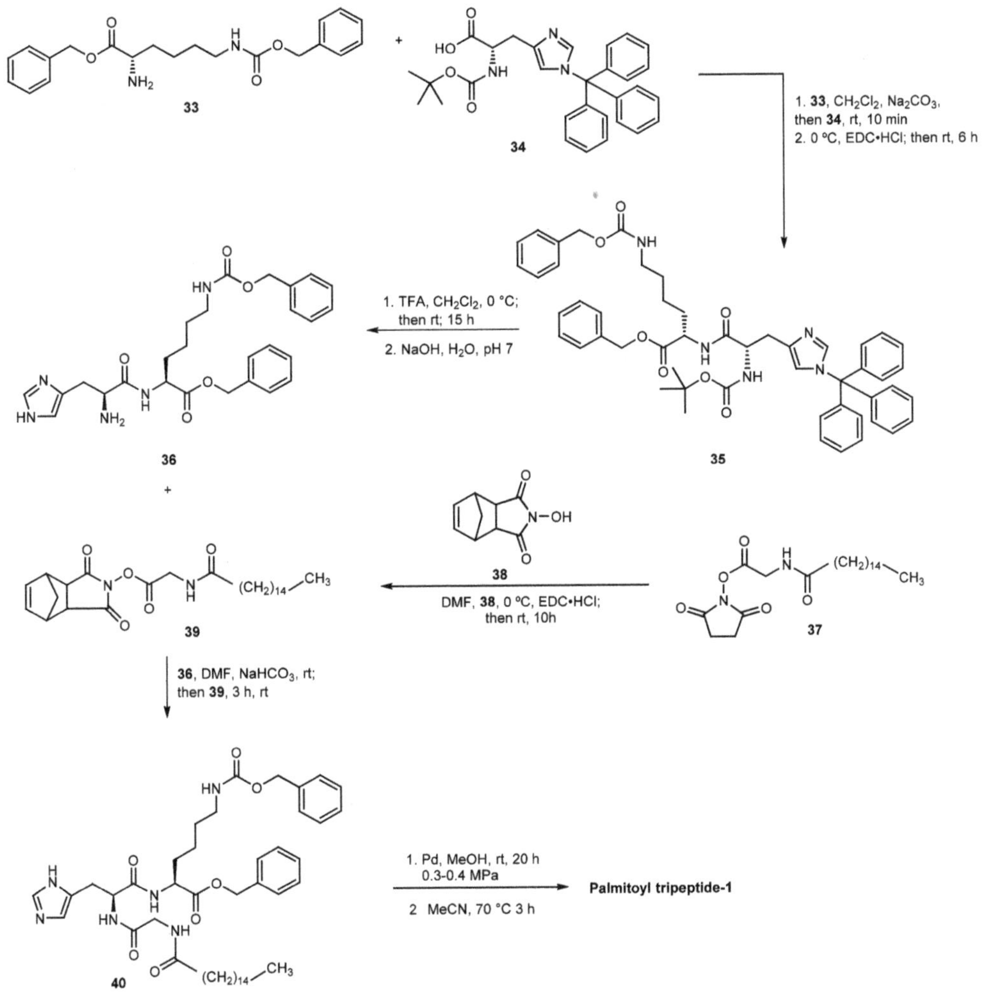

Scheme 7. Liquid-phase synthesis of palmitoyl tripeptide-1. NHS: N-Hydroxysuccinimide. Adapted from [73].

Palmitoyl hexapeptide-12 is an elastin fragment presenting the sequence N-(1-oxohexadecyl)-L-valyl-glycyl-L-valyl-L-alanyl-L-prolyl-glycine (Figure 1). The methodology for the preparation of this peptide is similar to the previously reported for other liquid-phase peptide syntheses, consisting in a series of deprotection/coupling reactions starting with the initial coupling of Boc-L-proline (50) and benzyl glycinate (51), followed by the coupling of Boc-L-alanine (53), Boc-L-valine (54), Boc-glycine (55), and Boc-L-valine (54) (Scheme 9) [74].

Although both peptides have been used in anti-aging cosmetics [22], neither studies or patents revealing the use of palmitoyl oligopeptide, palmitoyl tripeptide-1, or palmitoyl hexapeptide-12 for reducing sensitive skin symptoms were found.

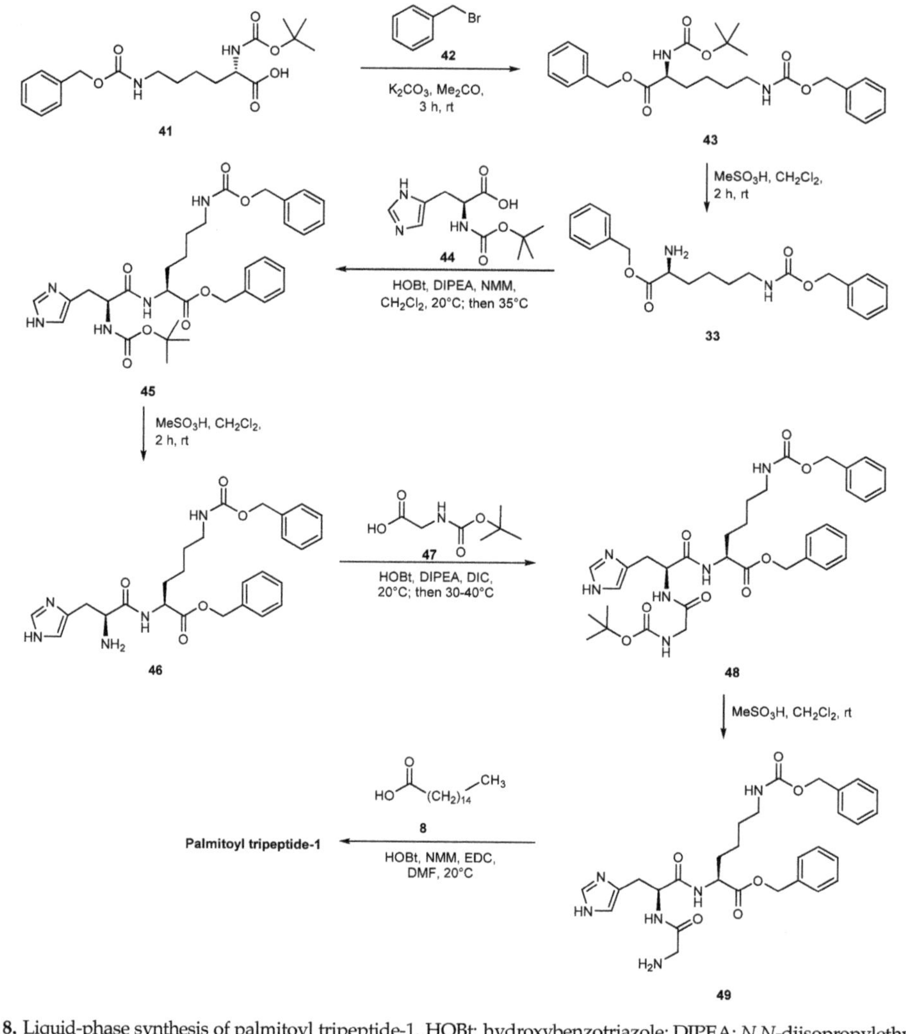

Scheme 8. Liquid-phase synthesis of palmitoyl tripeptide-1. HOBt: hydroxybenzotriazole; DIPEA: *N,N*-diisopropylethylamine; *N*-methylmorpholine; DIC: *N,N'*-diisopropylcarbodiimide; EDC: 1-ethyl-3-carbodiimide hydrochloride. Adapted from [72].

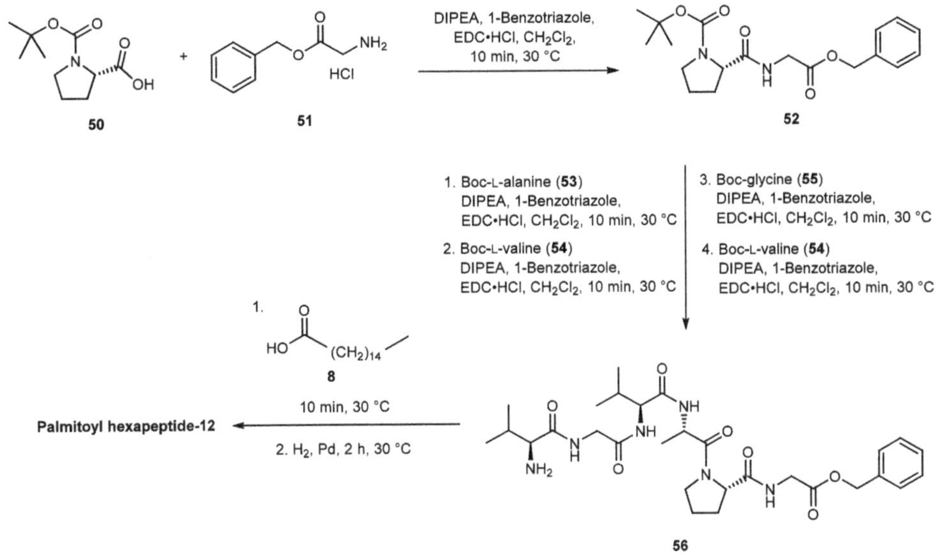

Scheme 9. Synthesis of palmitoyl hexapeptide-12. Adapted from [74].

3.2.8. Highlights in the Usage of Synthetic Peptides in Cosmetics for Sensitive Skin

Peptides disclosed in the composition of the facial cosmetics for sensitive skin investigated are all from synthetic origin. In contrast to natural peptides that have variable dimensions, reaching high molecular weights, and may contain allergenic moieties and their extraction can be costly [22], the synthesis of simpler peptides has the advantage of reaching both the pharmacophoric portion and improved bioavailability. Most peptides found in cosmetic products for sensitive skin are based on the pharmacologically active portions of endogenous molecules, whose low molecular weight and subsequent hydrophobization provide a better penetration through the stratum corneum. Noteworthy, acetyl dipeptide-1 cetyl ester, palmitoyl tripeptide-8, and acetyl tetrapeptide-15 are neurotransmitter-inhibiting peptides acting as agonists from cutaneous opioid system, such as μ receptor, which interacts with TRVP1 receptors through intracellular signaling. Therefore, these peptides reduce the activation of cutaneous nerve fibers, especially through TRVP1, thus preventing the release from CGRP. Conversely, acetyl hexapeptide-49 also reduces CGRP release after TRVP1 activation as well as the pro-inflammatory cytokine production by a signaling pathway involving proteinase activated receptor 2 (PAR-2). Interestingly, no peptide acting directly on TRVP1 receptors has been found. There are several TRVP1 antagonists reported in the scientific literature, but no peptides have been described, possibly due to the specificity of the receptor's binding site [75].

Concerning synthetic methodologies, these usually proceed via a linear approach, although a few convergent approaches were also described. Additionally, the fine-tuning of the chemistry associated with the synthetic methodology, ranging from standard coupling procedures (EDC/NHS) to new and improved methodologies, such as greener methods (EDC/Oxyma) can also contribute for the marked increase of peptides in this industry. The main bottleneck of these procedures is related to the activation of one of the carboxylic groups before the occurrence of the coupling reaction. This activation step, along with the next coupling reaction can lead to a potential loss of chiral integrity at the carboxyl residue undergoing activation. Although the above-described methodologies for the preparation of these peptides are fully optimized, the development of new procedures might need to take this challenge into account. New stand-alone coupling reagents, such as HOAt, containing better leaving groups can be used to enhance coupling rates and reduce the risk of racemization. Oxyma, a highly efficient leaving group, is safer and less hazardous than HOAt. Oxyma exhibited the same efficiency as HOAt and greater performance than HOBt. Of similar importance, is also the use of protecting groups, which can be used to maximize the yield of the desired products, as well as to minimize undesirable side reactions such as polymerization of the amino acids, usual in the synthesis of complex peptide-based structures.

3.3. Applicability of the Described Synthetic Peptides in Pharmeceuticals

Synthetic peptides whose mechanism of action has been previously act indirectly on TRPs. These receptors are associated with several skin diseases [76]. Pruritus, also known as itch, can be idiopathic or secondary to different pathologies, such as atopic dermatitis, psoriasis, urticaria, chronic renal failure, or liver diseases [77]. TRP channels, namely transient receptor potential cation channel subfamily A member 1 (TRPA1) and TRPV1 have shown to be greatly involved in itch development both under both physiological and pathological conditions. Rosacea is also aggravated by neuroinflammation, and the overexpression of TRPV1, TRPV2, TRPV4, and TRPV4 receptors has been proved in distinct subtypes from the disease [78]. Palmitoyl tripeptide-8, which is present in the formulation of a cosmetic product with proven efficacy for the treatment of rosacea patients, may be a prime candidate for the development of pharmaceuticals aimed at alleviating the signs and symptoms of this condition [39]. Moreover, the overexpression of TRPV1, TRPV4, and TRPV6 has been associated to nonmelanoma skin cancer, but their carcinogenic effect remains unknown [76].

Therefore, the recognition of the effectiveness from these synthetic and the further improvement of their molecular structures, may be useful for modulating TRP associated pathways, providing alternative treatments and/or symptom management for multiple diseases. These compounds are known to be safe for topical application, and their toxicity has been assessed both by the manufacturer, through material safety data sheets, and by the independent committee Cosmetic Ingredient Review supporting the Federal drug Administration in the US [79].

4. Conclusions

This study characterizes the usage of peptides in cosmetic products for sensitive skin for the first time. These ingredients were present in about 17% of the facial cosmetics for sensitive skin analyzed in 2019. Seven distinct peptides were found, namely acetyl dipeptide-1 cetyl ester, palmitoyl tripeptide-8, acetyl tetrapeptide-15, palmitoyl tripeptide-5, and acetyl hexapeptide-49, for which experimental data is reported to support use in cosmetics for sensitive skin, along with palmitoyl terrapeptide-7 and palmitoyl oligopeptide (the old name for the peptides palmitoyl tripeptide-1 and palmitoyl hexapeptide-12), whose efficacy is only documented for anti-aging cosmetics. Most of the available information regarding these ingredients is not reported in peer-reviewed scientific journals, but rather in patents and supplier brochures. Additionally, the small number of randomized clinical studies, and especially the fact that only one study included volunteers with sensitive skin (acetyl dipeptide-1 cetyl ester), hinders a robust evidence of the in vivo efficacy of these peptides. More clinical studies with good methodological quality are needed to provide sound evidence of the peptide's efficacy in sensitive skin care.

From a chemical perspective, the increasingly use of these peptides as ingredients in the cosmetic industry can be explained as a result of the development of solid-phase syntheses and automated methodologies. Finally, the implementation of reverse-phase high-performance liquid chromatography for peptide purification, in combination with the previously mentioned factors, has allowed the production of complex peptides in multi-kilogram amounts that was impossible to envisage only a few decades ago and that contributed to a boost in the use of peptides as ingredients in cosmetic formulations.

Peptides are useful ingredients in cosmetics for sensitive skin that may also be relevant for medical devices or medicines intended to treat or prevent the symptoms of diseases in which neurogenic inflammation plays an important role, such as rosacea and atopic or seborrheic dermatitis. Given the worldwide prevalence of sensitive skin and the growing interest in peptides by the cosmetic industry, it is foreseeable that the market for these products may increase in the coming years, fostering the design of new and more effective compounds. In the future, it is possible to see a further exploration of signaling pathways involving cutaneous opioid receptors, through the development of peptides acting upstream in this pathway, or as agonists of opioid receptors. Moreover, the development of peptides acting directly on TRVP1 receptors, either extra or intracellularly, could provide promising results.

Author Contributions: Conceptualization: I.F.A. and E.S.; Data collection and analysis: M.S.F.; Writing-Original draft preparation, and final manuscript: M.S.F. and D.I.S.P.R.; Supervision: J.M.S.-L.; Writing-Reviewing and Editing: I.F.A. and E.S., Funding acquisition J.M.S.-L. and E.S. All authors have read and agreed to the published version of the manuscript.

Funding: This work was financially supported by the Applied Molecular Biosciences Unit-UCIBIO which is financed by national funds from FCT/MCTES (UID/Multi/04378/2020), UIDB/04423/2020, UIDP/04423/2020, and under the project PTDC/SAUPUB/28736/2017 (reference POCI-01-0145-FEDER-028736), co-financed by COMPETE 2020, Portugal 2020 and the European Union through the ERDF and by FCT through national funds, as well as CHIRALBIOACTIVE-PI-3RL-IINFACTS-2019.

Institutional Review Board Statement: Not applicable.

Informed Consent Statement: Not applicable.

Data Availability Statement: Data sharing not applicable.

Acknowledgments: This work received support and help from FCT regarding M.S.F. doctoral grant (ref. SFRH/BD/144864/2019).

Conflicts of Interest: The authors declare no conflict of interest the funders had no role in the design of the study; in the collection, analyses, or interpretation of data; in the writing of the manuscript, or in the decision to publish the results.

Limitations: This study was performed for the Portuguese cosmetic market, which is dominated by multinational cosmetic brands. Therefore, this may result in discrepancies when comparing with other markets. Many ingredients found in cosmetic products from the market lack scientific literature regarding their efficacy. Therefore, some of the information used in this study was collected in technical documents and patents from suppliers.

Abbreviations and Acronyms

α-MSH	α-melanocyte stimulating hormone
ACTH	adrenocorticotropin
ADC	adenylate cyclase
AQP3	aquaporin 3
Boc	*tert*-Butyloxycarbonyl
Bu	Butyl
Bz	benzoyl
Cbz	benzyloxycarbonyl
CLA	conjugated linoleic acid
DCC	N,N'-dicyclohexylcarbodiimide
DIC	N,N'-diisopropylcarbodiimide
DIPEA	N,N-diisopropylethylamineDMAP
DMAP	4-(dimethylamino)pyridine
DMF	N,N-dimethylformamide
EDC	1-ethyl-3-(3-dimethylaminopropyl)carbodiimide
FLG	filaggrin
Fmoc	fluorenylmethyloxycarbonyl
h	hours
HBTU	N,N,N',N'-tetramethyl-O-(1H-benzotriazol-1-yl)uronium hexafluorophosphate

HOBt	hydroxybenzotriazole
INCI	international nomenclature of cosmetic ingredients
MMP's	metalloproteases
MSHs	melanocyte-stimulating hormones
NF-κβ	nuclear factor κ-β
NHS	N-hydroxysuccinimide
NMP	N-methyl-2-pyrrolidone
Pal	Palmitic acid
PAR-2	proteinase activated receptor 2
PKA	protein kinase A
POMC	pro-opiomelanocortin
PTSA	p-toluenesulfonic acid
RAM	Rink amide
rt	room temperature
SDS	sodium dodecyl sulfate
Smoc	2,7-disulfo-9-fluorenylmethoxycarbonyl
Su	succinimide
TEA	triethylamine
TFA	trifluoroacetic acid
THF	tetrahydrofuran
TIS	triisopropylsilane
Trt	Trityl
TRPV	Transient Receptor Potential Cation Channel Subfamily V
TPPA	Transient Receptor Potential Cation Channel Subfamily A
TSP-1	thrombospondin I

References

1. Misery, L. Sensitive skin, reactive skin. *Ann. Dermatol. Venereol.* **2019**, *146*, 585–591. [CrossRef] [PubMed]
2. Berardesca, E.; Farage, M.; Maibach, H. Sensitive skin: An overview. *Int. J. Cosmet. Sci.* **2013**, *35*, 2–8. [CrossRef]
3. Chen, W.; Dai, R.; Li, L. The prevalence of self-declared sensitive skin: A systematic review and meta-analysis. *J. Eur. Acad. Dermatol. Venereol.* **2020**, *34*, 1779–1788. [CrossRef]
4. Farage, M.A. The Prevalence of Sensitive Skin. *Front. Med. (Lausanne)* **2019**, *6*, 98. [CrossRef]
5. Farage, M.A.; Jiang, Y.; Tiesman, J.P.; Fontanillas, P.; Osborne, R. Genome-Wide Association Study Identifies Loci Associated with Sensitive Skin. *Cosmetics* **2020**, *7*, 49. [CrossRef]
6. Verhoeven, E.W.; de Klerk, S.; Kraaimaat, F.W.; van de Kerkhof, P.C.; de Jong, E.M.; Evers, A.W. Biopsychosocial mechanisms of chronic itch in patients with skin diseases: A review. *Acta Derm. Venereol.* **2008**, *88*, 211–218. [CrossRef]
7. Zheng, Y.; Liang, H.; Li, Z.; Tang, M.; Song, L. Skin microbiome in sensitive skin: The decrease of Staphylococcus epidermidis seems to be related to female lactic acid sting test sensitive skin. *J. Dermatol. Sci.* **2020**, *97*, 225–228. [CrossRef]
8. Misery, L.; Weisshaar, E.; Brenaut, E.; Evers, A.W.M.; Huet, F.; Stander, S.; Reich, A.; Berardesca, E.; Serra-Baldrich, E.; Wallengren, J.; et al. Pathophysiology and management of sensitive skin: Position paper from the special interest group on sensitive skin of the International Forum for the Study of Itch (IFSI). *J. Eur. Acad. Dermatol. Venereol.* **2020**, *34*, 222–229. [CrossRef] [PubMed]
9. Misery, L. Neuropsychiatric factors in sensitive skin. *Clin. Dermatol.* **2017**, *35*, 281–284. [CrossRef]
10. Misery, L.; Loser, K.; Stander, S. Sensitive skin. *J. Eur. Acad. Dermatol. Venereol.* **2016**, *30* (Suppl. 1), 2–8. [CrossRef]
11. Buhe, V.; Vie, K.; Guere, C.; Natalizio, A.; Lheritier, C.; Le Gall-Ianotto, C.; Huet, F.; Talagas, M.; Lebonvallet, N.; Marcorelles, P.; et al. Pathophysiological Study of Sensitive Skin. *Acta Derm. Venereol.* **2016**, *96*, 314–318. [CrossRef]
12. Richters, R.; Falcone, D.; Uzunbajakava, N.; Verkruysse, W.; van Erp, P.; van de Kerkhof, P. What is sensitive skin? A systematic literature review of objective measurements. *Skin Pharmacol. Physiol.* **2015**, *28*, 75–83. [CrossRef] [PubMed]
13. Ferrer-Montiel, A.; Camprubí-Robles, M.; García-Sanz, N.; Sempere, A.; Valente, P.; Nest, W.V.D.; Carreño, C. The contribution of neurogenic inflammation to sensitive skin: Concepts, mechanisms and cosmeceutical intervention. *Int. J. Cosmet. Sci.* **2009**, *11*, 311–315. [CrossRef]
14. Gutte, B. *Peptides: Synthesis, Structures, and Applications*; Elsevier: Amsterdam, The Netherlands, 1995.
15. Lintner, K.; Peschard, O. Biologically active peptides: From a laboratory bench curiosity to a functional skin care product. *Int. J. Cosmet. Sci.* **2000**, *22*, 207–218. [CrossRef]
16. Sato, A.K.; Viswanathan, M.; Kent, R.B.; Wood, C.R. Therapeutic peptides: Technological advances driving peptides into development. *Curr. Opin. Biotechnol.* **2006**, *17*, 638–642. [CrossRef]
17. Gorouhi, F.; Maibach, H.I. Role of topical peptides in preventing or treating aged skin. *Int. J. Cosmet. Sci.* **2009**, *31*, 327–345. [CrossRef]

18. Ahsan, H. The biomolecules of beauty: Biochemical pharmacology and immunotoxicology of cosmeceuticals. *J. Immunoass. Immunochem.* **2019**, *40*, 91–108. [CrossRef]
19. Hruby, V.J. Designing peptide receptor agonists and antagonists. *Nat. Rev. Drug Discov.* **2002**, *1*, 847–858. [CrossRef]
20. Kobiela, T.; Milner-Krawczyk, M.; Pasikowska-Piwko, M.; Bobecka-Wesolowska, K.; Eris, I.; Swieszkowski, W.; Dulinska-Molak, I. The Effect of Anti-aging Peptides on Mechanical and Biological Properties of HaCaT Keratinocytes. *Int. J. Pept. Res. Ther.* **2018**, *24*, 577–587. [CrossRef]
21. Schagen, S.K. Topical Peptide Treatments with Effective Anti-Aging Results. *Cosmetics* **2017**, *4*, 16. [CrossRef]
22. Ferreira, M.S.; Magalhães, M.C.; Sousa-Lobo, J.M.; Almeida, I.F. Trending Anti-Aging Peptides. *Cosmetics* **2020**, *7*, 91. [CrossRef]
23. Greff, D. Synthetic Peptides and Their Use in Cosmetic or Dermopharmaceutical Compositions. WO9807744A1, 26 February 1998.
24. Millington, G.W. Proopiomelanocortin (POMC): The cutaneous roles of its melanocortin products and receptors. *Clin. Exp. Dermatol.* **2006**, *31*, 407–412. [CrossRef]
25. Harno, E.; Gali Ramamoorthy, T.; Coll, A.P.; White, A. POMC: The Physiological Power of Hormone Processing. *Physiol. Rev.* **2018**, *98*, 2381–2430. [CrossRef]
26. Loing, E. Reaching a Zen-Like State in Skin: Biomimetic Peptide to Balance Skin. Available online: https://www.cosmeticsandtoiletries.com/testing/sensory/Reaching-a-Zen-like-State-in-Skin-Biomimetic-Peptide-to-Balance-Sensitivity-420538914.html (accessed on 7 May 2021).
27. Calmosensine Skin Pacified, Face Relaxed. Available online: https://www.ulprospector.com/documents/1003852.pdf?bs=11024&b=335122&st=20&r=la&ind=personalcare (accessed on 25 November 2020).
28. Khmaladze, I.; Österlund, C.; Smiljanic, S.; Hrapovic, N.; Lafon-Kolb, V.; Amini, N.; Xi, L.; Fabre, S. A novel multifunctional skin care formulation with a unique blend of antipollution, brightening and antiaging active complexes. *J. Cosmet. Dermatol.* **2020**, *19*, 1415–1425. [CrossRef]
29. Sulzberger, M.; Worthmann, A.C.; Holtzmann, U.; Buck, B.; Jung, K.A.; Schoelermann, A.M.; Rippke, F.; Stäb, F.; Wenck, H.; Neufang, G.; et al. Effective treatment for sensitive skin: 4-t-butylcyclohexanol and licochalcone A. *J. Eur. Acad. Dermatol. Venereol.* **2016**, *30* (Suppl. 1), 9–17. [CrossRef] [PubMed]
30. Calmosensine Sensual Healing. Available online: https://www.ulprospector.com/documents/1003854.pdf?bs=1240&b=44014&st=20&r=na&ind=personalcare (accessed on 7 May 2021).
31. Schoelermann, A.M.; Jung, K.A.; Buck, B.; Grönniger, E.; Conzelmann, S. Comparison of skin calming effects of cosmetic products containing 4-t-butylcyclohexanol or acetyl dipeptide-1 cetyl ester on capsaicin-induced facial stinging in volunteers with sensitive skin. *J. Eur. Acad. Dermatol. Venereol.* **2016**, *30* (Suppl. 1), 18–20. [CrossRef] [PubMed]
32. Archambault, J.-C.; Franchi, J.; Korichi, R. Cosmetic Composition Containing an Extract from Lotus and Method of Cosmetic Care Using said Composition. U.S. 20090148544A1, 11 June 2009.
33. Liu, Y. Nano-Encapsulated Skin Repair Agent Containing Blue Copper Peptide and its Preparation Method. CN111840125A, 30 October 2020.
34. Ding, W.; Lv, Q. Polypeptide Composition with Antiallergic Effect. CN106176274A, 7 December 2016.
35. Potin, A. Use of a Combination of Tyrosine-Arginine Dipeptide and Niacinamide as a Substance P Antagonist. FR2894142A1, 12 June 2009.
36. Potin, A. Use of Tyrosine-Arginine Dipeptide in a Cosmetic Composition for the Treatment of Cutaneous Redness. FR2894144A1, 12 June 2009.
37. Martinez, J.; Verdie, P.; Dubs, P.; Pinel, A.M.; Subra, G. Tripeptide-Carboxylic acid Conjugates as α-MSH Agonists and Their Therapeutic and Cosmetic Use. FR2870243A1, 19 November 2010.
38. NEUTRAZEN™ Active Ingredients Soothing Neurocosmetic. Available online: https://www.ulprospector.com/documents/1045285.pdf?bs=4499&b=125061&st=20&r=eu&ind=personalcare (accessed on 10 May 2021).
39. Baldwin, H.; Berson, D.; Vitale, M.; Yatskayer, M.; Chen, N.; Oresajo, C. Clinical effects of a novel topical composition on persistent redness observed in patients who had been successfully treated with topical or oral therapy for papulopustular rosacea. *J. Drugs Dermatol.* **2014**, *13*, 326–331.
40. Zhao, C.; Zhu, W.; Song, X.; Hui, Y.; Sun, L. Anti-Allergic Repair Mask Containing Polypeptide. CN110251415A, 20 September 2019.
41. Zhang, J.; Zhi, Q.; Yue, Z.; Song, X.; Liao, M.; Zhu, W. Polypeptide Composition Granules for Cosmetics with Anti-Inflammatory and Repairing Effects and Preparation Method Thereof. CN111514055A, 11 August 2020.
42. Liu, S.; Li, N.; Qiu, J.; Yin, Q.; Xiang, W.; Xiao, X. Preparation of Soothing and Anti-Allergic Cosmetic Composition. CN111494266A, 26 January 2021.
43. Chenevard, Y.; Fargeon, V. Soothing Cosmetic Composition Helicrysum Italicum and Glycyrrhizinic Acid Derivative. FR2965729A1, 13 April 2012.
44. BASF. Skinasensyl™ The Neurocosmeceutical Soother. Available online: https://www.carecreations.basf.com/product-formulations/products/products-detail/SKINASENSYL%20PW%20LS%209852/30537033 (accessed on 13 May 2021).
45. Yang, F.; Zheng, J. Understand spiciness: Mechanism of TRPV1 channel activation by capsaicin. *Protein Cell* **2017**, *8*, 169–177. [CrossRef]
46. Kapuscinska, A.; Olejnik, A.; Nowak, I. The conjugate of jasmonic acid and tetrapeptide as a novel promising biologically active compound. *New J. Chem.* **2016**, *40*, 9007–9011. [CrossRef]

47. Improving Skin Comfort Via Nervous System Modulation. Available online: https://personalcaremagazine.com/story/6258/improving-skin-comfort-via-nervous-system-modulation (accessed on 13 May 2021).
48. Tetrapeptide for Neurosensitive Skin. Available online: https://www.cosmeticsandtoiletries.com/formulating/function/antiirritant/35799934.html (accessed on 13 May 2021).
49. Zhang, L.; Falla, T.J. Cosmeceuticals and peptides. *Clin. Dermatol.* **2009**, *27*, 485–494. [CrossRef]
50. Yu, G.; Li, J.; Lin, Z.; Bian, F.; Si, C.; Liu, C. Liquid Phase Synthesis Method of Palmitoyl Tripeptide-5. CN111004306A, 25 September 2020.
51. Tao, Y.; Chen, J.; Wang, X. Method for Efficiently Preparing Palmitoyl Tripeptide-5 Based on Activated Ester. CN110423264A, 1 December 2020.
52. Ziegler, H.; Heidl, M.; Imfeld, D. Tripeptides and their derivatives for cosmetic applications for improving skin structure. WO2004099237A1, 18 November 2004.
53. DSM Launches Regu-Cea. Available online: https://www.happi.com/contents/view_breaking-news/2009-04-22/dsm-launches-regu-cea/ (accessed on 7 May 2021).
54. Milanello, S. REGU®-CEA: Approccio multifunzionale contro i sintomi della rosacea. *Kosmetica* **2009**, 56–58.
55. Liao, Y.; He, L.; Liu, X.; Liu, Y. Composition for Repairing Sensitive Skin, and its Application in Cosmetic. CN109010113A, 18 December 2018.
56. Lu, X.; Dai, C.; He, H.; Liang, J.; Tang, Z. Multifunctional Toning Lotion and its Preparation Method. CN111632001A, 8 September 2020.
57. Ji, X.; Lin, J.; Wang, L. Preparation Method of Dried Facial Mask Comprising Ceramide-2 and Traditional Chinese Medicine Extract for Caring Skin. CN109350591A, 25 May 2021.
58. Avcil, M.; Akman, G.; Klokkers, J.; Jeong, D.; Çelik, A. Efficacy of bioactive peptides loaded on hyaluronic acid microneedle patches: A monocentric clinical study. *J. Cosmet. Dermatol.* **2020**, *19*, 328–337. [CrossRef]
59. Delisens: Protect Skin, Reduce Discomfort. Available online: https://www.happi.com/issues/2013-10/view_features/protect-skin-reduce-discomfort/ (accessed on 7 May 2021).
60. Ding, W.; Peng, Y.; Huang, C. Polypeptide Composition with Soothing and Anti-Allergic Effects Containing Palmitoyl Tripeptide-8 In Water-In-Oil System for Preparing Skin Care Product. CN110833515A, 25 February 2020.
61. Ding, W. Polypeptide for Repairing Facial Steroid Dependent Dermatitis. CN109125107A, 4 January 2019.
62. Mondon, P.; Hillion, M.; Peschard, O.; Andre, N.; Marchand, T.; Doridot, E.; Feuilloley, M.G.; Pionneau, C.; Chardonnet, S. Evaluation of dermal extracellular matrix and epidermal-dermal junction modifications using matrix-assisted laser desorption/ionization mass spectrometric imaging, in vivo reflectance confocal microscopy, echography, and histology: Effect of age and peptide applications. *J. Cosmet. Dermatol.* **2015**, *14*, 152–160. [CrossRef]
63. Knauer, S.; Koch, N.; Uth, C.; Meusinger, R.; Avrutina, O.; Kolmar, H. Sustainable peptide synthesis enabled by a transient protecting group. *Angew. Chem., Int. Ed.* **2020**, *59*, 12984–12990. [CrossRef] [PubMed]
64. Mi, P.; Pan, J.; Liu, J. Preparation Method of Polypeptide. CN112110984A, 22 December 2020.
65. Hahn, H.J.; Jung, H.J.; Schrammek-Drusios, M.C.; Lee, S.N.; Kim, J.H.; Kwon, S.B.; An, I.S.; An, S.; Ahn, K.J. Instrumental evaluation of anti-aging effects of cosmetic formulations containing palmitoyl peptides, Silybum marianum seed oil, vitamin E and other functional ingredients on aged human skin. *Exp. Ther. Med.* **2016**, *12*, 1171–1176. [CrossRef]
66. Draelos, Z.D.; Kononov, T.; Fox, T. An open label clinical trial of a peptide treatment serum and supporting regimen designed to improve the appearance of aging facial skin. *J. Drugs Dermatol.* **2016**, *15*, 1100–1106. [PubMed]
67. Johnson, B.Z.; Stevenson, A.W.; Prele, C.M.; Fear, M.W.; Wood, F.M. The Role of IL-6 in Skin Fibrosis and Cutaneous Wound Healing. *Biomedicines* **2020**, *8*, 101. [CrossRef] [PubMed]
68. Zhu, Y. A Mild Polypeptide Repairing and Lightening Facial Mask for Facial Ulcers and Sensitive Skin. CN110302124A, 8 October 2019.
69. Zhang, X.; Zhong, W. Skin Care Composition Containing Plant Extract and Preparation Method Thereof. CN110279646A, 14 August 2020.
70. Husein El Hadmed, H.; Castillo, R.F. Cosmeceuticals: Peptides, proteins, and growth factors. *J. Cosmet. Dermatol.* **2016**, *15*, 514–519. [CrossRef] [PubMed]
71. Lintner, K. Cosmetic or Dermopharmaceutical Compositions Containing the n-palmytoyl-gly-hys-lys Tripeptide. WO2001043701A2, 21 June 2001.
72. Huang, Y.; Xing, H.; Wang, Z.; Yu, X. Liquid Phase Synthesis Method of Palmitoyl Tripeptide-1. CN 112409444A, 26 February 2021.
73. Zheng, Q. Preparation of Palmitoyl Tripeptide-1 by Liquid-Phase Peptide Synthesis Method. CN 108218956 A, 29 June 2018.
74. Su, X.; Yang, Y.; Bian, Y.; Cui, Y. Preparation of Palmitoyl Hexapeptide with Microchannel Modular Reaction Device. CN 109879936A, 14 June 2019.
75. Escelsior, A.; Sterlini, B.; Murri, M.B.; Serafini, G.; Aguglia, A.; da Silva, B.P.; Corradi, A.; Valente, P.; Amore, M. Red-hot chili receptors: A systematic review of TRPV1 antagonism in animal models of psychiatric disorders and addiction. *Behav. Brain Res.* **2020**, *393*, 112734. [CrossRef] [PubMed]
76. Caterina, M.J.; Pang, Z. TRP Channels in Skin Biology and Pathophysiology. *Pharmaceuticals (Basel)* **2016**, *9*, 77. [CrossRef]
77. Xie, Z.; Hu, H. TRP Channels as Drug Targets to Relieve Itch. *Pharmaceuticals (Basel)* **2018**, *11*, 100. [CrossRef]

78. Sulk, M.; Seeliger, S.; Aubert, J.; Schwab, V.D.; Cevikbas, F.; Rivier, M.; Nowak, P.; Voegel, J.J.; Buddenkotte, J.; Steinhoff, M. Distribution and expression of non-neuronal transient receptor potential (TRPV) ion channels in rosacea. *J. Investig. Dermatol.* **2012**, *132*, 1253–1262. [CrossRef]
79. Johnson, W., Jr.; Bergfeld, W.F.; Belsito, D.V.; Hill, R.A.; Klaassen, C.D.; Liebler, D.C.; Marks, J.G., Jr.; Shank, R.C.; Slaga, T.J.; Snyder, P.W.; et al. Safety Assessment of Tripeptide-1, Hexapeptide-12, Their Metal Salts and Fatty Acyl Derivatives, and Palmitoyl Tetrapeptide-7 as Used in Cosmetics. *Int. J. Toxicol.* **2018**, *37*, 90S–102S. [CrossRef]

MDPI
St. Alban-Anlage 66
4052 Basel
Switzerland
www.mdpi.com

Pharmaceuticals Editorial Office
E-mail: pharmaceuticals@mdpi.com
www.mdpi.com/journal/pharmaceuticals

Disclaimer/Publisher's Note: The statements, opinions and data contained in all publications are solely those of the individual author(s) and contributor(s) and not of MDPI and/or the editor(s). MDPI and/or the editor(s) disclaim responsibility for any injury to people or property resulting from any ideas, methods, instructions or products referred to in the content.

www.ingramcontent.com/pod-product-compliance
Lightning Source LLC
LaVergne TN
LVHW070734100526
838202LV00013B/1233